Springer Theses

Recognizing Outstanding Ph.D. Research

For further volumes:
http://www.springer.com/series/8790

Aims and Scope

The series "Springer Theses" brings together a selection of the very best Ph.D. theses from around the world and across the physical sciences. Nominated and endorsed by two recognized specialists, each published volume has been selected for its scientific excellence and the high impact of its contents for the pertinent field of research. For greater accessibility to non-specialists, the published versions include an extended introduction, as well as a foreword by the student's supervisor explaining the special relevance of the work for the field. As a whole, the series will provide a valuable resource both for newcomers to the research fields described, and for other scientists seeking detailed background information on special questions. Finally, it provides an accredited documentation of the valuable contributions made by today's younger generation of scientists.

Theses are accepted into the series by invited nomination only and must fulfill all of the following criteria

- They must be written in good English.
- The topic should fall within the confines of Chemistry, Physics, Earth Sciences, Engineering and related interdisciplinary fields such as Materials, Nanoscience, Chemical Engineering, Complex Systems and Biophysics.
- The work reported in the thesis must represent a significant scientific advance.
- If the thesis includes previously published material, permission to reproduce this must be gained from the respective copyright holder.
- They must have been examined and passed during the 12 months prior to nomination.
- Each thesis should include a foreword by the supervisor outlining the significance of its content.
- The theses should have a clearly defined structure including an introduction accessible to scientists not expert in that particular field.

Guofeng Shen

Emission Factors of Carbonaceous Particulate Matter and Polycyclic Aromatic Hydrocarbons from Residential Solid Fuel Combustions

Doctoral Thesis accepted by
College of Urban and Environmental Sciences,
Peking University, Beijing, China

Springer

Author
Dr. Guofeng Shen
Institute of Atmospheric Sciences
Jiangsu Academy of Environmental Science
Nanjing
People's Republic of China

Supervisor
Prof. Shu Tao
College of Urban and Environmental Sciences
Peking University
Beijing
China

ISSN 2190-5053 ISSN 2190-5061 (electronic)
ISBN 978-3-662-52477-0 ISBN 978-3-642-39762-2 (eBook)
DOI 10.1007/978-3-642-39762-2
Springer Heidelberg New York Dordrecht London

Softcover reprint of the hardcover 1st edition 2014

Printed on acid-free paper

Springer is part of Springer Science+Business Media (www.springer.com)

Parts of the thesis have been published in the following publications and reused with permission:

(1) **SHEN, G.F.**; Yang, Y.; Wang, W.; Tao, S.;* Zhu, C.; Min, Y.; Xue, N.; Ding, J.; Wang, B.; Wang, R.; Shen, H.; Li, W.; Wang, X.; Russell, A. Emission factors of particulate matter and elemental carbon for crop residues and coals burned in typical household stoves in China. **Environmental Science & Technology** 2010, 44, 7157–7162. ***(REUSE WITH PERMISSION)***

(2) **SHEN, G.F.**; Wang, W.; Yang, Y.; Zhu, C.; Min, Y.; Xue, M.; Ding, J.; Li, W.; Wang, B.; Shen, H.; Wang, R.; Wang, X.; Tao, S.*. Emission factors and particulate matter size distribution of polycyclic aromatic hydrocarbons from residential coal combustions in rural Northern China. **Atmospheric Environment** 2010, 44, 5237–5243. ***(REUSE WITH PERMISSION)***

(3) **SHEN, G.F.**; Wang, W.; Yang, Y.; Ding, J.; Xue, M.; Min, Y.; Zhu, C.; Shen, H.; Li, W.; Wang, B.; Wang, R.; Wang, X.; Tao, S.;* Russell, A. Emissions of PAHs from indoor crop residue burning in a typical rural stove: emission factors, size distributions and gas-particle partitioning. **Environmental Science & Technology** 2011, 45, 1206–1212. ***(REUSE WITH PERMISSION)***

(4) **SHEN, G.F.**; Tao, S.;* Wang, W.; Yang, Y.; Ding, H.; Xue, M.; Mi, Y.; Zhu, C.; Shen, H.; Li, W.; Wang, B.; Wang, R.; Wang, W.; Wang, X.; Russell, A. Emissions of Oxygenated Polycyclic Aromatic Hydrocarbons from indoor solid fuel combustion. **Environmental Science & Technology** 2011, 45, 3459–3465. ***(REUSE WITH PERMISSION)***

(5) **SHEN, G.F.**; Wei, S.; Zhang, Y.; Wang, R.; Wang, B.; Li, W.; Shen, H.; Huang, Y.; Chen, Y.; Chen, H.; Wei, W.; Tao, S.* Emissions of oxygenated polycyclic aromatic hydrocarbons from biomass pellet burning in a modern burner for cooking in China. **Atmospheric Environment** 2012. 60, 234–237. ***(REUSE WITH PERMISSION)***

(6) **SHEN, G.F.**; Tao, S.*; Wei, S.; Zhang, Y.; Wang, R.; Wang, B.; Li, W.; Shen, H.Z.; Huang, Y.; Yang, Y.; Wang, W.; Wei, W.; Wang, X.; Liu, W.; Wang, X.; Simonich, S. Reductions in emissions of carbonaceous particulate matter and polycyclic aromatic hydrocarbons from combustion of biomass pellets in comparison with raw fuel burning. **Environmental Science & Technology** 2012, 46, 6409–6416. ***(REUSE WITH PERMISSION)***

(7) **SHEN, G.F.**; Tao, S.*; Wei, S.; Zhang, Y.; Wang, R.; Wang, B.; Li, W.; Shen, H.Z.; Huang, Y.; Yang, Y.; Wang, W.; Wang, X.; Simonich, S. Retene emission from residential solid fuels in China and evaluation of retene as a unique marker for soft wood combustion. **Environmental Science & Technology** 2012, 46, 4666–4672. ***(REUSE WITH PERMISSION)***

(8) **SHEN, G.F.**; Wei, S.; Wei, W.; Zhang, Y.; Min, Y.; Wang, B.; Wang, R.; Li, W.; Shen, H.; Huang, Y.; Yang, Y.; Wang, W.; Wang, X.; Wang, X.; Tao, S.* Emission factors, size distributions and emission Inventories of carbonaceous particulate matter from residential wood combustion in rural China. **Environmental Science & Technology** 2012, 46, 4207–4214. ***(REUSE WITH PERMISSION)***

(9) **SHEN, G.F.**; Tao, S.*; Wei, S.; Zhang, Y.; Wang, R.; Wang, B.; Li, W.; Shen, H.; Huang, Y.; Chen, Y.C.; Chen, H.; Yang, Y.; Wang, W.; Wang, X.; Liu, W.; Simonich, S. Emissions of parent, nitro, and oxygenated polycyclic aromatic hydrocarbons from residential wood combustion in rural China. **Environmental Science & Technology** 2012. 46, 8123–8130. ***(REUSE WITH PERMISSION)***

(10) **SHEN, G.F.**; Wei, S.; Zhang, Y.; Wang, B.; Wang, R.; Shen, H.; Li, W.; Huang, Y.; Chen, Y.C.; Chen, H.; Tao, S.*Emission and size distribution of particle-bound polycyclic aromatic hydrocarbons from residential wood combustion in rural China. **Biomass and Bioenergy** 2013, 55. 141–147. ***(REUSE WITH PERMISSION)***

(11) **SHEN, G.F.**; Tao, S.;* Wei, S.; Chen, Y.; Zhang, Y.; Shen, H.; Huang, Y.; Zhu, D.; Yuan, C.; Wang, H.; Wang, Y.; Pei, L.; Lioa, Y.; Duan, Y.; Wang, B.; Wang, R.; Lv, Y.; Li, W.; Wang, X.; Zheng, X. Field measurement of emission factors of PM, EC, OC, parent, nitro- and oxy-polycyclic aromatic hydrocarbons for residential briquette, coal cake and wood in rural Shanxi, China. **Environmental Science & Technology** 2013, 47, 2998–3005. ***(REUSE WITH PERMISSION)***

(12) Wei, S.; **SHEN, G.F.**; Zhang, Y.; Xue, M.; Xie, H.; Lin, P.; Chen, Y.; Wang, X.; Tao, S.* Field measurement on the emissions of PM, OC, EC, and PAHs from indoor crop straw burning in rural China. **Environmental Pollution** 2014, 184, 18–24. ***(REUSE WITH PERMISSION)***

Supervisor's Foreword

Residential solid fuel combustion is a major source of incomplete combustion pollutants including particulate matter (PM), black carbon (BC), organic carbon (OC), polycyclic aromatic hydrocarbons (PAHs), and many more. This is particularly true for developing countries like China. Until recently, a majority of data on emission factors (EFs) of these pollutants from the residential sector had been measured in developed countries. However, residential stoves are very different between developed and developing countries. Because of the shortage of data on EFs for developing countries, emission inventories of pollutants generated by solid fuel combustion used for residential heating and cooking are associated with relatively large uncertainties, leading to biases in air quality modeling, exposure assessment and health analyses.

To fill the data gap, Guofeng's doctorial thesis focused on the measurement of EFs for solid fuel combustion from the residential sector in China. A simulated kitchen with real stoves commonly used in China was built particularly for this purpose. He tested a variety of residential solid fuels including coal, crop residues, and firewood in the measurements during a year-long experimental study. In addition, field measurements were conducted at selected rural sites to confirm the laboratory measurements.

As a result, a large volume of EF data of PM, BC, OC, and PAHs have been generated, which help fill a major data gap in the field. Soon after the publication, much of the data have been adopted for use in updating global emission inventories of BC, PAHs, and PM. Moreover, EFs of derivative PAHs including nitro-PAHs and oxy-PAHs for residential solid fuels and EFs of various pollutions for biomass pellet fuels, which are really scarce, were reported.

In addition to helping to fill the data gap, factors affecting EFs have been carefully investigated. It was found that modified combustion efficiency and fuel moisture are the most influential factors influencing EFs, and the knowledge gained in this work can be used to help quantify EFs for individual fuels. The models developed in this study can provide us with a better understanding of the generation mechanism of air pollutants during combustion.

Although the majority of the data collected in his study have been published in a series of papers internationally, we hope that an English version of this thesis with a collection of all measurements and key findings can help readers to use these results more efficiently.

Beijing, January 2014 Prof. Shu Tao

Foreword

Both emission inventories and ambient measurements indicate that residential solid fuel combustion is a major contributor to emissions of incomplete combustion products, particularly in developing countries. In the recent World Health Organization's (WHO) Global Burden of Disease study, of the over 60 factors quantified, indoor exposures to such pollutants is the second leading cause of premature death in developing countries, and the International Agency for Research on Cancer has determined such combustion products to be a Group 1 carcinogen. Further, the soot derived from incomplete solid fuel combustion absorbs radiation and can potentially exacerbate global change. Given the importance of this source to human and the environment health, it is important to improve our understanding of the emission characteristics of residential solid fuel combustion and to use that knowledge to develop more accurate estimate emissions from that source. This involves characterizing the physical and chemical properties of the emissions, emission factors, and the associated activity levels. However, unlike more concentrated point sources, conducting the appropriated analyses is difficult due to the variable nature of the sources, the need to conduct detailed laboratory analyses, and collect and analyze information from a large and diverse literature. This is a particular issue in developing countries, which is also where such information is so valuable.

Dr. Guofeng Shen, as described in this thesis, took on a great challenge to help provide such information. He conducted a series of measurements on emissions of carbonaceous particulate matter and polycyclic aromatic hydrocarbons from residential solid fuel combustion in rural China. This work provides a firsthand data of emission factors from a large number of tests that can be used in the development of more reliable inventories. In addition, the work identified key factors affecting pollutant emissions from the combustion process which can provide information for pollution control strategy development. Not only did he look at more traditional solid fuel use, he considered pollutant emissions from biomass pellets which are considered as a cleaner, alternative fuel to replace traditional solid fuels.

The results of his work have appeared in international journals, including Environmental Science and Technology and Atmospheric Environment. Given my interest in air pollution and health, and the important issues addressed by Dr. Shen, the publication of this thesis provides valuable information on pollutant emissions from residential solid fuel combustion. The outcome can be useful for emission inventory, health and climate impact analysis, and also the development of effective pollution control strategies. Further, this thesis can provide a foundation for future research in this area.

Georgia, Atlanta, November 18, 2013 Armistead G. Russell

Special Thanks and Acknowledgments

The work was supported by many people in field sampling, laboratory measurement, and data analysis. I appreciate their inspiration, support and kind suggestions in the whole process.

THANKS to Prof. Shu Tao who gave the detailed instruction in experimental design, data analysis, and the writing of manuscript and final report throughout my 5-year study in PKU. All my good progress is under his tirelessly tutelage. Prof. Tao inspired me to do things right, good, and challenging but also meaningful. He is absolutely an honorable scientist.

THANKS to Yu Liu, Bingjun Meng, and Xiaofang Fu from the research group. They helped me in the laboratory measurements of massive samples and quality control. It is their delicate work to achieve a large number of high quality and useful results.

THANKS to the colleagues in Peking University. I spent very happy time during my 5-year study in the research group. Even in the onerous, sometimes irksome work, we enjoyed the life though hard up. There are many colleagues to thank. It is a little pity that I cannot list all their names here. I appreciated their kind help and wish them forever.

THANKS to my family. My grandparents and parents in my hometown and my beloved wife Vicky Xue in Beijing support and encourage me all the time. Without their help, it was impossible to get into the university, and more important, graduate with satisfied harvest.

2012.07 @ PKU

In the publication of this English version, special thanks to the editors from the Springer office. They provide the opportunity to publish the thesis so that many friends all over the world can access it. It is my pleasure if the result can be helpful in their work. Wishes!

Guofeng Shen @ Nanjing

Contents

Figures

Tables

Chapter 1
Introduction

1.1 Background

1.1.1 Carbonaceous Particulate Matter

Particulate matter (PM) affects local/regional air quality, human health and climate change significantly, and thus of widespread interest nowadays. There is a growing desire for clean air worldwide, especially in developing countries including China. Increased mass loading of PM, especially fine PM like $PM_{2.5}$ (PM with diameter less than 2.5 μm) is speculated to be the main cause for the decreased visibility. The formation of regional haze is often believed to be strongly related to ambient $PM_{2.5}$. Exposure to ambient PM causes serious adverse impacts on human health, particularly for the women and children (Eaazti et al. 2002; Pope et al. 2002, 2009; Russell and Brunekreef 2009; Shannon et al. 2004; WHO 2002a, b, 2011a, b; Zhang and Smith 2007). It had been reported that exposure to ambient PM is strongly correlated with the increased risks of some respiratory and cardiovascular diseases (Dockery et al. 1993; Englert 2004; Pope et al. 2009). In comparison with coarse PM, fine PM such as $PM_{2.5}$ and $PM_{1.0}$ ($Da < 1.0$ μm) usually have more serious impacts on human health as they may penetrate deeper into the lung area (Englert 2004; Kumar et al. 2010; Russell and Brunekreef 2009; Samet et al. 2000).

PM compositions like sulfate, nitrate, organic carbon (OC) and elemental carbon (EC, sometimes known as black carbon) have significant effects on the local/regional climate change (Chow et al. 2011; Menon et al. 2002; McConnell et al. 2007; IPCC 2007). It has been generally accepted that EC or BC has a positive radiative forcing of 0.2–1.0 W/m^2, while OC, sulfate and nitrate may cool the atmosphere by light scattering (Chow et al. 2011). The net effect of PM depends on the amounts of these compositions, the mixing status, like the internal, external and coated mixture, particle size and form, and so on (Cheng et al. 2009; Martins et al. 1998; Jacobson 2000, 2001). Once emitted into the atmosphere, the amounts and mixing status of some compositions change dramatically because of chemical reaction, changes in gas-particle partitioning and particle size during the aging process (Cheng et al. 2009; Rogge et al. 1996; Martins et al. 1998; Venkataraman

G. Shen, *Emission Factors of Carbonaceous Particulate Matter and Polycyclic Aromatic Hydrocarbons from Residential Solid Fuel Combustions*, Springer Theses, DOI: 10.1007/978-3-642-39762-2_1, © Springer-Verlag Berlin Heidelberg 2014

and Friedlander 1994; Wang et al. 2010). As a result, the potential impacts of these compositions are usually complicated. In terms of the climate effect of PM, BC rather than EC is commonly used. The former, which is widely used for climate modelers, is often optically measured, implying that the aerosols have strong absorption ability, whereas the latter often operational definition based on the thermal stability of carbon and is often used in air quality and source apportionment studies (Bond et al. 2004; Chen et al. 2005, 2006; Huntzicker et al. 1982).

PM could be emitted from both natural and anthropogenic sources. In nature, dust storm is one main source of coarse PM, and the processes like volcanic eruptions are important natural source of PM. The anthropogenic sources of PM include the processes of residential combustion, vehicle and transportation etc. In addition to the primary sources, ambient fine PM can be formed from the secondary formation in atmosphere through chemical reactions of some precursors (Chan et al. 2009; Docherty et al. 2008; Odum et al. 1996; Turpin and Huntzicker 1995; Virtanen et al. 2010). Current researches on PM cover the studies on of its source and fate, temporal and spatial distributions, the impacts on human health and climate change, and the development on environmental criteria and standards. Different from that in most developed countries where there has been a much long history of researches on the fate and impacts of PM based on a systematic monitoring network, there are many gaps in the understanding of fate and impacts of PM in China. There are dramatically distinct temporal and spatial variations in the mass loads, chemical compositions and subsequent health and climate impacts of PM in China, which requiring much more basic studies.

1.1.2 Polycyclic Aromatic Hydrocarbons

Polycyclic Aromatic Hydrocarbons (PAHs) are a group of organic compounds with two or more aromatic rings. They are ubiquitous in various environment media like atmosphere, soil, water and food. PAHs are of worldwide interest because it can undergo long range transport and the exposure to PAHs would cause many adverse health outcomes like lung cancer and neural tube defects (Lang et al. 2007, 2008; Li et al. 2005; Liu et al. 2003; MacDonald et al. 2000; Prevedouros et al. 2004; Tao et al. 2006; Van Jaarsveld et al. 1997; Zhang et al. 2006, 2009). In most studies, 16 PAHs listed in the U.S. EPA priority compound list are often reported (U.S. EPA 1979). Recently, some parent PAHs not in the list are also of growing interest since they are found to affect the human health significantly and the development of advanced analytical instruments and methods provide the opportunity to detect these non-priority PAHs in environment (Jia et al. 2011; Layshock 2010; Wei et al. 2012). In addition the parent PAHs, more and more studies focus on the fates and impacts of PAH derivatives such as nitrated, oxygenated, and alkylated PAHs (Bolton et al. 2000; Durant et al. 1996; Ding et al. 2011; Lundstedt et al. 2007; Wang et al. 2011; Walgraeve et al. 2010). In comparison with parent PAHs, these derivatives usually have more detrimental toxic properties without the formation of intermediate products. Because of the relatively lower volatility, PAHs derivatives usually have more tendencies to be present in fine PM and can stay longer in air and penetrate deeper into the lung area. It is proposed that PAH derivatives may have more serious health impacts than the parent ones (Bolton et al. 2000; Durant et al. 1996; Li et al. 2003; Walgraeve et al. 2010).

The source of PAHs can be natural or anthropogenic. The natural source includes the processes like natural fire and volcanic eruptions, and the anthropogenic PAHs source includes the residential solid fuel combustion, industrial produces, vehicle emissions and so on (Mastral and Callean 2000; Tsibulsky et al. 2001). In general, anthropogenic source overwhelming the natural source contributes largely to the total PAHs emitted into the environment. Once emitted into the ambient air, PAHs re-distributed in air, soil, and water through the processes including dry/wet deposition, soil-air exchange, transport in air and water (Becker et al. 2006; Halsall et al. 2001; Kaneyasu and Takada 2004; Lang et al. 2007; MacDonald et al. 2000; Mackay 2001; Primbs et al. 2007; Prevedouros et al. 2004; Sehili and Lammel 2007; Tao et al. 2006; Van Jaarsveld et al. 1997). People are often exposed to PAHs in environment via dermal contact, inhalation and dietary ingestion, and the later two approaches are the main exposure pathways for most residents (Chuang et al. 1999; Duan et al. 2004; Jia and Batterman 2010; Jin et al. 2006; Li et al. 2009; Liu et al. 2007a, b; Xia et al. 2010; Yang 2010; Zhang et al. 2009).

There are very scarce studies on the fate and exposure risk of PAHs derivatives in comparison with that on parent PAHs. The derivatives can be produced from the primary incomplete combustion, and/or formed secondarily through the chemical reactions between parent PAHs and free radicals (Allen et al. 1997; Albinet et al. 2007, 2008a, b; Walgraeve et al. 2010; Wang 2010), but the relative contribution of primary sources and secondary formation of PAHs derivatives is not very clear at this stage and inconstant in the literature varying in sites and sampling periods (Andreou and Rapsomanikis 2009; Eiguren-Fernandez et al. 2008; Kojima et al. 2010; Walgraeve et al. 2010).

1.1.3 Residential Solid Fuel Combustion

To understand the fates and subsequent impacts of pollutants in environment, most studies often follow the "source to sink" roadmap of emission, transport and influence. The characterization of source and the development of emission inventory are important in understanding the fate and influence of pollutants in environment. Emission inventory is usually developed from the emission factor (EF) and the activity level. The former is defined as the mass of target emitted per mass or per energy unit of fuels. There are some inventories of PM and PAHs developed so far in global and national scales (Andreae and Rosenfeld 2008; Bond et al. 2004, 2007; Bond 2007; Zhang and Tao 2009), however, because of limited data about the energy consumption and the EFs, especially much little studies in developing countries, there are very large variations and uncertainties in current inventories (Bond et al. 2004, 2007; Lei et al. 2011; Streets et al. 2001; Xu et al. 2006; Zhang et al. 2007).

Residential solid fuel combustion, including the burning of crop residue, woody material and coal, is one major source of PM and PAHs. This is particularly true in developing countries since large amounts of solid fuels are burned inefficiently for daily cooking and heating because of a lack of cleaner fuels and easy access of traditional solid fuels in these areas. It was estimated that residential solid fuel combustion contributed about 34.6, 25.4, 15.4, 46.4, and 81.8 % of total national emissions of $PM_{2.5}$, PM_{10}, TSP, EC, and OC in China (Lei et al. 2011). For PAHs, the contribution of residential combustions of crop residue, woody material and coal made up to 34.6, 21.1 and 6.8 % of the national total (Xu et al. 2006).

Nowadays, more than half of world total populations rely on solid fuels for cooking and heating (WHO 2011a, b), and about three billion people in developing countries are using coal, wood and crop straw. It has is predicted that the population using solid fuels would increase in the years to come (IEA 2011). Inefficient solid fuel combustion in the residential sector produces large amounts of incomplete pollutants leading to serious indoor and outdoor air pollution and subsequently obvious impacts on human health and local/regional climate change. Globally, over 2 million people died due to exposure to smoke from the residential solid fuel combustion (WHO 2011a, b), and in China, residential solid fuel combustion is the largest environmental risk factor causing about 4.2 million deaths (Zhang and Smith 2007).

When it is realized that the wide use of solid fuels under inefficient residential combustion has significantly adverse impacts on air quality and human health, the deployment of renewable and cleaner fuels is one of research hotspots. It has been stated in the National Medium- and Long-Term Strategy Plan for Renewable Energy Development that the development of advanced bio-energy in future is to use cleaner and high quality energy resources such as electricity, biogas, liquid fuel and biomass pellets transformed from the raw biomass fuels under high efficient combustion. In China, the target goal of biomass pellet use is 50 million tons of annual consumption by 2020 (Chen et al. 2009). It could be expected that under the support of policy and financial support, the deployment of advanced renewable biomass fuels will benefit the environment obviously.

1.2 Main Objectives

- To establish methods for the emission characterization of residential sources, including the methods of sampling, laboratory analysis and calculation of emission factors.
- To obtain first-hand emission factors of carbonaceous particulate matter (CPM) and PAHs for crop residues, wood materials and coals burned in the residential stoves.

- To analyze the cause of variations in measured emission factors, and quantify the impacts of these influencing factors.
- To compare the emission characterization including size distribution of PM, gas-particle partitioning of organics, composition profiles, and isomer ratios among various solid fuels.
- To conduct some field measurements on emissions from residential solid fuel combustion in rural China.
- To analyze the emission characterization of CPM and PAHs from the burning of biomass pellets and compare with that for raw biomass fuel.

1.3 Thesis structure

Chapter 1. Introduction, including a brief introduction of the pollutants in the thesis study, the research objectives and structure of the thesis.

Chapter 2. Research background. In this section, a detailed review on the sources and impacts of CPM and PAHs is conducted. The importance of residential solid fuel combustion, particularly the measurements of emission factors is addressed

Chapter 3. Method. The simulation of residential solid fuel combustion, the chose of fuels, sampling and analytical methods of targets and the calculation of emission factors are described in detail in this section.

Chapter 4. Emission of CPM. Emission factors of CPM for solid fuels including crop residue, coal and wood are discussed in this section. The influence of fuel properties and combustion conditions is also discussed. In final, the emission characterization of CPM among different types of solid fuels is compared.

Chapter 5. Emission of PAHs. Similar to the structure in Chap. 4, we discussed the emission characterization of PAHs for coal, crop residue and wood in this section. Based on the measured results, the composition profiles, isomer ratios, gas-particle partitioning and size distribution of particle-bound PAHs are discussed and compared among different fuel types.

Chapter 6. Emission of PAH derivatives. Emission factors of PAHs derivatives including nitrated and oxygenated PAHs for residential solid fuels are measured for the first time in China. We analyzed the relationship between EFs of PAHs derivatives and EFs of corresponding PAHs, and the difference in the characterization such as gas-particle partitioning and size distribution.

Chapter 7. Field measurement. It is realized the emission in the simulated kitchen is different from that in field. In this section, we discussed the results from two field measurement campaigns in rural Jiangsu and rural Shanxi provinces.

Chapter 8. Biomass pellet. Emission characterization of two types of biomass pellets burned in a modern burner was measured, and compared with those for raw biomass fuels combusted in a traditional cooking stove. Based on the experimental data, the reduction in emission after the deployment of biomass pellets is analyzed.

Chapter 9. Conclusion and limitation. We summary the main conclusions of this thesis, and more important, the limitation of the present study is addressed.

References

Albinet, A., Leoz-Garziandia, E., Budzinski, H., & Villenave, E. (2007). Polycyclic aromatic hydrocarbons (PAHs), nitrated PAHs and oxygenated PAHs in ambient air of the Marseilles area (South of France): Concentrations and sources. *Science of Total Environment, 384*, 280–292.

Albinet, A., Leoz-Garziandia, E., Budzinski, H., Villenave, E., & Jaffrezo, J. (2008a). Nitrated and oxygenated derivatives of polycyclic aromatic hydrocarbons in the ambient air of two French alpine valleys Part 1: Concentrations, sources and gas/particle partitioning. *Atmospheric Environment, 42*, 43–54.

Albinet, A., Leoz-Garziandia, E., Budzinski, H., Villenave, E., & Jaffrezo, J. (2008b). Nitrated and oxygenated derivatives of polycyclic aromatic hydrocarbons in the ambient air of two French alpine valleys Part 2: Particle size distribution. *Atmospheric Environment, 42*, 55–64.

Allen, J. O., Dookeran, N. M., Taghizadeh, K., Lafleur, A. L., Smith, K. A., & Sarofim, A. F. (1997). Measurement of oxygenated polycyclic aromatic hydrocarbons associated with a size-segregated urban aerosol. *Environmental Science and Technology, 31*, 2064–2070.

Andreae, M. O., & Rosenfeld, D. (2008). Aerosol–cloud–precipitation interactions. Part 1. The nature and sources of cloud-active aerosols. *Earth-Science Review, 89*, 13–41.

Andreou, G., & Rapsomanikis, S. (2009). Polycyclic aromatic hydrocarbons and their oxygenated derivatives in the urban atmosphere of Athens. *Journal of Hazardous Material, 172*, 363–373.

Becker, S., Halsall, C. J., & Tych, W. (2006). Resolving the long-term trends of polycyclic aromatic hydrocarbons in the Canadian Arctic atmosphere. *Environmental Science and Technology, 40*, 3217–3222.

Bolton, J. L., Trush, M. A., Penning, T. M., Dryhurst, G., & Monks, T. J. (2000). Role of quinones in toxicology. *Chemical Research Toxicology, 13*, 135–160.

Bond, T. C., Streets, D. G., Yarber, K. F., Nelson, S. M., Woo, J., & Klimont, Z. (2004). A technology-based global inventory of black and organic carbon emissions from combustion. *Journal of Geophysical Research, 109*, D14203. doi:10.1029/2003JD003697

Bond, T. C. (2007). Testimony for the hearing on black carbon and climate change. House committee on oversight and government reform. United states house of representative. (October 18).

Bond, T. C., Bhardwaj, E., Dong, R., Jogani, R., Jung, S., Roden, C., et al. (2007). Historical emissions of black and organic carbon aerosol from energy-related combustion, 1850-2000. *Global Biogeochemical Cycle, 21,* GB2018.

Chan, A. W. H., Kautzman, K. E., Chhabra, P. S., Surratt, J. D., Chan, M. N., Crounse, J. D., et al. (2009). Secondary organic aerosol formation from photooxidation of naphthalene and alkylnaphthalenes: Implications for oxidation of intermediate volatility organic compounds (IVOCs). *Atmospheric Chemistry and Physics, 9*, 3049–3060.

Chen, Y. J., Sheng, G. Y., Bi, X. H., Feng, Y. L., Mai, B. X., & Fu, J. M. (2005). Emission factors for carbonaceous particles and polycyclic aromatic hydrocarbons from residential coal combustion in China. *Environmental Science and Technology, 39*, 1861–1867.

Chen, Y. J., Zhi, G., Feng, Y., Fu, J., Feng, J., Sheng, G., et al. (2006). Measurement of emission factors for primary carbonaceous particles from residential raw-coal combustion in China. *Geophysical Research Letter, 33*, L20815. doi:10.1029/2006GL026966

Chen, L., Xing, L., & Han, L. (2009). Renewable energy from agro-residues in China: Solid biofuels and biomass briquetting technology. *Renewable and Sustainable Energy Review, 13*, 2689–2695.

Cheng, Y. F., Berghof, M., Garland, R. M., Wiedensohler, A., Wehner, B., Muller, T., et al. (2009). Influence of soot mixing state on aerosol light absorption and single scattering albedo during air mass aging at a polluted regional site in Northeastern China. *Journal of Geophysical Research, 114*, D00G10, doi:10.1029/2008JD010883

Chow, J. C., Watson, J. G., Lowenthal, D. H., Antony Chen, L.-W., & Motallebi, N. (2011). PM2.5 source profiles for black and organic carbon emission inventories. *Atmospheric Environment, 45*, 5407–5414.

Chuang, J. C., Callahan, P. J., Lyu, C. W., & Wilson, N. K. (1999). Polycyclic aromatic hydrocarbon exposures of children in low-income families. *Journal of Exposure Analysis and Environmental Epidemiology, 9*, 85–98.

Ding, J., Zhong, J., Yang, Y., Li, B., Shen, G., Su, Y., et al. (2011). Occurrence and exposure to polycyclic aromatic hydrocarbons and their derivatives in a rural Chinese home through biomass fuelled cooking. *Environmental Pollution*. doi:10.1016/j.envpol.2011.10.008

Docherty, K. S., Stone, E. A., Ulbrich, I. M., DeCarlo, P. F., Snyder, D. C., Schauer, J. J., et al. (2008). Apportionment of primary and secondary organic aerosols in southern California during the 2005 study of organic aerosols in riverside (SAQR-1). *Environmental Science and Technology, 42*, 7655–7662

Dockery, D. W., Pope III, C. A., Xu, P., Spengler, J. D., Ware, J. H., & Fay, M. E. (1993). An association between air pollution and mortality in six U.S. cities. *The New England Journal of Medicine, 329*, 1573–1579.

Duan, X., Wei, F., Yang, H., Zhang, J., & Xu, J. (2004). Daily total exposure amount to polycyclic aromatic hydrocarbons of crowd of different work environment. *China Environmental Science, 24*(5), 515–518. (In Chinese).

Durant, J. L., Busby, W. F., Lafleur, A. L., Penman, B. W., & Crespi, C. L. (1996). Human cell mutagenicity of oxygenated, nitrated and unsubstituted polycyclic aromatic hydrocarbons associated with urban aerosols. *Mutation Research, 371*, 123–157.

Eiguren-Fernandez, A., Miguel, A. H., Lu, R., Purvis, K., Grant, B., Mayo, P., et al. (2008). Atmospheric formation of 9, 10-phenanthraquinone in the Los Angeles air basin. *Atmospheric Environment, 42*, 2312–2319

Englert, N. (2004). Fine particles and human health-a review of epidemiological studies. *Toxicology Letters, 149*, 235–242.

Ezzati, M., Lopez, A. D., Rodgers, A., Hoorn, S. V., Murray, C. J. L., & The comparative risk assessment collaborating group. (2002). Selected major risk factors and global and regional burden of disease. *Lancet, 360*, 1347–1360.

Halsall, C. J., Sweetman, A. J., & Barrie, L. A. (2001). Modeling the behavior of PAHs during atmospheric transport from the UK to the Arctic. *Atmospheric Environment, 35*, 255–267.

Huntzicker, J. J., Johnson, R. L., Shah, J. J., & Cary R. A. (1982). Analysis of organic and elemental carbon in ambient aerosols by a thermal-optical method. In G. T. Wolff & R. L. Klimisch (Eds.), *Particulate carbon: Atmospheric life cycle* (pp. 79–88). New York: Plenum.

Intergovernmental Panel on Climate Change (IPCC). (2007). In: R. K. Pachauri & A. Reisinger (Eds.), *Fourth assessment report: Climate change 2007 (AR4)*. Geneva: IPCC.

International Energy Agency (IEA). (2011). Energy for all, financing access for the poor, special early excerpt of the world energy outlook 2011. http://www.iea.org/papers/2011/weo2011_energy_for_all.pdf

Jacobson, M. Z. (2000). A physically-based treatment of elemental carbon optics: Implications for global direct forcing of aerosols. *Geophysical Research Letters, 27*, 217–220.

Jacobson, M. Z. (2001). Strong radiative heating due to the mixing state of black carbon in atmospheric aerosols. *Nature, 409*, 695–697.

Jia, C., & Batterman, S. (2010). A critical review of naphthalene sources and exposures relavant to indoor and outdoor air. *International Journal of Environmental Research and Public Health, 7*, 2903–2939.

Jia, Y., Stone, D., Wang, W., Schrlau, J., Tao, S., & Simonich, S. L. M. (2011). Estimated reduction in cancer risk due to PAH exposure if source control measures during the 2008 Beijing Olympics were sustained. *Environmental Health Perspective, 119*, 815–820.

Jin, Y., Ma, X., Chen, X., Cheng, Y., Baris, E., & Ezzati, M. (2006). Exposure to indoor air pollution from household energy use in rural China: The interactions of technology, behavior, and knowledge in health risk management. *Social Science and Medicine, 62*, 3161–3176.

Kaneyasu, N., & Takada, H. (2004). Seasonal variations of sulfate, carbonaceous species (black carbon and polycyclic aromatic hydrocarbons), and trace elements in fine atmospheric aerosols collected at subtropical islands in the East China Sea. *Journal of Geophysical Research-Atmospheres, 109*(D6), D06211.

Kojima, Y., Inazu, K., Hisamatsu, Y., Okochi, H., Baba, T., & Nagoya, T. (2010). Influence of secondary formation on atmospheric occurrences of oxygenated polycyclic aromatic hydrocarbons in airborne particles. *Atmospheric Environment, 44*, 2873–2880.

Kumar, P., Robins, A., Vardoulakis, S., & Britter, R. (2010). A review of the characteristics of nanoparticles in the urban atmosphere and the prospects for developing regulatory controls. *Atmospheric Environment, 44*, 5035–5052.

Lang, C., Tao, S., Zhang, G., Fu, J. M., & Simonich, S. (2007). Outflow of polycyclic aromatic hydrocarbons from Guangdong, Southern China. *Environmental Science and Technology, 41*, 8370–8375.

Lang, C., Tao, S., Liu, W. X., Zhang, Y. X., & Simonich, S. (2008). Atmospheric transport and outflow of polycyclic aromatic hydrocarbons from China. *Environmental Science and Technology, 42*, 5196–5201.

Layshock, J. (2010). Beyond the 16 EPA priority pollutant PAHs: Environmental characterizations of oxygenated PAHs and dibenzopyrene isomers (Ph.D Thesis, Oregon State University).

Lei, Y., Zhang, Q., He, K. B., & Streets, D. G. (2011). Primary anthropogenic aerosol emission trends for China, 1990–2005. *Atmospheric Chemistry and Physics, 11*, 931–954.

Li, N., Hao, M. Q., Phalen, R. F., Hinds, W. C., & Nel, A. E. (2003). Particulate air pollutants and asthma-a paradigm for the role of oxidative stress in PM-induced adverse health effects. *Clinical Immunology, 109*, 250–265.

Li, X., Li, B., Tao, S., Guo, M., Cao, J., & Wang, X. (2005). Population exposure to PAHs in Tianjin area. *Acta Scientiae circumstantiae, 25*(7), 989–993. (In Chinese).

Li, X., Zhao, T., Yu, Y., Zhang, C., Li, P., & Li, S. (2009). Population exposure to PAHs and the health risk assessment in Beijing area. *Journal of Agro-Environmental Science, 28*(8), 1758–1765. (In Chinese).

Liu, H. Y., Jacob, D. J., Bey, I., Yantosca, R. M., Duncan, B. N., & Sachse, G. W. (2003). Transport pathways for Asian pollution outflow over the Pacific: interannual and seasonal variations. *Journal of Geophysical Research Atmospheres, 108*, D20.

Liu, Y. N., Tao, S., Dou, H., Zhang, W. T., Zhang, X. L., & Dawson, R. (2007a). Exposure of traffic police to Polycyclic aromatic hydrocarbons in Beijing, China. *Chemosphere, 66*, 1922–1928.

Liu, Y. N., Tao, S., Yang, Y. F., Dou, H., Yang, Y., & Coveney, R. M. (2007b). Inhalation exposure of traffic police officers to polycyclic aromatic hydrocarbons (PAHs) during the winter in Beijing, China. *Science of the Total Environment, 383*, 98–105.

Lundstedt, S., White, P., Lemieux, C., Lynes, K., Lambert, I., Öberg, L., et al. (2007). Sources, fate, and toxic hazards of oxygenated polycyclic aromatic hydrocarbons (PAHs) at PAH-contaminated sites. *Ambio, 36*, 475–485.

MacDonald, R. W., Barrie, L. A., Bidleman, T. F., Diamond, M. L., Gregor, D. J., Semkin, R. G., et al. (2000). Contaminants in the Canadian Arctic: 5 years of progress in understanding sources, occurrence and pathways. *Science and Total Environment*, *254*, 93–234.

Mackay, D. (2001). *Environmental multimedia models: The fugacity approach.* Boca Raton: Lewis Publishers.

Martins, J. V., Artaxo, P., Liousse, C., Reid, J. S., Hobbs, P. V., & Kaufman, Y. J. (1998). Effects of black carbon content, particle size, and mixing on light absorption by aerosols from biomass burning in Brazil. *Journal of Geophysical Research, 103*, 32041–32050.

Mastral, A. M., & Calleän, M. S. (2000). A review on polycyclic aromatic hydrocarbon (PAH) emissions from energy generation. *Environmental Science and Technology, 34*, 3051–3057.

McConnell, J. R., Edwards, R., Lok, G. L., Flanner, M. G., Zender, C. S., Saltzman, E. S., et al. (2007). 20th-century industrial black carbon emissions altered arctic climate forcing. *Science, 317*, 1381–1385.

Menon, S., Hansen, J., Nazarenko, L., & Luo, Y. (2002). Climate effects of black carbon aerosol in China and India. *Science, 297*, 2250–2254.

Odum, J. R., Hoffmann, T., Bowman, F., Collins, D., Flagan, R. C., & Seinfeld, J. H. (1996). Gas/particle partitioning and secondary organic aerosol yields. *Environmental Science and Technology, 30*, 2580–2585.

Pope, C. A, I. I. I., Ezzati, M., & Dockery, D. W. (2009). Fine-particulate air pollution and life expectancy in the United States. *The New England Journal of Medicine, 360*, 376–386.

Pope III, C. A., Burnett, R. T., Thun, M. J., Calle, E. E., Krewski, D., Ito, K., et al. (2002). Lung cancer, cardiopulmonary mortality, and long-term exposure to fine particulate air pollution. *The Journal of the American Medical Association, 287*, 1132–1141.

Prevedouros, K., Brorström-Lundén, E., Halsall, C. J., Jones, K. C., Lee, R. G. M., & Sweetman, A. J. (2004). Seasonal and long-term trends in atmospheric PAH concentrations: Evidence and implications. *Environmental Pollution, 128*, 17–27.

Primbs, T., Simonich, S., Schmedding, D., Wilson, G., Jaffe, D., Takami, A., et al. (2007). Atmospheric outflow of anthropogenic semivolatile organic compounds from East Asia in spring 2004. *Environmental Science and Technology, 41*, 3551–3558.

Rogge, W. F., Hildemann, L. M., Mazurek, M., Cass, G. R., & Simoneit, B. R. T. (1996). Mathematical modeling of atmospheric fine particle-associated primary organic compound concentrations. *Journal of Geophysical Research, 101*, 19379–19394.

Russell, A. G., & Brunekreef, B. (2009). A focus on particulate matter and health. *Environmental Science and Technology, 43*, 4620–4625.

Samet, J. M., Dominici, F., Curriero, F. C., Coursac, I., & Zeger, S. L. (2000). Fine particulate air pollution and mortality in 20 U.S. cities. *The New England Journal of Medicine, 343*, 1742–1749.

Sehili, A. M., & Lammel, G. (2007). Global fate and distribution of polycyclic aromatic hydrocarbons emitted from Europe and Russia. *Atmospheric Environment, 41*, 8301–8315.

Shannon, M. W. et al., Committee on Environmental Health 2003–2004. (2004). Ambient air pollution: Health hazards to children. *Pediatrics, 114*, 1699–1707.

Streets, D. G., Gupta, S., Waldhof, S. T., Wang, M. Q., Bond, T. C., & Yi, B. (2001). Black carbon emissions in China. *Atmospheric Environment, 35*, 4281–4296.

Tao, S., Li, X. R., Yang, Y., Coveney, R. M., Lu, X. X., Chen, H. T., et al. (2006). Dispersion modeling of polycyclic aromatic hydrocarbons from combustion of biomass and fossil fuels and production of coke in Tianjin, China. *Environmental Science and Technology, 40*, 4586–4591.

Tsibulsky, V., Sokolovsky, V., & Dutchak, S. (2001). MSC-E contribution to the HM and POP emission inventories, Technical Note 7/2001. http://www.msceast.org/publications.html

Turpin, B. J., & Huntzicker, J. J. (1995). Idenfication of secondary organic aerosol episodes and quantification of primary and secondary organic aerosol concentrations during SCAQS. *Atmospheric Environment, 29*, 3527–3544.

United States, Environmental Protection Agency (EPA). (1979). *Water-related environmental fate of 129 priority pollutants*. EPA Publication EPA-440/4-79-029a (Volume 1) and EPA-440/4-79-029b (Volume II). Washington, Office of Water Planning and Standards. Available from NTIS, Springfield.

Van Jaarsveld, J. A., Van Pul, W. A. J., & De Leeuw, F. A. A. M. (1997). Modelling transport and deposition of persistent organic pollutants in the European region. *Atmospheric Environment, 31*, 1011–1024.

Venkataraman, C., & Friedlander, S. (1994). Size distributions of polycyclic aromatic hydrocarbons and elemental carbon. 2. ambient measurement and effects of atmospheric processes. *Environmental Science and Technology, 28*, 563–572.
Virtanen, A., Joutsensaari, J., Koop, T., Kannosto, J., Yli-Pirila, P., Leskinen, J., et al. (2010). An amorphous solid state of biogenic secondary organic aerosol particles. *Nature, 467*, 824–827.
Walgraeve, C., Demesstere, K., Dewulf, J., Zimmermann, R., & van Langenhove, H. (2010). Oxygenated polycyclic aromatic hydrocarbons in atmospheric particulate matter: Molecular characterization and occurrence. *Atmospheric Environment, 44*, 1831–1846.
Wang, W. (2010). Distribution and soil-air exchange of polycyclic aromatic hydrocarbons and derivatives in the Beijing-Tianjin region (Ph.D. Thesis, Peking University) (In Chinese).
Wang, R., Tao, S., Wang, B., Yang, Y., Lang, C., Zhang, Y., et al. (2010). Source and pathways of polycyclic aromatic hydrocarbons transported to Alert, the Canadian High Arctic. *Environmental Science and Technology, 44*, 1017–1022.
Wang, W., Jariyasopit, N., Schrlau, J., Jia, Y., Tao, S., Yu, T., et al. (2011). Concentration and photochemistry of PAHs, NPAHs, and OPAHs and toxicity of PM2.5 during the Beijing Olympic Games. *Environmental Science and Technology, 45*, 6887–6895.
Wei, S., Liu, M., Huang, B., Bi, X., Sheng, G., & Fu, J. (2012). Polycyclic aromatic hydrocarbons with molecular weight 302 in PM2.5 at two industrial sites in South China. *Journal of Environmental Monitoring, 13*, 2568–2574.
World Health Organization (WHO). (2002a). The health effects of indoor air pollution exposure in developing countries; WHO/SDE/OEH/02.05.
World Health Organization (WHO). (2002b). *The world health report: Reducing risks, promoting healthy life*. Geneva: WHO.
World Health Organization (WHO). (2011a). Indoor air pollution and health, Fact sheet N292, 2011. Retrieved Oct 2011 from http://www.who.int/mediacentre/factsheets/fs292/en/index.html
World Health Organization (WHO). (2011b). Database: Outdoor air pollution in cities. Retrieved April 2012 from http://www.who.int/phe/health_topics/outdoorair/databases/en/index.html
Xia, Z., Duan, X., Qiu, W., Liu, D., Wang, B., Tao, S., et al. (2010). Health risk assessment on dietary exposure to polycyclic aromatic hydrocarbons (PAHs) in Taiyuan, China. *Science of the Total Environment, 408*, 5331–5337.
Xu, S. S., Liu, W. X., & Tao, S. (2006). Emission of polycyclic aromatic hydrocarbons in China. *Environmental Science and Technology, 40*, 702–708.
Yang, Y. (2010). Indoor pollution of particulate matter and polycyclic aromatic hydrocarbons in rural Northern China (Master Thesis, Peking University) (In Chinese).
Zhang, W., Dou, H., Liu, Y., Zhang, X., & Tao, S. (2006). Exposure to traffic police in Haidian District to PAHs. *Environmental Science, 27*, 224–227. (In Chinese).
Zhang, J., & Smith, K. R. (2007). Household air pollution from coal and biomass fuels in China: Measurements, health impacts, and interventions. *Environmental Health Perspective, 115*, 848–855.
Zhang, Y. X., Tao, S., Cao, J., & Coveney, R. M. (2007). Emission of polycyclic aromatic hydrocarbons in China by county. *Environmental Science and Technology, 41*, 683–687.
Zhang, Y. X., & Tao, S. (2009). Global atmospheric emission inventory of polycyclic aromatic hydrocarbons (PAHs) for 2004. *Atmospheric Environment, 43*, 812–819.
Zhang, Y. X., Tao, S., Shen, H. Z., & Ma, J. M. (2009). Inhalation exposure to ambient polycyclic aromatic hydrocarbons and lung cancer risk of Chinese population. *Proceedings of the National Academy of Sciences, 106*, 21063–21067.

Chapter 2
Research Background

2.1 Environmental Impact

2.1.1 Carbonaceous Particulate Matter

It has been widely realized that PM affects the local/regional air quality, human health and climate change significantly and is of worldwide interest. The exposure to ambient PM caused about 0.8 million death per year (WHO 2002a, b). In addition, the light extinction of PM could have obvious impacts on the ecosystem such as the yield of crop (Chameides et al. 2009; Russell et al. 1999). There are many ways to characterize the physic-chemical properties of PM, such as surface area, size, light extinction efficiency, chemical compositions of inorganic and organics, and toxic effects. One widely reported physical property is the PM aerodynamic size. Particles with different diameters have different transportation abilities and distinct health impacts (Ansmann and Müller 2005; Dockery et al. 1993; Englert 2004; Pope et al. 2009; Veselovskii et al. 2004). Fine PM may undergo longer air transport in environment and penetrate deeper into the end of bronchus and alveolus area, causing serious impacts on human health. Because of relatively large surface area of fine PM, there are mamy toxic organics bound in the fine PM, and hence increased the health risks of fine PM. In our daily lives, several commonly used terms include Total Suspended Particle (TSP, PM with diameter <100 μm), PM_{10} (PM with diameter <10 μm), $PM_{2.5}$ (PM with diameter <2.5 μm), $PM_{1.0}$ (PM with diameter <1.0 μm), and ultra fine $PM_{0.1}$ (PM with diameter <0.1 μm). Though a large number of researches had highlighted the high toxic potentials of fine PM, some studies also arose the attention on the toxic of coarse PM (Russell and Brunekreef 2009; Pope et al. 2009). The size distribution of PM can be separated into three modes, including Aitken nuclei mode ($Dp < 0.05$ μm), accumulated mode (0.05–2 μm) and coarse particle mode ($D_P > 2$ μm) (Tang et al. 2006; Whitby 2007). Aitken nuclei mode can be found in some new formed particles. PM in accumulated mode can be either coagulated from the Aitken nuclei mode, or directly produced from the condensation of vapors during the combustion process. PM in the emissions from the solid fuel

G. Shen, *Emission Factors of Carbonaceous Particulate Matter and Polycyclic Aromatic Hydrocarbons from Residential Solid Fuel Combustions*, Springer Theses,
DOI: 10.1007/978-3-642-39762-2_2,

combustion is mainly in accumulated mode (Bond et al. 2004; Chen et al. 2005; Rogge et al. 1996; Reid et al. 2005). In general, secondarily formed aerosols are mainly in Aitken nuclei and accumulated modes. The size distribution of PM in the air is affected by a number of factors including ambient temperature and relatively humidity (RH) (McMurry and Stolzenburg 1989; Shi et al. 2007; Tan et al. 2009; Vasconcelos et al. 1994). In most cases, the PM size distribution in air follows the log transformed normal distribution (Tang et al. 2006; Hinds 1999).

In most developing countries, the air quality has been deteriorating and there are many programs studying the cause and potential impacts of severe air pollution (Balakrishnan et al. 2002, 2004; Cao et al. 2003a, b; Chan and Yao 2008; Kim Oanh and Zhang 2004; Kim Oanh et al. 2006; Pandey et al. 2006, 2008), however, current knowledge about the air pollution and its environmental impact is still lack, particularly when compared with the long history studies in developed countries (Ansmann and Müller 2005; Cercasov and Wulfmeyer 2008; Chow et al. 2008, 2011; Dockery et al. 1993; Englert 2004; Keeler 2004; Laden et al. 2000; Pope et al. 2009; Veselovskii et al. 2004). Because of the distinct sources and meteorological conditions, the emissions, ambient pollution levels as well as the potential impacts of PM have obviously temporal and spatial variations. In China, the fast economical development and large consumption of energy lead to severe air pollution during the last several decades (Cao et al. 2003a, b; Chan and Yao 2008; Davis and Guo 2000; Deng et al. 2008; Feng et al. 2006; Liu et al. 2008a, b; Lin et al. 2010; Remer et al. 2008; Sun et al. 2004; Wu et al. 2006; Zhao et al. 2009). The migration of population from the rural area to urban/sub-urban areas, particularly after the 1980s, leads to high population densities and serious environmental issues in the large cities. In China, a total number of city is about 660, among which about 170 ones are megacities with population over 1 million (Chan and Yao 2008). The urban population increased from 19.6 to 40.5 % during 1980–2005, and there would be about 350 million people living in the urban by 2025 (Zhu et al. 2011). Most of available studies are mainly in large cities, like Beijing, Shanghai and cities in the Pearl River Delta (Chan and Yao 2008; Hao et al. 2005; Zhang et al. 2009a). For example, it was reported that the main air pollutant is PM_{10}, and there were about 40 % days annually with 24 h average PM_{10} concentrations exceed the national standard even though the annual average concentration showed a decreasing trend from 1999 to 2005. The emission of PM in Beijing decreased from 2.70 million tones in 1994 to 58 thousand tones in 2005, but the industrial dust increased from 59 thousand tones to 100 thousand tones. Besides PM_{10}, $PM_{2.5}$ is of wide and increasing interest, especially in the last several years (van Donkelaar et al. 2010; WHO 2011; Yuan et al. 2012). According to the plan from the Ministry of Environmental Protection, some cities and pilot regions would start to monitor in daily air pollutant monitoring, and by 2015, all county cities should monitor and report $PM_{2.5}$ (Yuan et al. 2012). Since $PM_{2.5}$ is not only emitted from the primary sources but also formed during the secondary transformation, increased emissions of some precursors, for example NOx and volatile organic compounds from vehicle emissions, would deteriorate the air quality and increase the technical and political difficulty in air pollution control. It is expected that with the efforts of effective source control strategies and the implication of pollution control techniques, the emission and pollution level of PM would decrease in future. Even so, PM would still be the main air pollutant in China, and the severe air pollution levels should be arisen high attention by both scientists and policy makers (Chan and Yao 2008; Lei et al. 2011; Lin et al. 2010). One important and large source of PM in China is the consumption of traditional solid fuels including coals, wood and crop residues. In rural area, a large amount of solid fuels is used by residents for daily cooking and heating. The inefficient combustions of these fuels often cause serious air pollution in the region. It was reported that the PM_{10} concentration in the kitchen during the cooking time might be 6 times of that during the non-cooking period, and the personal exposure level was three times for the cooker to that for the non-cooker (Jiang and Bell 2008). In addition to the use in residential household, biomass is also widely burned in open field, especially after the harvest. It was estimated that about 110–158 Tg biomass fuels were burned annually in open field (Streets et al. 2003; Cao et al. 2006).

Carbon fraction, mainly including organic carbon (OC) and elemental carbon (EC), is an important chemical composition of PM. Sometimes, the term "black carbon (BC)" is used. In general, the climate modelers prefer to use BC which often refers to the carbon or substance having light absorbing ability, whereas most people in the field of aerosol chemistry use the term EC. The analytical methods of EC and BC are usually different. For the former, thermal or thermal-optical methods are used in which the carbon was oxidized to CO_2 and quantified. The most widely used instruments to measure EC are EC/OC analyzer from Sunset, or DRI. BC is measured using the optical method at a specific spectrum, like 840 nm. Because the EC and BC data often correlated with one anther and in many cases are comparable, they are sometimes considered as the same mass when the other measurement was unavailable (Bond et al. 2004; Chen et al. 2005; Chow et al. 2011). The sampling and analytical methods have significant influences on the obtained OC and EC results. However, current work on how to eliminate the influence of sampling and analytical methods on the analyzed carbon results is limited. It is expected that there would be more studies on the issues like how to estimate and control the absorption of gaseous organics, the vaporization of volatile organics in PM, and how to correct the difference or bias in different analytical temperature protocols and charring correction methods (Cheng et al. 2009a, b; Duan et al. 2007; Mader et al. 2001).

OC can be emitted from the primary combustion, and/or formed from the atmospheric reactions secondarily. Different from OC, EC is produced from the primary combustion sources such as the industrial processes, vehicle emission and coal combustions. Because of the enhanced secondary formation of OC, the ratio of OC to EC is usually higher in summer in comparison with that in winter (Chen et al. 2002; Yu et al. 2004). The ratio of OC and EC can be used to estimate the relatively contribution of secondary formation of OC in atmosphere (Tang et al. 2006; Tan et al. 2009; Turpin and Huntzicker 1995; Na et al. 2004). OC and EC, particularly the later, usually prefer to be present in fine PM (Chen et al. 1997; Tang et al. 2006; Tan et al. 2009; Venkataraman and Friedlander 1994;

Venkataraman et al. 1994), and absorb (for OC) or adsorb (for EC) organics causing the toxic of PM (Crutzen and Andreae 1990; Seinfeld and Pandis 2006; Allen et al. 1996; Venkataraman et al. 2002). OC and EC also influence the local/regional climate change through the light scattering and absorbing, respectively, leading to negative and positive radiative forcing, respectively (Yu et al. 2004; Chow et al. 2011). The net climate effect of carbon fraction in PM is often controlled by the ratio of OC to EC (Zhi et al. 2009). The positive radiative forcing of EC highly depends on the mixing status (inter- or external mixture). It was reported that the forcing of EC may range from 0.27 to 0.78 W/m^2 (Jacobson 2001). During the last three decades, the average temperature in the Arctic increased by about 1.5 °C, of which about 1.1 °C might be contributed by the warm effect of aerosol because of the light absorption of BC (IPCC 2001; Shindell and Faluvegi 2009). The emissions from open biomass burning may increase the ambient concentration in the Arctic by about 2 times (Koch and Hansen 2005; Stohl et al. 2006; Warneke et al. 2009, 2010). Since the PM in background air in the Arctic mainly consist sulfate and much small percents of nitrate, organics and BC (Law and Stohl 2007; Quinn et al. 2006), while PM from the biomass burning generally has relatively higher fractions of OC and EC but less sulfate (Warneke et al. 2009), the transport of PM from biomass burning to the Arctic may have much significantly impacts on the ambient concentrations of OC and EC in the Arctic (Warneke et al. 2010). The mass percentages of total carbon (the sum of OC and EC) vary in different sources (Chow et al. 2011; Duan et al. 2007; Seinfeld and Pandis 2006). For example, the mass percents of OC and EC in PM from residential wood combustion were reported to be around 50 and 12 %, and in the emission from residential coal combustion at 70 and 26 %, but in the industrial coal combustion, they might vary from 2–34 to 1–8 % (Chow 1995; Chow et al. 2011; Duan et al. 2007).

In addition to the impacts on air quality and climate change, OC and EC have significant impacts on human health since they often absorb and/or adsorb many toxics and can penetrate deeper into the lung area and bronchus (Bond et al. 2004; EPA 2012; McCracken et al. 2010; Zhi et al. 2009). The monitoring of particle carbon fraction has been included into some research programs and observation campaign from the 1980s (Qin et al. 2001). These programs provided important data on the concentration levels and dynamic change patterns of BC. In China, the air pollution of BC is of growing interest, especially after the large increase of vehicles and consumption of coals both of which are large emitters of BC. During the 1980s, there are some, though very limited observation of BC, and after the 1990s, more and more field BC observation studies come out. In general, we have very limited studies on the source, ambient level and impact of BC in China, and there are often large biases and uncertainties in the inventories and environmental impacts of BC (Gu et al. 2006; Zhang et al. 2009b; Zhi et al. 2009). For example, it was once reported that the flood in the south China and drought in the north China may be strongly related to the emissions of BC from residential biomass and coal combustion (Menon et al. 2002), but in the another study, the results were found to be hard to be predicted when including the emissions of BC from residential solid fuel combustion into the atmospheric model (Gu et al. 2006; Zhang and Tao 2009).

2.1.2 Polycyclic Aromatic Hydrocarbons and Derivatives

Polycyclic aromatic hydrocarbons (PAHs) are a group of hydrocarbon organic pollutants with two or more benzene rings (Kanaly and Harayama 2000; Mumtaz et al. 1996; Neff 1979). They have been ubiquitously found in air, soil, water and organism. Because of the carcinogenicity and mutagenic toxicities, PAHs have been widely studied in many areas around the world (Fernandes et al. 1997; Gaspari et al. 2003; IARC 1987; Perera et al. 2002; Pufulete et al. 2004; Xue and Warshawsky 2004; Zhang et al. 2009c). There are many PAH isomers in environment, and they are usually a mixture in environment. The physical and chemical properties of PAH isomers differ obviously. PAHs with the same rings and molecular weight often have the comparable physic-chemical properties. In general, with the increase of aromatic rings and molecular weight, the solubility, vapor pressure and volatility decrease, but the hydrophobic effect increases (Smith and Harrison 1996). PAHs are included in the priority list in many countries and organizations. In the middle 1970s, 129 compounds are listed by U.S. EPA as priority control pollutants, among which there are 16 PAHs. PAHs are also the Regionally-based Assessment of Persistent Toxic Substances in the United Nations Environment Program, 2002. Because of the ability of volatility and long range transport (Becker et al. 2006; Halsall et al. 2001; Kaneyasu and Takada 2004; Lang et al. 2007; MacDonald et al. 2000; Mackay 2001; Primbs et al. 2007; Prevedouros et al. 2004; Sehili and Lammel 2007; Tao et al. 2006; Van Jaarsveld et al. 1997), PAHs are also included into the Convention on Long-range Tansboundary Air Pollution by the United Nations Economic Commission for Europe, 1998. In China, some PAHs including naphthalene, fluoranthene, benzo[b[fluoranthene, benzo[k]fluoranthene, benzo[a]pyrene, benzo[ghi]pyrene, and benzo[-ghi]perylene, are included into the priority control pollutant list (Zhou et al. 1990). Benzo[a]pyrene, as an example of toxic PAHs, is included in the environmental air quality standard (MEP 1996).

Pollution level and characterization of PAHs pollution in environment have obviously temporal and spatial variations due to distinct emission sources and the influence of meteorological conditions (Garban et al. 2002; Hafner et al. 2005; Liu et al. 2007; Liu 2008; Wang 2010; Zhang 2010). Once emitted into the air, PAHs often redistribute between the gaseous and particulate phases (Bidleman 1988; Goss and Schwarzenbach 1998; Lohmann and Lammel 2004; Pankow 1987), and undergo the complicated chemical reactions and physical processes like dry and wet deposition (Behymer and Hites 1988; Reisen and Arey 2002, 2005; Garban et al. 2002; Wang et al. 2010, 2011a, b). Due to the long range transport ability, PAHs can redistribute in the areas like some background sites and the Arctic area (Daly and Wania 2005; Ding et al. 2007; Garban et al. 2002; MacDonald et al.

2000; Van Jaarsveld et al. 1997; Wania and Mackay 1993; Wang et al. 2010). PAHs in air can enter into the soil and water through the deposition and washout process, and PAHs in soil could also be reemitted into the air or water (Cousins et al. 1999; Meijer et al. 2003; Wang et al. 2011a; Wild and Jones 1995). As a result, the PAH content in soil varies in site and time, which is heavily associated with the emission densities, physic-chemical properties of PAHs and structure and properties of soil (Heywood et al. 2006; Wilcke and Amelung 2000).

PAHs exposure would cause serious adverse human health outcomes through direct (inhalation or dietary exposure) or indirect exposure (the absorption of animals and plants) pathways (An et al. 2005; Duan and Wei 2002; Duan et al. 2008; Li et al. 2005, 2009a, b; Menzie et al. 1992; McClean et al. 2004, 2007; Tao et al. 2006; Wu et al. 2010; Xia et al. 2010). The relative contributions of inhalation and dietary exposures may vary in site and time. In the regions with serious ambient PAHs pollution, the inhalation exposure would contribute largely to the total exposure (Duan et al. 2008; Li et al. 2009a, b; Tao et al. 2006; Xia et al. 2010). For example, it was reported that for the non-smokers, the inhalation exposure made up to 90 % of total PAHs exposure (Suzuki and Yoshinaga 2007; Vyskocil et al. 2000, 2004). Meanwhile, in some regions where the air levels of PAHs are very low, personal exposed PAHs may be mainly from the ingestion of high molecular weight PAHs in food (Van Rooij et al. 1994; Vyskocil et al. 1997). PAHs would form carcinogenic PAHs-DNA adducts in vivo after exposure (Boysen and Hecht 2003; Cheng 1998; Jacob and Seidel 2002; Denissenko et al. 1996; Ramesh et al. 2004). The formation of DNA adduct is an important process of PAHs-induced toxicity. The risk of many diseases, like lung cancer after PAHs exposure has been reported in many epidemiologic studies (Karlehagen et al. 1992; Romundstad et al. 2000; Simioli et al. 2004; Vyskocil et al. 2004). It was reported that after the adjusting of gene susceptibility and spatial distributions of ambient PAHs and population, abut 1.6 % of lung cancer cases (0.65×10^{-5}) could be attributed to the exposure to ambient PAHs in China (Zhang et al. 2009c).

Based on the information of source, pollution level and dynamic changes of PAHs, pollution control strategy on PAHs could be conducted effectively (Halsall et al. 1994; Buehler and Hites 2002; Wild and Jones 1995; Venier and Hites 2010). The pollution of PAHs may be expected to decrease after the effective implication of the regulations and laws, which actually occurs in many developed countries (Coleman et al. 1997; Jacob et al. 1997; Katsoyiannis et al. 2011; Menichini et al. 1999). Due to the emission sources of industrial and vehicle emissions, the PAHs pollution levels in urban areas in these developed countries are generally higher than those in the rural areas (Allen et al. 1996; Cotham and Bidleman 1995; Gigliotti et al. 2005; Motelay-Massei et al. 2005; Smith and Harrison 1996). In China, most studies on PAHs so far are in urban area (Duan et al. 2006; Guo et al. 2003a, b; Wan et al. 2006; Wang and Kawamura 2005; Wang et al. 2006a, b, 2007a, b; Wu et al. 2005, 2006; Xie et al. 2008), and much higher pollution levels and adverse health outcomes in these densely population regions have been identified in comparison with those in developed countries (Wang et al. 2006a; Wu et al. 2006; Zhang and Tao 2009; Zhou et al. 2005). Because of large consumption of solid fuels under ineffective residential combustions in rural area, and the deployment of central heating system (Guo et al. 2004; Liu et al. 2007; Liu 2008; Mumford et al. 1987, 1993; Wang et al. 2009a; Wu et al. 2006; Wang 2010; Wang et al. 2011a, b; Zhang et al. 2007a, 2008a, b; Zhang and Tao 2009), the pollution of PAHs in the urban area is more or less controlled but the pollution in rural area is still very serious leading to comparable, even slight higher pollution levels of PAHs in the rural China. There are some research programs, although limited due to technical difficulty and expensively finical cost, in China focusing on the pollution characterization of PAHs in the last several years. In an extensive field study in the Beijing-Tianjin region, it has been found that the ambient PAHs pollution in the area was much higher than that in the developed countries (Okuda et al. 2006; Wu et al. 2005, 2006; Zhou et al. 2005), and there was no significant difference between the rural and urban areas (Liu et al. 2007). Significant seasonal distribution was revealed with high contamination level in the winter, followed by autumn and relatively low levels in the summer and spring. The distribution of PAHs was strongly correlated with the factors like emission density, population and GDP (Liu et al. 2007; Liu et al. 2008a). The source apportionment result showed that the main sources of PAHs in the Beijing-Tianjin region were industrial process, crop residue burning, coal combustion and coke production.

Besides parent PAHs, PAHs derivatives, such as nitrated, oxygenated, hydroxyl-PAHs, are of growing concern recently. In comparison with parent PAHs, PAHs derivatives often have more direct and strong toxic (Allen et al. 1996, 1997; Miller 1978; Bolton et al. 2000; Cheeseman et al. 1985; Palmer and Paulson 1997; Walgraeve et al. 2010). Some derivatives, like oxygenated PAHs (oPAHs) including ketones and quinones, are important intermediary metabolites of many carcinogens (Bolton et al. 2000). The mechanism(s) of these derivatives is not very clear, but it is widely accepted that this was related to the formation of free radicals, and the produce of reactive oxygen species (ROS). PAHs derivatives can be emitted from the primary sources including the incomplete combustions of biomass and coal, and formed secondarily through the reaction of parent PAHs and free radicals (Nielson 1984; Arey et al. 1989; Atkinson et al. 1990; Albinet et al. 2007, 2008a, b; Walgraeve et al. 2010). Nitrated and oxygenated PAHs had been measured in the smoke from the emissions of vehicle, residential coal and biomass burning, and affected by the factors like combustion temperature, air supply condition and relatively humidity (Gullett et al. 2003; Sidhu et al. 2005; Fitzpatrick et al. 2007). The secondary formed PAHs highly vary in sampling site and period (Eiguren-Fernandez et al. 2008; Kojima et al. 2010; Walgraeve et al. 2010). Because of limited studies available, it is not very clear the relative contributions of primary and secondary sources nowadays. .

Ambient measurements on PAHs derivatives are limited, and mainly in developed countries and regions (Allen et al. 1997; Lundstedt et al. 2007; Reisen and Arey 2002, 2005; Tang et al. 2005; Wilson et al. 1995). Some limited studies in China included the emission characterization of nitrated PAHs (nPAHs) from the vehicle emission (Xu 1984) and in some environmental samples (Hattori et al. 2007; Tang et al. 2005). Based on the literature reported datas, oPAHs concentrations are comparable with the parent PAHs within the same order of magnitude, while nitrated PAHs are generally 2–3 orders of magnitude lower than the level of parent PAHs. It was reported that some quinones are chemically steady and can stay longer in atmosphere after emission (Church and Pryor 1985; McFerrin et al. 2008; Pryor et al. 1983). Similar to the parent PAHs, PAHs derivatives can be either in gaseous or particulate phases, but because of relatively low vapor pressure, these derivatives have mode tendencies to be present in particulate phase, especially fine PM (Angelo et al. 1999; Youhei et al. 2004; Walgraeve et al. 2010).

2.2 Emission Inventory

2.2.1 Inventory of CPM

PM can be emitted from both natural sources like soil dust and volcano eruption, and anthropogenic sources. Fine PM can be also formed from the secondary chemical reactions of some precursors (Odum et al. 1996; Andreae and Rosenfeld 2008). It was estimated that globally the anthropogenic emission of PM was about 210 Tg/year, and secondarily formed aerosol was about 139 Tg/year (Andreae and Rosenfeld 2008). In Europe, annual emission of PM was about 1.83 Tg, of which 0.52 and 0.23 Tg were from the indoor burning and agricultural activities (Amann et al. 2005). The annual emission of PM in Asia was about 30 Tg, of which 17 Tg was emitted from the open biomass burning (Hays et al. 2005; Streets et al. 2003). In regard of the carbon fraction, based on the fuel consumption data from the International Energy Agency (IEA) and reported EFs in the literature, it was estimated that total emission of BC in 1996 was about 7.95 Tg, among which open biomass burning, residential biomass burning and coal combustion contributed 42, 18 and 6 %, respectively (Bond et al. 2004, 2007). The historical emissions of BC and OC from 1850 to 2000 were found to increase from 1.0 to 4.4, and from 4.1 to 8.7 Tg, respectively (Bond et al. 2007). By updating the data for biomass burning, the estimated emission of BC in 1996 in the second estimation was lower than the previously estimated 7.95 Tg (Bond et al. 2007; Fernandes et al. 2007). In the latest report on BC, global total emission of BC in 2007 was about 7.6 Tg, among which open biomass burning, residential combustion, transportation and industrial sources contributed 35.5, 25.1, 19.0 and 19.3 %, respectively (EPA 2012). It was thought that because of increased use of fossil fuels, the emission of BC increased during the last several decades (Junker and Liousse 2008; Novakov et al. 2000), however, due to improved techniques in combustion and control technologies, the emission of BC has decreased after the 1980s (Junker and Liousse 2008). There are very large uncertainties in these inventories. Even though different methods reported similar temporal and spatial distributions, the estimated amounts varied dramatically because of the distinct data of EFs and fuel consumption.

China is one large emitter of BC (Cooke et al. 1999; Bond et al. 2004). It was reported that anthropogenic emissions of BC and OC in Asia were 2.54 and 10.4 Tg, respectively, among which 1.05 and 3.4 Tg were from China (Streets et al. 2003). It is obvious that fuel type and burning condition in China differed dramatically from those in other countries and regions, and hence the lack of information of direct measurements on the emission characterization of the specific fuel-stove combinations in China would lead to considerable bias in the developed inventories. Based on the literature reported EFs of BC or PM, the emission of BC in China was estimated at 1.34 Tg, and about 605.4, 512, 82.5, and 36.9 Gg from the burning of residential coal, biomass, industrial coal and diesel, respectively (Streets et al. 2001). It was also predicted that with the development of approaches like new and clean techniques and the deployment of briquettes, the emission of BC would decrease to 1.22 Tg by 2020, and the residential biomass and coal combustions were still the two dominated sources contributing 535 and 387 Gg, respectively (Streets et al. 2001). Recently, Lei et al. (2011) estimated the emissions of BC and OC in China from 1995 to 2000. In the study, EFs of BC and OC were calculated from EF of PM and the mass fractions of EC and OC in PM, respectively. The national total emissions of $PM_{2.5}$, PM_{10}, TSP, BC and OC were 9.28, 13.5, 24.9, 1.13 and 2.87 Tg in 1990, and by 2005, they increased to 12.95, 18.83, 34.26, 1.51, and 3.19 Tg, respectively (Lei et al. 2011). The emission of PM increased from 1990 to 1996, and after 1996 there was a short decrease, and then increased again after 2000. The emissions of BC and OC increased after 2000. Residential solid fuel combustion is a dominated source of CPM in China. Based on the estimated consumption of biomass fuels and literature reported EFs, it was estimated that the total emissions of PM and BC increased from 2000 to 2007 at a rate of 2.6 %, and at 3.62 and 0.43 million tones in 2007. Due to limited domestic measurements, in most emission inventories EFs reported for other countries and cities are often adopted, which often bias the estimated amounts. In general, the emission of BC in China during 1995 to 2001 was 1.37–1.71 Tg (Streets et al. 2001; Bond et al. 2004; Cao et al. 2006; Streets, et al. 2003; Zhang and Smith 2007). Residential combustions of coal and biomass fuels are the dominated sources, and the open burning of biomass burning after the harvest also contributes largely to the total anthropogenic emission. It is realized that when there were still very large emissions from residential solid fuel combustion, with the increased emissions from the gasoline and diesel powered vehicles (Zhi et al. 2009; Wang et al. 2012), anthropogenic emission of BC would be an important air pollutant of concern and requires further studies.

2.2.2 Inventory of PAHs

PAHs are mainly produced from the incomplete combustion processes (Mastral and Calleän 2000; Tsibulsky et al. 2001), and are ubiquitous in environment after the emissions of both anthropogenic and natural sources. Generally, the sources of

PAHs include the industrial, transport, domestic burning, agricultural activity and natural emissions (Tsibulsky et al. 2001; Zhang 2010). Volcano eruption, forest fire, and the biological formation are all able to produce PAHs. Though it is different to estimate the activity level and the EFs of PAHs from the natural sources, it is widely accepted that anthropogenic emissions contribute largely to the total PAHs in environment. PAHs from the industrial process are highly related to the techniques and pollution control technologies, and PAHs from the transport sources are related to the type of vehicles, usage, loading and operation modes, as well as fuel types. Residential combustion sources of PAHs include the burnings of biomass fuels and coals. In most developing countries, these solid fuels are widely used for daily cooking and heating by rural resident. Because of relatively low burning efficiency, residential solid fuel combustions often yield high emissions of incomplete pollutants, and it is the main source of PAHs globally and particularly in developing countries.

There are some inventories on the emissions of PAHs at the country and regional levels (Berdowski et al. 1997; Galarneau et al. 2006; Pacyna et al. 2003; Tsibulsky et al. 2001; Wenborn et al. 1999; Zhang and Tao 2009). For instance, it had been estimated that global total emission of BaP from anthropogenic source during 1966–1969 was about 5 Gg/y (Suess 1976). Zhang and Tao (2009) developed a global emission inventory of PAHs in 2004, and the differences among countries and different sources were addressed. The results showed that global total emission of PAHs was 0.52 million tones in 2004, and there was 0.29 million tones from Asia. China, India and U.S. were three large emitters with the national total emissions of 114, 90 and 32 thousand tones, respectively. In terms of the source contribution, biomass burning contributed over 57 % of the total, and the emission from industrial source only made up to 10 % the global total. The source profiles varied among different countries. In general, vehicle emission is the main source in developed countries, and in most developing countries, residential solid fuel combustion was the largest emitter. In comparison with less work on global emission estimation, there are more inventories on the national or regional scales (Tsibulsky et al. 2001; Berdowski et al. 1997; Pacyna et al. 2003; Wenborn et al. 1999). For example, Tsibulsky et al. (2001) estimated the emissions of 6 PAHs in the Soviet Union during 1990 to 1997 based on the literature reported EFs. Wenborn et al. (1999) reported a total emission of 1.7 Gg/year from U.K. and most of PAHs were from the industrial activities while the residential combustion was a minor source. Berdowski et al. (1997) estimated the emissions of FLA, BaP, BbK, BkF, BghiP and IcdP from 23 European countries, and pointed out that the emissions from residential source would increase while that from the industrial source will decrease gradually. The historical emission of BaP from Europe was reported by Pacyna et al. (2003). According to the report by European Monitoring and Evaluation program, total emissions of BaP, BbF, BkF and IcdP in Europe were 2.4 and 1.3 Gg in 1990 and 2003, respectively. It was reported that the emissions of 16 PAHs and 7 carcinogenic PAHs from the U.S. in 1990 were 26.5 and 2.0 Gg, respectively, among which firewood combustion contributed about 33 % of the total (U.S. EPA 1998). Afterwards, the U.S. EPA updated the emission inventory of 7 carcinogenic PAHs between 1970 and 1995, and more important, a database was established so as to share the results for the scientists all over the world (U.S. EPA 2001). The emission of BaP in the Great Lake region and the Ontario was 26.8 Mg/year, and manufacturing industry, indoor firewood burning and open biomass burning were three main sources contributing 33, 28 and 13 % of the total emission (GLC 2007). It was pointed out that with the effective control on vehicle emission, the implement of regulations and control strategies, and improved combustion efficiencies of indoor cooking stoves, regional PAHs emission decreased obviously.

Emission inventories of PAHs in developing countries are rarely developed so far. To our knowledge, in addition to the global inventory by Zhang and Tao (2009) in which PAHs emissions were estimated at a national scale, one widely used inventory in developing country is that developed by Tao group (Xu et al. 2006; Zhang et al. 2007b, 2008a) for PAHs emission in China. The total emissions were 25.3 Gg/year, and residential wood, crop residue and coal combustion contributed 34, 26 and 20 %, respectively. The contribution of coking production was 16 %. In comparison with the composition profile in other developed countries or regions, the fractions of high molecular weight PAHs most of which are toxic ones were generally higher in the emissions from China. This might be due to relatively high contributions of emissions from low efficient residential biomass and coal combustions. Xu et al. (2006) analyzed the temporal-spatial and historical distribution of PAHs in China. After that, Zhang et al. (2007b) developed a high resolution emission inventory of PAHs at a county level. The estimated emission was 28.8 Gg, and the contributions of biomass burning, residential coal combustion and coking were 59, 23 and 15 %, respectively. The composition profiles varied significantly among different counties. The study on historical change suggested that with the development of social economic and energy consumption, the emissions of PAHs increased simultaneously. From 1950 to 1979, the total PAHs emission increased from 19.0 to 67.0 thousand tones, and between 1950 to 1961, because of the fast increase of coal consumption in the residential sector and widespread indigenous coking, PAHs emissions increased by about 1.4 times. After the middle of 1970s, because of the urbanization and family planning policy, the natural increase rate of rural population decreased and subsequent residential consumption of biomass stopped the increase. During the period, indigenous coking was the main cause of PAHs increase in China. In 2003, the coal low was issued and implemented, and most indigenous coking was banned. After that, the PAHs emission decreased accordingly.

2.2.3 Uncertainty in Inventory

In almost all inventories developed so far, there are considerable biases and uncertainties (Bond et al. 2004; Streets et al. 2001; Zhang et al. 2009c). Emission inventory is developed based on the activity level and corresponding emission factor (EF). The information of activity level can be obtained from the field survey or statistical reports. EFs defined as the pollutant mass for per fuel mass or energy (unit: mg/kg or mg/kJ) is the basic input for the development of emission inventory. The lack of data on either activity level or EFs would lead to large biases in developed inventories. Nowadays, the information on the fuel consumption of major stationary sources is available and accessible for most studies. However, the information for the minor or scattered pointed sources is lack since they are usually difficult to count out. In terms of the EFs, in addition to the lack of sufficient data, variations in reported data due to influencing factors such as fuel properties, burning conditions and even experimental methods are more responsible for the uncertainties in the developed inventories. EFs reported in the literature can vary in orders of magnitude (Streets et al. 2001). The influences of many factors are completed and usually complicated. For example, the pollutant emission would increase under high excess air and subsequent low burning temperature and thermal efficiency (Fan et al. 2010; Johansson et al. 2004). The burning of high moisture fuel would require an amount of heat to vaporize water and hence lead to incomplete combustion of fuel and emissions of incomplete burning pollutant, but during the combustion of low moisture fuel, the fast burning would cause an oxygen deficient atmosphere in the stove chamber which may also lead to increased emissions of incomplete pollutants (Rogge et al. 1998; Simoneit 2002). Generally, the EFs are lower under high burning efficiencies, and the later is affected by a number of factors such as fuel moisture, fuel loading and air supply. It was reported that from 1990 to 2005, because of the implement of pollution control devices and technology improvement, the EFs of $PM_{2.5}$ and TSP from the combustion sources in China decreased obviously, by about 7–69 and 18–80 %, respectively (Lei et al. 2011). EFs can be measured directly from the field or laboratory experiments, or calculated from the other related pollutants (Bond et al. 2004; Zhang et al. 2000). For example, EFs of particle-bound PAHs can be estimated from the EFs of PM and PAHs mass fraction in PM. In a study on global carbonaceous carbon inventory, Bond et al. (2004) quantitatively analyzed a number of factors influencing the emissions and identified the most significant factor. For instance, in the coking production, three most significant factors are technology, EF of PM, and the carbon fraction in PM. In the coal combustion, EF of PM was the main influencing factor, followed by the technology and carbon fraction. For most sources, EF is the most important factor or of relatively high importance in comparison with technology level and energy consumption data (Bond et al. 2004). It has been previously reported that relative variations in PAHs emissions from the aluminum electrolysis with pre-baked anode, non-transportation petroleum, industrial coal and gasoline and kerosene consumption in transportation were 314, 228, 161, 173 and 179 %, respectively. The calculated variations for the EFs (16.1 %) were significantly higher than that of 5.41 % for the activity level based on the Monte Carlo simulation, which indicated that EFs were the main source of uncertainty in developed inventory (Zhang and Tao 2009).

2.3 Residential Solid Fuel Combustion

2.3.1 Solid Fuels

Solid fuels mainly include natural combustible materials like crop straw, firewood and coal. Because of easy access, high abundance and relatively low cost, solid fuels are widely used in rural households for daily cooking and heating. There are over 3.0 billion people all over the world using solid fuels in daily lives, and most of them live in developing countries. Biomass fuels are one most widely used solid fuel in rural area. The consumption of biomass contributed about 14 % of global total energy consumption, and in developing countries, it could make up to about 40 % (IEA 2011a, b). During the burning of biomass fuel, combustible fractions such as lignin, cellulose and hemi-cellulose, were burned and emitted vaporized organics some of which would participate into the formation of PM and other incomplete pollutants. The process can be separated into two phases of flaming and smoldering with or without obvious fire. Low efficient biomass burning often produces large amounts of incomplete particles and gaseous pollutants (Andreae 1991; Andreae and Merlet 2001; Simoneit 2002). In China, biomass fuel contributed about 24 % of national total energy consumption. The burnings of crop residues and wood materials are the major sources of pollutants like PM, volatile organic compounds and SO_2 (Cao et al. 2005, 2007; Crutzen and Andreae 1990; Tian et al. 2011).

Coal is the most abundant and widely used fossil fuel. The consumption of coal in China is very large, and it can be expected that in the years to come, coal would still be the main energy source in the country. The coal consumption increased from 603 Mtce in 1980 to 1389 Mtce by 1996, and after that although there was a short decrease, it raised up to 2656 Mtce in 2007. Coal consumption comprises up to 70–76 % of national total energy consumption (Chen and Xu 2010). Coal combustion process covers the phases of pre-heating, burning of volatile fractions and combustion of the other solid materials. Pollutant emissions from the first two phases were about 50 times of that in the last burning period (Bond et al. 2002; Butcher and Ellenbecker 1982). But it is noted that these three phases are not absolutely independent (Bond et al. 2002, 2004). It is accepted that at the initial burning period, a large number of pollutant was released and increased obviously, and then in the smoldering phase without obvious fire, pollutant emissions usually decreased.

The combustions of biomass and coal under relatively low burning efficiencies often produce various pollutants, and subsequently affect the indoor and outdoor air quality and human health. It was reported that the indoor $PM_{2.5}$ concentration in the kitchen using briquette coal was 13 and 30 times of those using natural gas and liquid petroleum gas, respectively (Peng et al. 2005). Exposure to these toxic pollutants is associated with increased risks of various diseases (Chapman et al. 2005; Hu and Guo 2007; Tian et al. 2009; He et al. 2005; Lin et al. 2005; Mumford et al. 1987; Smith 1993; Zhang and Smith 2007). The occurrence of COPD in population using solid fuels for daily cooking was 16–19 %, significantly higher than that of 8.7 % in the population using liquid petroleum gas (Wen et al. 2006). In China, exposure to smoke from residential solid fuel combustion caused about 0.42 million premature death, which was about 40 % higher than that of 0.30 million caused by the outdoor air pollution (Zhang and Smith 2007).

Residential solid fuel combustion is a major source for many pollutants like PM and PAHs, especially in developing countries. It is necessary to develop reliable inventories of these pollutants from solid fuel combustion so as to estimate the impacts on air quality, human health and also regional climate change of these pollutants of widespread concern.

2.3.2 EF Measurements

EFs can be affected by a number of factors and usually vary in orders of magnitude (Streets et al. 2001; Xu et al. 2006). Beside the use in the development of emission inventory, EFs are also widely used in the development and implement of many environment pollution control strategies. For instance, the emission standard is one most important term used to regulate the pollutant emission behaviors of most industrial sources. Emission permits, standards and control strategies based on the inadequate or even wrong EFs might ignore some important sources of target pollutants, and even result in wrong emission reduction plans (U.S. EPA 2006; Pouliot et al. 2012).

When the mass or energy of combusted fuel and the total mass of emitted pollutants are known, the EF can be calculated directly. This is one most widely used approach to measure the EFs, especially in laboratory chamber study. The fuel burning process differs significantly in field in comparison with that in laboratory chamber. Field measurement might be closer to the real practice and more reliable as an input for the study on emission and transport behaviors of these pollutants, however, it is realized that it is difficulty to collect all emitted pollutants in field. Alternatively, the calculation of EFs based on the carbon mass balance method is developed and commonly adopted in the measurement of EFs for fuel combustion. The method follows the assumption that carbon in fuel would be burned and released into the atmosphere as the forms of gaseous CO, CO_2, total hydrocarbon carbon (THC) and particle-bound carbon. The results calculated based on the carbon mass balance method were found to be comparable to the direct measured ones within 20 % (Dhammapala et al. 2006).

In most EF measurement experiment, sampling dilution system or cooling system is adopted to avoid the potential impacts of high temperature and relatively humidity (RH) in the emission smoke on sampling and EF measurements. In a dilution system, high purity inert gas or clean air is used to dilute the emission exhaust and cool the smoke temperature. In the system, the dilution ratio, dilution rate and temperature of diluted gas affect the mass and size distribution of PM significantly (Purvis et al. 2000; Lipsky and Robinson 2006). For example, it was reported that when the dilution ratio increased from 20:1 to 350:1, the EF of $PM_{2.5}$ decreased by about 50 %. The dilution also changes the gas-particle partitioning of volatile organics (Lipsky and Robinson 2006). In practice, it is necessary to test and control the dilution ratio and rate, and monitor the change of gas temperature and RH. The dilution system is often used in laboratory study (Hildemann et al. 1989; Hays et al. 2005; Schimidl et al. 2011; Schauer et al. 2001; Rogge et al. 1993a, b, c, d, 1994). The cooling system with a cooling device after the exhaust exit to cool the smoke temperature is another widely used sampling system (Bond et al. 2004), for example in the emission source measurement study by U.S. EPA. In addition, in some studies, especially field measurements, the sampling was done directly at the chimney exit (Roden et al. 2006, 2009). In this case, the exhaust is diluted naturally by ambient air and assumed that smoke temperature decreased significantly after the mixing with cold ambient air once emitted.

There has been a long research on the emission measurements of particle and gaseous pollutants in developed countries (Gupta et al. 1998; Jetter and Kariher 2009). For example, by using a dilution system (Hildemann et al. 1989), Schauer and Rogge and the colleagues investigated the EFs of PM and particle-bound pollutants for a variety of sources like residential wood combustion, vehicle emission and open biomass burning (Simoneit et al. 1993; Schauer et al. 1999a, b, 2001, 2002a, b; Rogge et al. 1991, 1993a, b, c, d, 1994, 1997a, b, 1998). They also analyzed the influence of fuel type and burning conditions on the EFs and composition profiles. Jenkins et al. (1993, 1996a, b) simulated open burning of biomass fuels in a laboratory tunnel, and measured EFs of PM and PAHs at 5.05–7.27 g/kg and 5.04–683 mg/kg, respectively, varying under different burning conditions. Dhammapala et al. (2006, 2007a, b) measured EFs of PM, OC, EC and organics from the burning of crop straw. They also compared the results from direct calculation and that using the carbon mass balance method, and found that the results were comparable (Dhammapala et al. 2006).

Although there are lots of tests on emissions from biomass burning in the literature for developing countries, the studies mainly focused on the open burning rather than the combustion in residential sector. In most developing countries and regions, like Asia and Africa, more attention should be paid to the residential solid fuel combustion since it produces largely to the total pollutant emission. Venkataraman and Kim Oanh and the colleagues measured EFs of PM, CO, VOCs and OC for a variety of solid fuels burned in different stoves that were often used in Asian countries like India, Thailand and Malaysia by using a hood and dilution system in laboratory (Bhattacharya et al. 2002; Kim Oanh et al. 1999, 2002; Venkataraman and Rao 2001; Venkataraman et al. 2002). Roden et al. (2009) conducted a series of measurements on PM emission from wood combustion in Honduras and found that the EF of PM in field measurement was 6.1 g/kg, about 4 times of that of 1.5 g/kg measured in laboratory study. For the same fuel and stove, the EF of PM was significantly different between the field (6.6 g/kg) and laboratory studies (1.8 g/kg).

In China, much less work has been done. Moreover, most studies so far are in the laboratory chamber. Zhang and colleagues measured EFs of CO, CO_2, NO_2,

SO_2, VOCs and PM for a variety of fuels like coal, crop residue, wood, kerosene and natural gas burned in different stove types (Edwards et al. 2004; Tsai et al. 2003; Zhang et al. 1999, 2000; Zhang and Smith 1999). The results for a large fuel/stove combination are one most widely used database for the development of emission inventory in China. The research group leaded by Dr. Hao designed a sampling system and measured the EFs of CPM and gases for residential biomass burning in rural China (Li et al. 2007a, b, 2009a, b; Wang et al. 2009b). Chen and colleagues reported the EFs of PM, OC and EC for a variety of coals with different maturity (Chen 2004; Chen et al. 2004, 2005, 2006, 2009; Zhi et al. 2008, 2009). In their study, a cooling system was developed to avoid the influence of high temperature on sampling. They found that PM emission varied significantly among different coal types. The emission for anthracite was about 6–15 times lower than that for the bituminous. PM emitted from residential coal combustion was fine with high abundance of sub-micron particle. Among these limited measurement studies, PAHs emission was much less (Dou et al. 2007; Li et al. 2008; Zhang et al. 2008c, 2011), and because of different methods used, variety in fuel and stove types, these reported results varied dramatically. Lu et al. (2009) compared EFs of PAHs from the rice straw burning under different fuel moistures, burning temperatures and air supply amounts. The results showed that the EFs of PAHs increased with the increase of combustion temperature ranging from 200 to 700 °C. With the increase of air supply, the EFs of PAHs increased and decreased with a maximum at air supply of 40 %. It is thought that adequate air supply is necessary to achieve low pollutant emission. For PAHs emissions from wood combustion, Zhang et al. (2003) reported PAHs EFs for four wood fuels burned in different conditions, but the study was done in Austria using local fuel and local purchased stove. PAHs EFs for coal reported in the literature ranged from 6.0 to 882 mg/kg, depending on coal property, form and burning conditions (Cui et al. 1993; Chen et al. 2004, 2005, 2006; Dou et al. 2007; Liu et al. 2009). Emissions from the burning of anthracite with low volatile matter content are generally lower than those for sub-bituminous and bituminous coals. Li et al. (2003) measured PAHs in the emissions from the burning of four different coal types in a laboratory tunnel, and found that PAHs in the exhaust were about 30–60 times of that in the coal indicating the formation and emission of PAHs under high temperature pyrolysis and pyrosynthesis. Total EFs of 16 PAHs ranged from 0.010 to 0.225 mg/kg, or 0.001–0.009 mg/KJ. Dou et al. (2007) reported PAHs EFs for bituminous coals from Shanxi and Beijing were 882 and 880 mg/kg, and for anthracite were 104 and 71 mg/kg, respectively. The emission was lower for the briquette compared with the raw chunk, and the emissions from the coal stove with one chimney were about 4–12 times of that burned in two-chimney stove (Liu et al. 2009). Chen et al. (2005) measured EFs of 14 priority PAHs for five different coals. The EF for anthracite was about 3 orders of magnitude lower than that for the bituminous.

The main objective of this thesis is to measure emissions of CPM (EF_{CPM}) and PAHs (EF_{PAHs}) from the residential combustions of different solid fuels. Among various sources, we focused on the residential solid fuel combustion since it is one major source of these pollutants in China, particularly in rural area, and there are limited data available so far. It is hoped that the study can provide important first-hand data for the future development of CPM and PAHs inventories in China, and also as a part of input in environmental and ecological analysis modeling, the results can provide helpful information in the analysis of the impacts of these pollutants on human health and climate change.

References

Albinet, A., Leoz-Garziandia, E., Budzinski, H., & Villenave, E. (2007). Polycyclic aromatic hydrocarbons (PAHs), nitrated PAHs and oxygenated PAHs in ambient air of the Marseilles area (South of France): Concentrations and sources. *Science of Total Environment, 384*, 280–292.

Albinet, A., Leoz-Garziandia, E., Budzinski, H., Villenave, E., & Jaffrezo, J. (2008a). Nitrated and oxygenated derivatives of polycyclic aromatic hydrocarbons in the ambient air of two French alpine valleys Part 1: Concentrations, sources and gas/particle partitioning. *Atmospheric Environment, 42*, 43–54.

Albinet, A., Leoz-Garziandia, E., Budzinski, H., Villenave, E., & Jaffrezo, J. (2008b). Nitrated and oxygenated derivatives of polycyclic aromatic hydrocarbons in the ambient air of two French alpine valleys Part 2: Particle size distribution. *Atmospheric Environment, 42*, 55–64.

Allen, J. O., Dookeran, N. M., Smith, K. A., Sarofim, A. F., Taghizadeh, K., & Lafleur, A. (1996). Measurement of polycyclic aromatic hydrocarbons associated with size-segregated atmospheric aerosols in Massachusetts. *Environmental Science and Technology, 30*, 1023–1031.

Allen, J. O., Dookeran, N. M., Taghizadeh, K., Lafleur, A. L., Smith, K. A., & Sarofim, A. F. (1997). Measurement of oxygenated polycyclic aromatic hydrocarbons associated with a size-segregated urban aerosol. *Environmental Science and Technology, 31*, 2064–2070.

Amann, M., Bertok, I., Cofala, J., Gyarfas, F., Heyes, C., Klimont, Z., Schöpp, W., & Winiwarter, W. (2005). Final report: Baseline Scenarios for the Clean Air for Europe (CAFE) Programme. http://www.iiasa.ac.at/rains/CAFE_files/Cafe-Lot1_FINAL(Oct).pdf

An, S., Chen, J., & Chen, X. (2005). Research progress on the molecular toxicity of carcinogenicity of polycyclic aromatic hydrocarbons. *Foreign Medical Sciences (Section Hygiene), 32*, 10–13. (In Chinese).

Andreae, M. O. (1991). Biomass burning: Its history, use, and distribution and its impact on environmental quality and global climate. In J. S Levine (Ed.), *Global Biomass Burning, Atmospheric, Climatic, and Biospheric Implications* (pp. 3–21). Cambridge, MA: MIT Press.

Andreae, M. O., & Merlet, P. (2001). Emission of trace gases and aerosols from biomass burning. *Global Biogeochemistry Cycle, 15*, 955–966.

Andreae, M. O., & Rosenfeld, D. (2008). Aerosol–cloud–precipitation interactions. Part 1. The nature and sources of cloud-active aerosols. *Earth-Science Review, 89*, 13–41.

Angelo, C., Fabio, M., & Patrizia, D. F. (1999). Distribution of n-alkanes, polynuclear aromatic hydrocarbons and nitrated polynuclear aromatic hydrocarbons between the fine and coarse fractions of inhalable atmospheric particulates. *Journal of Chromatography A, 846*, 255–264.

Ansmann, A., & Müller, D. (2005). Lidar and atmospheric aerosol particles. In C. Weitkamp (Ed.), *Lidar range-resolved optical remote sensing of the atmosphere*. Singapore: Springer.

Arey, J., Zielinska, B., Atkinson, R., & Aschmann, S. M. (1989). Nitroarene products from the gas-phase reactions of volatile polycyclic aromatic hydrocarbons with the OH radical and N_2O_5. *International Journal of Chemical Kinetics, 21*, 775–799.

Atkinson, R., Arey, J., Zielinska, B., & Aschmann, S. M. (1990). Kinetics and nitro-products of gas-phase OH and NO_3 radical-initiated reactions of naphthalene, fluoranthene and pyrene. *International Journal of Chemical Kinetics, 22*, 999–1014.

Balakrishnan, K., Sankar, S., Parikh, J., Padmavathi, R., Srividya, K., Venugopal, V., et al. (2002). Daily average exposures to respirable particulate matter from combustion of biomass fuels in rural households of southern India. *Environmental Health Perspective, 110*, 1069–1075.

Balakrishnan, K., Sambandam, S., Ramaswamy, P., Mehta, S., & Smith, K. R. (2004). Exposure assessment for respirable particulates associated with household fuel use in rural districts of Andhra Pradesh, India. *Journal of Exposure Science and Environmental Epidemiology, 14*, S14–S25.

Becker, S., Halsall, C. J., & Tych, W. (2006). Resolving the long-term trends of polycyclic aromatic hydrocarbons in the Canadian Arctic atmosphere. *Environmental Science and Technology, 40*, 3217–3222.

Behymer, T. D., & Hites, R. A. (1988). Photolysis of polycyclic aromatic hydrocarbons adsorbed on fly ash. *Environmental Science and Technology, 22*, 1311–1319.

Berdowski, J. J. M., Baas, J., Bloos, J. P. J., Visschedijk, A. J. H., & Zandweld, P. Y. J. (1997). *The European emission inventory of heavy metals and persistent organic pollutants for 1990*. TNO Institute of Environmental Sciences, Energy Research and Process Innovation.

Bhattacharya, S. C., Albina, D. O., & Salam, P. A. (2002). Emission factors of wood and charcoal-fired cookstoves. *Biomass and Bioenergy, 23*, 453–469.

Bidleman, F. (1988). Atmospheric process. *Environmental Science and Technology, 21*, 361–367.

Bolton, J. L., Trush, M. A., Penning, T. M., Dryhurst, G., & Monks, T. J. (2000). Role of quinones in toxicology. *Chemical Research Toxicology, 13*, 135–160.

Bond, T. C., Covert, D. S., Kramlich, J. C., Larson, T. V., & Charlson, R. J. (2002). Primary particle emissions from residential coal burning: Optical properties and size distributions. *Journal of Geophysical Research-Atmospheres, 107*(D21), 8347. doi:10.1029/2001JDO00571

Bond, T. C., Streets, D. G., Yarber, K. F., Nelson, S. M., Woo, J., & Klimont, Z. (2004). A technology-based global inventory of black and organic carbon emissions from combustion. *Journal of Geophysical Research, 109*, D14203. doi:10.1029/2003JD003697

Bond, T. C., Bhardwaj, E., Dong, R., Jogani, R., Jung, S., Roden, C., Streets, D., & Trautmann, N. M. (2007). Historical emissions of black and organic carbon aerosol from energy-related combustion, 1850–2000. *Global Biogeochemical Cycle, 21*, GB2018.

Boysen, G., & Hecht, S. S. (2003). Analysis of DNA and protein adducts of benzo[a]pyrene in human tissues using structure-specific methods. *Mutation Research, 543*, 17–30.

Buehler, S. S., & Hites, R. A. (2002). The Great Lakes' integrated atmospheric deposition network. *Environmental Science and Technology, 36*, 354–359.

Butcher, S. S., & Ellenbecker, M. J. (1982). Particulate emission factors for small wood and coal stoves. *Journal of the Air Pollution Control Association, 32*, 380–384.

Cao, J. J., Lee, S. C., Ho, K. F., Zhang, X. Y., Zou, S. C., & Fung, K. (2003a). Characteristics of carbonaceous aerosol in Pearl River Delta Region China during 2001 winter Period. *Atmospheric Environment, 37*, 1451–1460.

Cao, J.J., Lee, S.C., Ho, K.F., Zou, S.C., Zhang, X.Y., & Pan, J. G. (2003b). Spatial and seasonal distributions of atmospheric carbonaceous aerosols in Pearl River Delta region, China. *China Particuology, 1*, 33–37.

Cao, G., Zhang, X., Wang, D., & Zheng, F. (2005). Inventory of atmospheric pollutants discharged from biomass burning in China continent. *China Environmental Science, 25*(4), 389–394. (In Chinese).

Cao, G., Zhang, X., & Zheng, F. (2006). Inventory of black carbon and organic carbon emissions from China. *Atmospheric Environment, 40*, 6516–6527.

Cao, G., Zhang, X., Wang, Y., & Zheng, F. (2007). Emission inventory of air pollutants from open burning of crop straw in China. *Chinese Science Bulletin, 52*, 1826–1831. (In Chinese).

Cercasov, V., & Wulfmeyer, V. (2008). Trends in airborne particulates in Stuttgart, Germany: 1972–2005. *Environmental Pollution, 152*, 304–313.

Chameides, W. L., Yu, H., Liu, S. C., Bergin, M., Zhou, X., Mearns, L., et al. (2009). Case study of the effects of atmospheric aerosols and regional haze on agriculture: An opportunity to enhance crop yields in China through emission controls? *Proceedings of the National Academy of Sciences, 96*, 13626–13633.

Chan, C. K., & Yao, X. (2008). Air pollution in mega cities in China. *Atmospheric Environment, 32*, 1–42.

Chapman, R. S., He, X., Blair, A. E., & Lan, Q. (2005). Improvement in household stoves and risk of chronic obstructive pulmonary disease in Xuanwei, China: Retrospective cohort study. *BMJ, 331*, 1050–1055. doi:10.1136/bmj.38628.676088.55

Cheeseman, K. H., Albano, E. F., Tomasi, A., & Slater, T. F. (1985). Biochemical studies on the metabolic activation of halogenated alkanes. *Environment Health Perspective, 64*, 85–101.

Chen, S., Liao, S., Jian, W., & Lin, C. C. (1997). Particle size distribution of aerosol carbons in ambient air. *Environment International, 23*, 475–488.

Chen, L. W. A., Bruce, G. D., Russell, R. D., Chow, J. C., & Henry, R. C. (2002). Origins of fine aerosol mass in the Baltimore-Washington corridor: Implications from observation, factor analysis, and ensemble air parcel back trajectories. *Atmospheric Environment, 36*, 4541–4554.

Chen, Y. (2004). *Emission characteristics of carbonaceous particulate and polycyclic aromatic hydrocarbons from residential honeycomb-coal-briquette combustion*. Ph.D thesis, University of Chinese Academy of Sciences (In Chinese).

Chen, Y. J., Bi, X. H., Mai, B. X., Sheng, G. Y., & Fu, J. M. (2004). Emission characterization of particulate/gaseous phases and size association for polycyclic aromatic hydrocarbons from residential coal combustion. *Fuel, 83*, 781–790.

Chen, Y. J., Sheng, G. Y., Bi, X. H., Feng, Y. L., Mai, B. X., & Fu, J. M. (2005). Emission factors for carbonaceous particles and polycyclic aromatic hydrocarbons from residential coal combustion in China. *Environmental Science and Technology, 39*, 1861–1867.

Chen, Y. J., Zhi, G., Feng, Y., Fu, J., Feng, J., Sheng, G., et al. (2006). Measurement of emission factors for primary carbonaceous particles from residential raw-coal combustion in China. *Geophysical Research Letter, 33*, L20815. doi:10.1029/2006GL026966

Chen, Y., Zhi, G., Feng, Y., Liu, D., Zhang, G., Li, J., et al. (2009). Measurements of black and organic carbon emission factors for household coal combustion in China: Implication for emission reduction. *Environmental Science and Technology, 43*, 9495–9500.

Chen, W., & Xu, R. (2010). Clean coal technology development in China. *Energy Policy, 38*, 2123–2130.

Cheng, Y. (1998). *Carcinogenic polycyclic aromatic hydrocarbons* (pp. P163–P183). People's medical publishing house Co. Ltd. (In Chinese).

Cheng, Y., He, K. B., Duan, F. K., Zheng, M., Ma, Y. L., & Tan, J. H. (2009a). Measurement of semivolatile carbonaceous aerosols and its implications: A review. *Environment International, 35*, 674–681.

Cheng, Y., He, K. B., Duan, F. K., Zheng, M., Ma, Y. L., & Tan, J. H. (2009b). Positive sampling artifact of carbonaceous aerosols and its influence on the thermal-optical split of OC/EC. *Atmosphere Chemistry and Physics, 9*, 7243–7256.

Chow, J. C. (1995). Measurement methods to determine compliance with ambient air quality standards for suspended particles. *Journal of the Air and Waste Management Association, 45*, 320–382.

Chow, J. C., Watson, J. G., Lowenthal, D. H., & Magliano, K. L. (2008). Size resolved aerosol chemical concentrations at rural and urban sites in Central California, USA. *Atmosphere Research, 90*, 243–252.

Chow, J. C., Watson, J. G., Lowenthal, D. H., Antony Chen, L. -W., & Motallebi, N. (2011). PM2.5 source profiles for black and organic carbon emission inventories. *Atmospheric Environment, 45*, 5407–5414.

Church, D., & Pryor, W. (1985). Free-redical chemistry of cigarette smoke and its toxicological implication. *Environmental Health Perspective, 64*, 111–126.

Coleman, P. J., Lee, R. G. M., Alcock, R. E., & Jones, K. C. (1997). Observations on PAH, PCB, and PCDD/F trends in UK urban air, 1991–1995. *Environmental Science and Technology, 31*, 2120–2124.

Cooke, W. F., Liousse, C., Cachier, H., & Feichter, J. (1999). Construction of a 1 × 1 fossil fuel emission data set for carbonaceous aerosol and implementation and radiative impact in the ECHAM4 model. *Journal of Geophysical Research, 104*, 22137–22162.

Cotham, W. E., & Bidleman, T. F. (1995). Polycyclic aromatic hydrocarbons and polychlorinated biphenyls in air at an urban and a rural site near Lake Michigan. *Environmental Science and Technology, 29*, 2782–2789.

Cousins, I. T., Gevao, B., & Jones, K. C. (1999). Measuring and modelling the vertical distribution of semivolatile organic compounds in soils. I: PCB and PAH soil core data A-6944–2008. *Chemosphere, 39*, 2507–2518.

Crutzen, P. J., & Andreae, M. O. (1990). Biomass burning in the tropics: Impact on atmospheric chemistry and biogeochemical cycles. *Science, 250*, 1669–1678.

Cui, W., Yao, W., & Xu, X. (1993). Emission characterization of polycyclic aromatic hydrocarbons from coal combustion. *Acta Scientiae Circumstantiae, 13*, 317–324. (In Chinese).

Daly, G. L., & Wania, F. (2005). Organic contaminants in mountains. *Environmental Science and Technology, 39*, 385–398.

Davis, B. L., & Guo, J. (2000). Airborne particulate study in five cities of China. *Atmospheric Environment, 34*, 2703–2711.

Deng, X., Tie, X., Wu, D., Zhou, X., Bi, X., Tan, H., et al. (2008). Long-term trend of visibility and its characterizations in the Pearl River Delta Region (PRD). *Atmospheric Environment, 42*, 1424–1435.

Denissenko, M. F., Pao, A., & Tang, M. (1996). Preferential formation of Benzo[a]pyrene adducts at lung cancer mutational hotspots in P53. *Science, 274*, 430–432.

Dhammapala, R., Claiborn, C., Corkill, J., & Gullett, B. (2006). Particulate emissions from wheat and Kentucky bluegrass stubble burning in eastern Washington and northern Idaho. *Atmospheric Environment, 40*, 1007–1015.

Dhammapala, R., Claiborn, C., Simpson, C., & Jimenez, J. (2007a). Emission factors from wheat and Kentucky bluegrass stubble burning: Comparison of field and simulated burn experiments. *Atmospheric Environment, 41*, 1512–1520.

Dhammapala, R., Claiborn,C., Jimenez, J., Corkill, J., Gullett, B., Simpson, C., & Paulsen, M. (2007b). Emission factors of PAHs, methoxyphenols, levoglucosan, elemental carbon and organic carbon from simulated wheat and Kentucky bluegrass stubble burns. *Atmospheric Environment, 41*, 2660–2669.

Ding, X., Wang, X. M., Xie, Z., Xiang, C., Mai, B., Sun, L., et al. (2007). Atmospheric polycyclic aromatic hydrocarbons observed over the North Pacific Ocean and the Arctic area: Spatial distribution and source identification. *Atmospheric Environment, 41*, 2061–2072.

Dockery, D. W., Pope, C. A, I. I. I., Xu, P., Spengler, J. D., Ware, J. H., Fay, M. E., et al. (1993). An association between air pollution and mortality in six U.S. cities. *The New England Journal of Medicine, 329*, 1573–1579.

Dou, H., Chang, B., Wei, Z., Qiu, W., Liu, S., & Liu, Y. (2007). Emission factors of PAHs in residential coal combustion in China. *Acta Scientiae Circumstantiae, 27*, 1783–1788. (In Chinese).

Duan, F., He, K., Liu, X., Dong, S., & Yang, F. (2007). Research progress on carbonaceous aerosol: Organic carbon and elemental carbon. *Chinese Journal of Environmental Engineering, 1*, 1–8. (In Chinese).

Duan, J., Bi, X., & Tan, J. (2006). The particle diameter distribution of polycyclic aromatic hydrocarbons (PAHs) in atmospheric particle during haze period in Guangzhou. *China Environmental Science, 26*, 6–10. (In Chinese).

Duan, X., & Wei, F. (2002). The environmental pollution caused by Benzo[a]pyrene, its harm to health and the research focuses on it. *World Science and Technology Research and Development, 24*, 11–17. (In Chinese).

Duan, X., Wei, F., Zhang, J., & Yang, H. (2008). Study on biomarkers for polycyclic aromatic hydrocarbons exposure. *Industrial Health and Occupational Diseases, 34*, 129–132. (In Chinese).

Edwards, R. D., Smith, K. R., Zhang, J., & Ma, Y. (2004). Implications of changes in household stoves and fuel use in China. *Energy Policy, 32*, 395–411.

Eiguren-Fernandez, A., Miguel, A. H., Lu, R., Purvis, K., Grant, B., Mayo, P., et al. (2008). Atmospheric formation of 9, 10-phenanthraquinone in the Los Angeles air basin. *Atmospheric Environment, 42*, 2312–2319.

Englert, N. (2004). Fine particles and human health-a review of epidemiological studies. *Toxicology Letters, 149*, 235–242.

Fan, X., Lv, Z., Li, D., & Yu, X. (2010). Performance of cooking stoves with biomass pellet fuel. *Transactions of the Chinese Society of Agricultural Engineering, 26*, 280–284. (In Chinese).

Feng, J., Hu, M., Chan, C. K., Lau, P. S., Fang, M., He, L., et al. (2006). A comparative study of the organic matter in PM2.5 from three Chinese megacities in three different climatic zones. *Atmospheric Environment, 40*, 3983–3994.

Fernandes, M. B., Sicre, M. A., Boireau, A., & Tronczynski, J. (1997). Polyaromatic Hydrocarbon (PAH) distributions in the Seine River and its estuary. *Marine Pollution Bulletin, 34*, 857–867.

Fernandes, S. D., Trautmann, N. M., Streets, D. G., Roden, C. A., & Bond, T. C. (2007). Global biofuel use, 1850–2000. *Global biogeochemical cycle, 21*, GB2019.

Fitzpatrick, E. M., Ross, A. B., Bates, J., Andrews, G., Jones, J. M., Phylaktou, H., et al. (2007). Emission of oxygenated species from the combustion of pine wood and its relation to soot formation. *Process Safety and Environmental Protection, 85*, 430–440.

Galarneau, E., Bidleman, T. F., & Blanchard, P. (2006). Seasonality and interspecies differences in particle/gas partitioning of PAHs observed by the Integrated Atmospheric Deposition Network (IADN). *Atmospheric Environment, 40*, 182–197.

Garban, B., Blanchoud, H., Motelay-Massei, A., Chevreuil, M., & Ollivon, D. (2002). Atmospheric bulk deposition of PAHs onto France: Trends from Urban to remote sites. *Atmospheric Environment, 36*, 5395–5403.

Gaspari, L., Chang, S. S., & Santella, R. M. (2003). Polycyclic aromatic hydrocarbon-DNA adducts in human sperm as marker of DNA damage and infertility. *Mutation Research, 535*, 155–160.

Gigliotti, C. L., Totten, L. A., Offenberg, J. H., Dachs, J., Reinfelder, J. R., Nelson, E. D., et al. (2005). Atmospheric concentrations and deposition of polycyclic aromatic hydrocarbons to the Mid-Atlantic east coast region. *Environmental Science and Technology, 39*, 5550–5559.

GLC (2007) Assessment of Benzo(a)pyrene Air Emissions In the Great Lakes Region, Great Lakes Regional Toxic Air Emissions Inventory Steering Committee.

Goss, K., & Schwarzenbach, R. P. (1998). Gas/solid and gas/liquid partitioning of organic compounds: Critical evaluation of the interpretation of equilibrium constants. *Environmental Science and Technology, 32*, 2025–2032.

Gu, Y., Liou, K. N., Xue, Y., Mechoso, C. R., Li, W., & Luo, Y. (2006). Climate effects of different aerosols types in China simulated by the UCLA general circulation model. *Journal of Geophysical Research, 111*, D15201.

Gullett, B. K., Touati, A., & Hays, M. D. (2003). PCDD/F, PCB, HxCBz, PAH, and PM emission factors for fireplaces and woodstove combustion in the San Francisco Bay region. *Environmental Science and Technology, 37*, 1758–1765.

Guo, H., Lee, S. C., Ho, K. F., Wang, X. M., & Zou, S. C. (2003a). Particle-associated polycyclic aromatic hydrocarbons in Urban air of Hong Kong. *Atmospheric Environment, 37*, 5307–5317.

Guo, Z.G., Sheng, L.F., Feng, J.L., & Fang, M. (2003b). Seasonal variation of solvent extractable organic compounds in the aerosols in Qingdao, China. *Atmospheric Environment, 37*, 1825–1834.

Guo, H., Wang, T., Simpson, I. J., Blake, D. R., Yu, X. M., Kwok, Y. H., et al. (2004). Source contributions to ambient VOCs and CO at a rural site in eastern China. *Atmospheric Environment, 38*, 4551–4560.

Gupta, S., Saksena, S., Shankar, V. R., & Joshi, V. (1998). Emission factors and thermal efficiencies of cooking biofuels from five countries. *Biomass and Bioenergy, 14*, 547–559.

Hafner, W. D., Carlson, D. L., & Hites, R. A. (2005). Influence of local human population on atmospheric polycyclic aromatic hydrocarbon concentrations. *Environmental Science and Technology, 39*, 7374–7379.

Halsall, C. J., Coleman, P. J., Davis, B. J., Burnett, V., Waterhouse, K. S., Harding-Jones, P., et al. (1994). Polycyclic aromatic hydrocarbons in U.K. Urban Air. *Environmental Science and Technology, 28*, 2380–2386.

Halsall, C. J., Sweetman, A. J., & Barrie, L. A. (2001). Modeling the behavior of PAHs during atmospheric transport from the UK to the Arctic. *Atmospheric Environment, 35*, 255–267.

Hao, J., Wang, L., Li, L., Hu, J., & Yu, X. (2005). Air pollutants contribution and control strategies of energy-use related sources in Beijing. *Science in China, Series D, 48*, 138–146.

Hattori, T., Tang, N., Tamura, K., Hokoda, A., Yang, X. Y., Igarashi, K., et al. (2007). Particulate polycyclic aromatic hydrocarbons and their nitrated derivatives in three cities in Liaoning province, China. *Environmental Forensics, 8*, 165–172.

Hays, M. D., Fine, P. M., Geron, C. D., Kleeman, M. J., & Gullett, B. K. (2005). Open burning of agricultural biomass: Physical and chemical properties of particle-phase emissions. *Atmospheric Environment, 39*, 6747–6764.

He, G., Ying, B., Liu, J., Gao, S., Shen, S., Balakrishnan, K., et al. (2005). Patterns of household concentrations of multiple indoor air pollutants in China. *Environmental Science and Technology, 39*, 991–998.

Heywood, E., Wright, J., Wienburg, C. L., Black, H., Long, S. M., Osborn, D., et al. (2006). Factors influencing the national distribution of polycyclic aromatic hydrocarbons and polychlorinated biphenyls in British soils E-9466-2010. *Environmental Science and Technology, 40*, 7629–7635.

Hildemann, L. M., Cass, C. R., & Markowski, G. R. (1989). A dilution stack sampler for collection of organic aerosol emissions: Design, characterization and field tests. *Aerosol Science and Technology, 10*, 193–204.

Hinds, W. C. (1999). *Aerosol technology, properties, behavior, and measurements of airborne particles*. New York: Wiley Interscience.

Hu, J., & Guo, X. (2007). Research progress in indoor air pollution and health effects caused by combustion of biomass fuels. *Journal of Environmental and Health, 24*, 827–829. (In Chinese).

IARC (1987) IARC monographs on the evaluation of the carcinogenic risk of chemicals to humans. Overall evaluation of carcinogenicity: An updating of IAPC monographs (Vol. 1–42, Suppl. 7). Lyon, France: International Agency for Research on Cancer.

Intergovernmental Panel on Climate Change (IPCC) (2001) Climate change 2001: The scientific basis. In J.T Houghton (Ed.), *Contribution of working group I to the third assessment report of the intergovernmental panel on climate change*. New York: Cambridge University Press.

International Energy Agency (IEA) (2011a) Energy for all, financing access for the poor, special early excerpt of the world energy outlook 2011. http://www.iea.org/papers/2011/weo2011_energy_for_all.pdf

International Energy Agency (IEA) (2011b) Key world energy statistics, 2011. http://www.iea.org. Assessed Dec.2011.

Jacob, J., Grimmer, G., & Hildebrand, A. (1997). Long-term decline of atmospheric and marine pollution by polycyclic aromatic hydrocarbons (PAHs) in Germany. *Chemosphere, 34*, 2099–2108.

Jacob, J., & Seidel, A. (2002). Biomonitoring of polycyclic aromatic hydrocarbons in human urine. *Journal of Chromatography B, 778*, 31–47.

Jacobson, M. Z. (2001). Strong radiative heating due to the mixing state of black carbon in atmospheric aerosols. *Nature, 409*, 695–697.

Jenkins, B. M., Kennedy, I. M., Turn, S. Q., Williams, R. B., Hall, S. G., Teague, S. V., et al. (1993). Wind tunnel modeling of atmospheric emissions from agricultural burning: Influence of operating configuration on flame structure and particle emission factor for a spreading-type fire. *Environmental Science and Technology, 27*, 1763–1775.

Jenkins, B. M., Jones, A. D., Turn, S. Q., & Williams, R. B. (1996a). Particle concentrations, gas-particle partitioning and species intercorrelations for polycyclic aromatic hydrocarbons (PAHs) emitted during biomass burning. *Atmospheric Environment, 30*, 3825–3835.

Jenkins, B. M., Jones, A. D., Turn, S. Q., & Williams, R. B. (1996b). Emission factors for polycyclic aromatic hydrocarbons from biomass burning. *Environmental Science and Technology, 30*, 2462–2369.

Jetter, J. J., & Kariher, P. (2009). Solid-fuel household cook stoves: Characterization of performance and emissions. *Biomass and Bioenergy, 33*, 294–305.

Jiang, R., & Bell, M. L., (2008). A comparison of particulate matter from biomass burning rural and non-biomass burning urban households in Northeastern China. *Environmental Health Perspectives, 116*, 907–914.

Johansson, L. S., Leckner, B., Gustavsson, L., Cooper, D., Tullin, C., & Potter, A. (2004). Emission characteristics of modern and old-type residential boilers fired with wood logs and wood pellets. *Atmospheric Environment, 38*, 4183–4195.

Junker, C., & Liousse, C. (2008). A global emission inventory of carbonaceous aerosol form historic records of fossil fuel and biofuel consumption for the period 1860–1997. *Atmospheric Chemistry and Physics, 8*, 1195–1207.

Kanaly, R. A., & Harayama, S. (2000). Biodegradation of high-molecular-weight polycyclic aromatic hydrocarbons by bacteria: A mini-review. *Journal of Bacteriology, 182*, 2059–2067.

Kaneyasu, N., & Takada, H. (2004). Seasonal variations of sulfate, carbonaceous species (black carbon and polycyclic aromatic hydrocarbons), and trace elements in fine atmospheric aerosols collected at subtropical islands in the East China Sea. *Journal of Geophysical Research-Atmospheres, 109*(D6), D06211.

Karlehagen, S., Andersen, A., & Ohlson, C. G. (1992). Cancer incidence among creosote-exposed workers. *Scandinavian Journal of Work, Environment and Health, 18*, 26–29.

Katsoyiannis, A., Sweetman, A., & Jones, K. C. (2011). PAH molecular diagnostic ratios applied to atmospheric sources: A critical evaluation using two decades of source inventory and air concentration data from the UK. *Environmental Science and Technology, 45*, 8897–8906.

Keeler, G. (2004). Characterization of ultrafine particle number concentration and size distribution during a summer campaign in southwest Detroit. *Journal of the Air and Waste Manage Association, 54*, 1079–1090.

Kim Oanh, N. T., Reutergardh, L. B., & Dung, N. T. (1999). Emission of polycyclic aromatic hydrocarbons and particulate matter from domestic combustion of selected fuels. *Environmental Science and Technology, 33*, 2703–2709.

Kim Oank, N. T., Nghiem, L. H., & Yin, L. P. (2002). Emission of polycyclic aromatic hydrocarbons, toxicity and mutagenicity from domestic cooking using sawdust briquettes, wood and kerosene. *Environmental Science and Technology, 36*, 833–839.

Kim Oanh, N. T., & Zhang, B. N. (2004). Photochemical smog modeling for assessment of potential impacts of different management strategies on air quality of the Bangkok Metropolitan Region, Thailand. *Journal of the Air and Waste Manage Association, 54*, 1321–1338.

Kim Oanh, N. T., Upadhyay, N., Zhuang, Y. H., Hao, Z. P., Murthy, D. V. S., Lestari, P., et al. (2006). Particulate air pollution in six Asian cities: Spatial and temporal distributions, and associated sources. *Atmospheric Environment, 40*, 3367–3380.

Koch, D., & Hansen, J. (2005). Distant origins of arctic black carbon: A goddard institute for space studies model E experiment. *Journal of Geophysical Research, 110*, D04204.

Kojima, Y., Inazu, K., Hisamatsu, Y., Okochi, H., Baba, T., & Nagoya, T. (2010). Influence of secondary formation on atmospheric occurrences of oxygenated polycyclic aromatic hydrocarbons in airborne particles. *Atmospheric Environment, 44*, 2873–2880.

Laden, F., Neas, L. M., Dockery, D. W., & Schwartz, J. (2000). Association of fine particulate matter from different sources with daily mortality in 6 U.S. cities. *Environmental Health Perspective, 108*, 941–947.

Lang, C., Tao, S., Zhang, G., Fu, J. M., & Simonich, S. (2007). Outflow of polycyclic aromatic hydrocarbons from Guangdong, Southern China. *Environmental Science and Technology, 41*, 8370–8375.

Law, K. S., & Stohl, A. (2007). Arctic air pollution: Origins and impacts. *Science, 315*, 1537–1540.

Lei, Y., Zhang, Q., He, K. B., & Streets, D. G. (2011). Primary anthropogenic aerosol emission trends for China, 1990–2005. *Atmospheric Chemistry and Physics, 11*, 931–954.

Li, X., Fu, G., You, X., Yan, J., (2003). A study of the PAHs emissions resulting from the combustion of different ranks of coal. *Journal of Engineering for Thermal Energy & Power, 18*, 125–127. (In Chinese).

Li, X., Li, B., Tao, S., Guo, M., Cao, J., & Wang, X. (2005). Population exposure to PAHs in Tianjin area. *Acta Scientiae Circumstantiae, 25*(7), 989–993. (In Chinese).

Li, X., Duan, L., Wang, S., Duan, J., Guo, X., Yi, H., et al. (2007a). Emission characteristics of particulate matter from rural household biofuel combustion in China. *Energy and Fuels, 21*, 845–851.

Li, X., Wang, S., Duan, L., Hao, J., Li, C., Chen, Y., & Yang, L. (2007b). Particulate and trace gas emissions from open burning of wheat straw and corn stover in China. *Environmental science and technology, 41*, 6052–6058.

Li, J., Dong, Y., & Cao, Z. (2008). Emission factors of PAHs from rice straw burning. *China Environmental Science, 28*, 23–36. (In Chinese).

Li, X., Wang, S., Duan, L., Hao, J., & Nie, Y. (2009a). Carbonaceous aerosol emissions from household biofuel combustion in China. *Environmental Science and Technology, 43*, 6076–6081.

Li, X., Zhao, T., Yu, Y., Zhang, C., Li, P., & Li, S. (2009b). Population exposure to PAHs and the health risk assessment in Beijing area. *Journal of Agro-environmental Science, 28*(8), 1758–1765 (In Chinese).

Liu, S.Z., Tao, S., Liu, W.X., Liu, Y.N., Dou, H., Zhao, J.Y., Wang, L.G., Wang, J.F., Tian, Z.F., & Gao, Y. (2007). Atmospheric polycyclic aromatic hydrocarbons in North China: A wintertime study. *Environmental Science and Technology, 41*, 8256–8261.

Liu, S. (2008). *Ambient polycyclic aromatic hydrocarbons in the western watershed of Bohai Sea*. Ph.D thesis, Peking University (In Chinese).

Liu, S. Z., Tao, S., Liu, W. X., Dou, H., Liu, Y. N., Zhao, J. Y., et al. (2008a). Seasonal and spatial occurrence and distribution of atmospheric polycyclic aromatic hydrocarbons (PAHs) in rural and Urban areas of the North Chinese Plain. *Environmental Pollution, 156*, 651–656.

Liu, Z., Liu, J., Wang, B., Lu, F., Huang, S., Wu, D., & Han, D. (2008b). Aerosol observation in Fengtai area, Beijing. *Particuology, 6*, 214–217.

Liu, W. X., Dou, H., Wei, Z. C., Chang, B., Qiu, W. X., Liu, Y., et al. (2009). Emission characteristics of polycyclic aromatic hydrocarbons from combustion of different residential coals in North China. *Science of the Total Environment, 407*, 1436–1446.

Lin, Y., Zhou, Z., He, G., Wei, H., Liu, J., Liu, F., et al. (2005). Geographical, spatial and temporal distributions of multiple indoor air pollutants in four Chinese provinces. *Environmental Science and Technology, 39*, 9431–9439.

Lin, J., Nielsen, C. P., Zhao, Y., Lei, Y., Liu, Y., & McElroy, M. B. (2010). Recent changes in particulate air pollution over China observed from space and the ground: Effectiveness of emission control. *Environmental Science and Technology, 44*, 7771–7776.

Lipsky, E. M., & Robinson, A. L. (2006). Effects of dilution on fine particle mass and partitioning of semivolatile organics in diesel exhaust and wood smoke. *Environmental Science and Technology, 40*, 155–162.

Lohmann, R., & Lammel, G. (2004). Adsorptive and absorptive contributions to the gas-particle partitioning of polycyclic aromatic hydrocarbons: State of knowledge and recommended parametrization for modeling. *Environmental Science and Technology, 38*, 3793–3803.

Lu, H., Zhu, L., & Zhu, N. (2009). Polycyclic aromatic hydrocarbon emissions from straw burning and the influence of combustion parameters. *Atmospheric Environment, 43*, 978–983.

Lundstedt, S., White, P., Lemieux, C., Lynes, K., Lambert, I., Öberg, L., et al. (2007). Sources, fate, and toxic hazards of oxygenated polycyclic aromatic hydrocarbons (PAHs) at PAH-contaminated sites. *Ambio, 36*, 475–485.

MacDonald, R. W., Barrie, L. A., Bidleman, T. F., Diamond, M. L., Gregor, D. J., Semkin, R. G., Strachan, W. M. J., Li, Y. F., Wania, F., Alaee, M., Alexeeva, L. B., Bachus, S. M., Bailey, R., Bewers, J.M., Gobeil, C., Halsall, C. J., Harner, T., Hoff, J. T., Jantunen, L. M. M., Lockhort, W. L., Mackay, D., Muir, D. C. G., Pudykiewicz, J., Reimer, K. J., Smith, J. N., Stern, G. A., Schroeder, W. H., Wagemann, R., & Yunker, M. B. (2000). Contaminants in the Canadian Arctic: 5 years of progress in understanding sources, occurrence and pathways. *Science and Total Environment, 254*, 93–234.

Mackay, D. (2001). *Environmental multimedia models: The fugacity approach*. Boca Raton: Lewis Publishers.

Mader, B. T., Flagan R, C., & Seinfeld, J. H. (2001). Sampling atmospheric carbonaceous aerosols using a particle trap impactor/denuder sampler. *Environmental Science and Technology, 35*, 4857–4867.

Mastral, A., & Calleän, M. S. (2000). A review on Polycyclic Aromatic Hydrocarbon (PAH) emissions from energy generation. *Environmental Science and Technology, 34*, 3051–3057.

McClean, M. D., Rinehart, R. D., & Ngo, L. (2004). Urinary 1-Hydroxypyrene and polycyclic aromatic hydrocarbon exposure among asphalt paving workers. *The Annals of Occupational Hygiene, 48*, 565–578.

McClean, M. D., Rinehart, R. D., & Sapkota, A. (2007). Dermal exposure and urinary 1-hydroxypyrene among asphalt roofing workers. *Journal of Occupational and Environmental Hygiene, 4*, 118–126.

McCracken, J., Baccarelli, A., Hoxha, M., Dioni, L., Melly, S., Coull, B., et al. (2010). Annual ambient black carbon associated with shorter telomeres in elderly men: Veterans affairs normative aging study. *Environmental Health Perspective, 118*, 1562–1570.

McFerrin, C. A., Hall, R. W., & Dellinger, B. (2008). Ab initio study of the formation and degradation reactions of semiquinone and phenoxyl radicals. *Journal of Molecular Structure, 848*, 16–23.

McMurry, P. H., & Stolzenburg, M. (1989). On the sensitivity of particle size to reactive humidity for Los Angeles aerosols. *Atmospheric Environment, 23*, 497–507.

Meijer, S. N., Ockenden, W. A., Steinnes, E., Corrigan, B. P., & Jones, K. C. (2003). Spatial and temporal trends of POPs in Norwegian and UK background air: Implications for global cycling. *Environmental Science and Technology, 37*, 454–461.

Menichini, E., Monfredini, F., & Merli, F. (1999). The temporal variability of the profile of carcinogenic polycyclic aromatic hydrocarbons in Urban air: A study in a medium traffic area in Rome, 1993–1998. *Atmospheric Environment, 33*, 3739–3750.

Menon, S., Hansen, J., Nazarenko, L., & Luo, Y. (2002). Climate effects of black carbon aerosol in China and India. *Science, 297*, 2250–2254.

Menzie, C. A., Potocki, B. B., & Santodonato, J. (1992). Exposure to carcinogenic PAHs in the environment. *Environmental Science and Technology, 7*, 1278–1284.

Miller, E. C. (1978). Some current perspectives on chemical carcinogenesis in humans and experimental animals: Presidential address. *Cancer Research, 38*, 1479–1496.

Ministry of Environmental Protection of the People's Republic of China. Ambient air quality standard. GB 3095-1996 (In Chinese).

Motelay-Massei, A., Harner, T., Shoeib, M., Diamond, M., Stern, G., & Rosenberg, B. (2005). Using passive air samplers to assess urban–rural trends for persistent organic pollutants and polycyclic aromatic hydrocarbons. 2. Seasonal trends for PAHs, PCBs, and organochlorine pesticides. *Environmental Science and Technology, 39*, 5763–5773.

Mumford, J. L., He, X. Z., Chapman, R. S., Cao, S. R., Harris, D. B., Li, X. M., et al. (1987). Lung cancer and indoor air pollution in Xuan Wei, China. *Science, 235*, 217–220.

Mumford, J. L., Lee, X., Lewtas, J., Young, T. L., & Santella, R. M. (1993). DNA adducts as biomarkers for assessing exposure to polycyclic aromatic hydrocarbons in tissues from Xuan-Wei women with high exposure to coal combustion emissions and high lung cancer mortality. *Environmental Health and Perspective, 99*, 83–87.

Mumtaz, M. M., George, J. D., & Gold, K. W. (1996). ATSDR evaluation of health effects of chemicals. IV. Polycyclic aromatic hydrocarbons (PAHs): Understanding a complex problem. *Toxicology and Industrial Health, 12*, 742–971.

Na, K., Sawant, A. A., & Song, C. (2004). Primary and secondary carbonaceous species in the atmosphere of western riverside county, California. *Atmospheric Environment, 38*, 1345–1355.

Neff, J. M. (1979). *Polycyclic aromatic hydrocarbons in the aquatic environment: Sources fates and biological effects*. London: Applied Science Publication.

Nielson, T. (1984). Reactivity of polycyclic aromatic hydrocarbons toward nitrating species. *Environmental Science and Technology, 18*, 157–163.

Novakov, T., Andreae, M. O., Gabriel, R., Kirchstetter, T., Mayol-Bracero, O., & Ramanathan, V. (2000). Origin of carbonaceous aerosols over the tropical Indian Ocean: Biomass burning or fossil fuels. *Geophysical Research Letter, 27*, 4061–4064.

Odum, J. R., Hoffmann, T., Bowman, F., Collins, D., Flagan, R. C., & Seinfeld, J. H. (1996). Gas/particle partitioning and secondary organic aerosol yields. *Environmental Science and Technology, 30*, 2580–2585.

Okuda, T., Naoi, D., Tenmoku, M., Tanaka, S., He, K. B., Ma, Y. L., et al. (2006). Polycyclic aromatic hydrocarbons (PAHs) in the aerosol in Beijing, China, measured by aminopropylsilane chemically-bonded stationary-phase column chromatography and HPLC/fluorescence detection. *Chemosphere, 35*, 427–435.

Pacyna, J. M., Breivik, K., Munch, J., & Fudala, J. (2003). European atmospheric emissions of selected persistent organic pollutants, 1970–1995. *Atmospheric Environment, 37*, 119–131.

Palmer, H. J., & Paulson, K. E. (1997). Reactive oxygen species and antioxidants in signal transduction and gene expression. *Nutrition Review, 55*, 353–361.

Pandey, K., Wheeler, D., Ostro, B., Deichmann, U., Hamilton, K., & Bolt, K. (2006). Ambient particulate matter concentrations in residential and pollution hotspot areas of world cities: New estimates based on the Global Model of Ambient Particulates (GMAPS) The World Bank Development Economics Research Group and the Environment Department Working Paper. Washington: The World Bank.

Pandey, S. K., Kim, K. H., Chung, S. Y., Cho, S. J., Kim, M. Y., & Shon, Z. H. (2008). Long-term study of NOx behavior at Urban roadside and background locations in Seoul, Korea. *Atmospheric Environment, 42*, 607–622.

Pankow, J. (1987). Review and comparative analysis of the theories on partitioning between the gas and aerosol particulate phases in the atmosphere. *Atmospheric Environment, 21*, 2275–2283.

Perera, F., Hemminki, K., Jedrychowski, W., & Whyatt, R. (2002). In utero DNA damage from environmental pollution is associated with somatic gene mutation in newborns. *Cancer Epidemiology, Biomarkers and Prevention, 11*, 1134–1137.

Peng, R., Liu, J., Pan, X., Liu, H., Wen, T., Jin, X., et al. (2005). The content and the proportion of the combustion particles PM2.5, PM10 and TSP in three main fuels. *Journal of Environment and Health, 22*, 13–15. (In Chinese).

Pope, C. A, I. I. I., Ezzati, M., & Dockery, D. W. (2009). Fine-particulate air pollution and life expectancy in the United States. *The New England Journal of Medicine, 360*, 376–386.

Pouliot, G., Wisner, E., Mobley, D., & Hunt, W, Jr. (2012). Quantification of emission factor uncertainty. *Journal of the Air and Waste Management Association, 62*, 287–298.

Prevedouros, K., Brorström-Lundén, E., Halsall, C. J., Jones, K. C., Lee, R. G. M., & Sweetman, A. J. (2004). Seasonal and long-term trends in atmospheric PAH concentrations: Evidence and implications. *Environmental Pollution, 128*, 17–27.

Primbs, T., Simonich, S., Schmedding, D., Wilson, G., Jaffe, D., Takami, A., et al. (2007). Atmospheric outflow of anthropogenic semivolatile organic compounds from East Asia in spring 2004. *Environmental Science and Technology, 41*, 3551–3558.

Pryor, W. A., Prier, D. G., & Church, D. F. (1983). Electron-spin resonance study of mainstream and sidestream cigarette smoke: Nature of the free radicals in gas-phase smoke and in cigarette tar. *Environmental Health Perspective, 47*, 345–355.

Pufulete, M., Battershill, J., Boobis, A., & Fielder, R. (2004). Approaches to carcinogenic risk assessment for polycyclic aromatic hydrocarbons: A UK perspective. *Regulatory Toxicology and Pharmacology, 40*, 54–66.

Purvis, C. R., McCrillis, R. C., & Kariher, P. H. (2000). Fine particulate matter (PM) and organic speciation of fireplace emissions. *Environmental Science and Technology, 34*, 1653–1658.

Qin, S., Tang, J., & Wen, Y. (2001). Black carbon and its importance in climate change studies. *Meteorological Monthly, 27*, 2–6. (In Chinese).

Quinn, P. K., Shaw, G., Andrews, E., Dutton, E. G., Ruoho-airola, T., & Gong, S. L. (2006). Arctic haze: Current trends and knowledge gaps. *Tellus, 59*, 99–114.

Ramesh, A., Walker, S. A., & Hood, D. B. (2004). Bioavailability and risk assessment of orally ingested polycyclic aromatic hydrocarbons. *International Journal of Toxicology, 23*, 301–333.

Reid, J. S., Koppmann, R., Eck, T. F., & Eleuterio, D. P. (2005). A review of biomass burning emissions part II: Intensive physical properties of biomass burning particles. *Atmospheric Chemistry and Physics, 5*, 799–825.

Reisen, F., & Arey, J. (2002). Reactions of hydroxyl radicals and ozone with acenaphthene and acenaphthylene. *Environmental Science and Technology, 36*, 4302–4311.

Reisen, F., & Arey, J. (2005). Atmospheric reactions influence seasonal PAH and Nitro-PAH concentrations in the Los Angeles Basin. *Environmental Science and Technology, 39*, 64–73.

Remer, I. A., Kleidman, R. G., Levy, R.C., Kaufman,Y. J., Tanre, D., Mattoo, S., Martins, H. V., Ichoku, C., Koren, I., Yu, H. B., & Holben, B. N. (2008). Global aerosol climatology from the MODIS satellite sensors. *Journal of Geophysical Research, 113*, D14S07.

Roden, C. A., Bond, T. C., Conway, S., & Pinel, A. B. O. (2006). Emission factors and real-time optical properties of particles emitted from traditional wood burning cookstoves. *Environmental Science and Technology, 40*, 6750–6757.

Roden, C. A., Bond, T. C., Conway, S., Pinel, A. B. O., MacCarty, N., & Still, D. (2009). Laboratory and field investigations of particulate and carbon monoxide emissions from traditional and improved cook stoves. *Atmospheric Environment, 43*, 1170–1181.

Rogge, W. F., Hildemann, L. M., Mazurek, M. A., Cass, G. R., & Simoneit, B. R. T. (1991). Sources of fine organic aerosol. 1. charbroilers and meat cooking operations. *Environmental Science and Technology, 25*, 1112–1125.

Rogge, W. F., Hildemann, L. M., Mazurek, M. A., Cass, G. R., & Simoneit, B. R. T. (1993a). Sources of fine organic aerosol. 2. noncatalyst and catalyst-equipped automobiles and heavy duty diesel trucks. *Environmental Science and Technology, 27*, 636–651.

Rogge, W. F., Hildemann, L. M., Mazurek, M. A., Cass, G. R., & Simoneit, B. R. T. (1993b). Sources of fine organic aerosol. 3. road dust, tire debris, and organ metallic brake lining dust: Roads as sources and sinks. *Environmental Science and Technology, 27*, 1892–1904.

Rogge, W. F., Hildemann, L. M., Mazurek, M. A., Cass, G. R., & Simoneit, B. R. T. (1993c). Sources of fine organic aerosol. 4. particulate abrasion products from leaf surfaces of Urban plants. *Environmental Science and Technology, 27*, 2700–2711.

Rogge, W. F., Hildemann, L. M., Mazurek, M. A., Cass, G. R., & Simoneit, B. R. T. (1993d). Sources of fine organic aerosol. 5. natural gas home appliances. *Environmental Science and Technology, 27*, 2736–2744.

Rogge, W. F., Hildemann, L. M., Mazurek, M. A., Cass, G. R., & Simoneit, B. R. T. (1994). Sources of fine organic aerosol. 6. cigarette smoke in the Urban atmosphere. *Environmental Science and Technology, 28*, 1375–1388.

Rogge, W. F., Hildemann, L. M., Mazurek, M., Cass, G. R., & Simoneit, B. R. T. (1996). Mathematical modeling of atmospheric fine particle-associated primary organic compound concentrations. *Journal of Geophysical Research, 101*, 19379–19394.

Rogge, W. F., Hildemann, L. M., Mazurek, M. A., Cass, G. R., & Simoneit, B. R. T. (1997a). Sources of fine organic aerosol. 7. hot asphalt roofing tar pot fumes. *Environmental Science and Technology, 31*, 2726–2730.

Rogge, W. F., Hildemann, L. M., Mazurek, M. A., Cass, G. R., & Simoneit, B. R. T. (1997b). Sources of fine organic aerosol. 8. boilers burning No. 2 distillate fuel oil. *Environmental Science and Technology, 31*, 2731–2737.

Rogge, W. F., Hildemann, L. M., Mazurek, M. A., Cass, G. R., & Simoneit, B. R. T. (1998). Sources of fine organic aerosol. 9. pine, oak and synthetic log combustion in residential fireplaces. *Environmental Science and Technology, 32*, 13–22.

Romundstad, P., Haldorsen, T., & Andersen, A. (2000). Cancer incidence and cause-specific mortality among workers in two Norwegian aluminium reduction plants. *American Journal of Industrial Medicine, 37*, 175–183.

Russell, P. B., Hobbs, P. V., & Stowe, L. L. (1999). Aerosol properties and radiative effects in the United States East Coast haze plume: An overview of the tropospheric aerosol radiative forcing observational experiment (TARFOX). *Journal of Geophysical Research, 104*, 2213–2222.

Russell, A. G., & Brunekreef, B. (2009). A focus on particulate matter and health. *Environmental Science and Technology, 43*, 4620–4625.

Schauer, J. J., Rogge, W. F., Hildemann, L. M., Mazurek, M. A., & Cass, G. L. (1996). Source apportionment of airborne particulate matter using organic compounds as tracers. *Atmospheric Environment, 30*, 3837–3855.

Schauer, J. J., Kleeman, M. J., Cass, G. R., & Simoneit, B. R. T. (1999a). Measurement of emissions from air pollution sources 1. C1 through C29 organic compounds from meat charbroiling. *Environmental Science and Technology, 33*, 1566–1577.

Schauer, J. J., Kleeman, M.J., Cass, G. R., & Simoneit, B. R. T. (1999b). Measurement of emissions from air pollution sources 2. C1 through C30 organic compounds from medium duty diesel trucks. *Environmental Science and Technology, 33*, 1578–1587.

Schauer, J. J., Kleeman, M. J., Cass, G. R., & Simoneit, B. R. T. (2001). Measurement of emissions from air pollution sources 3. C1 through C29 organic compounds from fireplace combustion of wood. *Environmental Science and Technology, 33*, 1716–1728.

Schauer, J. J., Kleeman, M. J., Cass, G. R., & Simoneit, B. R. T. (2002a). Measurement of emissions from air pollution sources 4. C1 through C27 organic compounds from cooking with seed oils. *Environmental Science and Technology, 36*, 567–575.

Schauer, J. J., Kleeman, M.J., Cass, G. R., & Simoneit, B. R. T. (2002b). Measurement of emissions from air pollution sources 5. C1-C32 organic compounds from gasoline-powered motor vehicles. *Environmental Science and Technology, 33*, 1169–1180.

Schimidl, C., Luisser, M., Padouvas, E., Lasselsberger, L., Rzaca, M., Curz, C. R., et al. (2011). Particulate and gaseous emissions from manually and automatically fired small scale combustion systems. *Atmospheric Environment, 45*, 7443–7454.

Sehili, A. M., & Lammel, G. (2007). Global fate and distribution of polycyclic aromatic hydrocarbons emitted from Europe and Russia. *Atmospheric Environment, 41*, 8301–8315.

Seinfeld, J. H., & Pandis, S. N. (2006). *Atmospheric chemistry and physics: Air pollution to climate*. New York: Wiley.

Shi, Z. B., He, K. B., Yu, X. C., Yao, Z. L., Yang, F., Ma, Y. L., et al. (2007). Diurnal variation of number concentration and size distribution of ultrafine particles in the Urban atmosphere of Beijing in winter. *Journal of Environmental Science, 19*, 933–938.

Shindell, D., & Faluvegi, G. (2009). Climate response to regional radiative forcing during the twentieth century. *Nature Geoscience, 2*, 294–300.

Sidhu, S., Gullett, B., Striebich, R., Klosterman, J., Contreras, J., & DeVito, M. (2005). Endocrine disrupting chemical emissions from combustion sources: Diesel particulate emissions and domestic waste open burn emissions. *Atmospheric Environment, 39*, 801–811.

Simioli, P., Lupi, S., & Gregorio, P. (2004). Non-smoking coke oven workers show an occupational PAH exposure-related increase in urinary mutagens. *Mutation Research-Genetic Toxicology and Environmental Mutagenesis, 562*, 103–110.

Simoneit, B. R. T., Rogge, W. F., Mazurek, M. A., Standley, L. J., Hildemann, L. M., & Cass, G. R. (1993). Lignin pyrolysis products, lignans, and resin acids as specific tracers of plant classes in emissions from biomass combustion. *Environmental Science and Technology, 27*, 2533–2541.

Simoneit, B. R. T. (2002). Biomass burning- a review of organic tracers for smoke from incomplete combustion. *Applied Geochemistry, 17*, 129–162.

Smith, K. R. (1993). Fuel combustion, air pollution exposure, and health: The situation in developing countries. *Annual Review of Environment and Resources, 18*, 529–566.

Smith, D. J. T., & Harrison, R. M. (1996). Concentrations, trends and vehicle source profile of polynuclear aromatic hydrocarbons in the U.K. atmosphere. *Atmospheric Environment, 30*, 2513–2525.

Stohl, A., Andrews, E., Burkhart, J. F., Forster, C., Herber, A., Hoch, S. W., et al. (2006). Pan-arctic enhancements of light absorbing aerosols concentrations due to North American boreal forest fires during summer 2004. *Journal of Geophysical Research, 111*, D22214.

Streets, D. G., Gupta, S., Waldhof, S. T., Wang, M. Q., Bond, T. C., & Yi, B. (2001). Black carbon emissions in China. *Atmospheric Environment, 35*(428), 1–4296.

Streets, D. G., Bond, T. C., Carmichael, G. R., Fernandes, S.D., Fu, Q., He, D., Klimont, Z., Nelson, S. M., Tsai, N.Y., Wang, M. Q., Woo, J. -H., & Yarber, K. F. (2003). An inventory of gaseous and primary aerosol emissions in Asia in the year 2000. *Journal of Geophysical Research, 108*, D21, 8809, 1–23.

Suess, M. J. (1976). The environmental load and cycle of polycyclic aromatic hydrocarbons. *Science of the Total Environment, 6*, 239–250.

Sun, Y., Zhuang, G., Wang, Y., Han, L., Guo, J., Dan, M., et al. (2004). The air-borne particulate pollution in Beijing-concentration, composition, distribution and sources. *Atmospheric Environment, 38*, 5991–6004.

Suzuki, K., & Yoshinaga, J. (2007). Inhalation and dietary exposure to polycyclic aromatic hydrocarbons and urinary 1-hydroxypyrene in non-smoking university students. *Institute of Occupational and Environmental Health, 81*, 115–121.

Tan, J., Duan, J., Zhao, J., Bi, X., Sheng, G., Fu, J., et al. (2009). The size distribution of organic carbon and elemental carbon during haze period in Guangzhou. *Environmental Chemistry, 28*, 267–271. (In Chinese).

Tang, N., Hattori, T., Taga, R., Igarashi, K., Yang, X. Y., Tamura, K., et al. (2005). Polycyclic aromatic hydrocarbons and nitropolycyclic aromatic hydrocarbons in Urban air particulates and their relationship to emission sources in the Pan-Japan Sea countries. *Atmospheric Environment, 39*, 5817–5826.

Tang, X., Zhang, Y., & Shao, M. (2006). *Atmospheric environmental chemistry*. High Education Press. (In Chinese).

Tang, X., Bi, X., Chen, Y., Sheng, G., & Fu, J. (2006b). Study on the size distribution of organic carbon and elemental carbon in the aerosol. *Research of Environmental Science, 19*, 104–108. (In Chinese).

Tao, S., Li, X. R., Yang, Y., Coveney, R. M., Lu, X. X., Chen, H. T., et al. (2006). Dispersion modeling of polycyclic aromatic hydrocarbons from combustion of biomass and fossil fuels and production of coke in Tianjin, China. *Environmental Science and Technology, 40*, 4586–4591.

Tian, L., Lan, Q., Yang, D., He, X., Yu, I. T. S., & Hammond, S. K. (2009). Effect of chimneys on indoor air concentrations of PM10 and benzo[a]pyrene in Xuan Wei, China. *Atmospheric Environment, 43*, 3352–3355.

Tsai, S. M., Zhang, J., Smith, K. R., Ma, Y., Rasmussen, R. A., & Khalil, M. A. K. (2003). Characterization of non-methane hydrocarbons emitted from various cookstoves used in China. *Environmental Science and Technology, 37*, 2869–2877.

Tsibulsky, V., Sokolovsky, V., & Dutchak, S. (2001). MSC-E contribution to the HM and POP emission inventories, Technical Note 7/2001. http://www.msceast.org/publications.html

Turpin, B. J., & Huntzicker, J. J. (1995). Identification of secondary organic aerosol episodes and quantification of primary and secondary organic aerosol concentrations during SCAQS. *Atmospheric Environment, 29*, 3527–3544.

Tian, H., Zhao, D., & Wang, Y. (2011). Emission inventories of atmospheric pollutants discharged from biomass burning in China. *Acta Scientiae Circumstantiae, 31*, 349–357. (In Chinese).

U.S. EPA (1998). (1990) Emissions Inventory Of Section 112(C) (6) Pollutants: Polycyclic Organic Matter (Pom), 2,3,7,8-Tetrachlorodibenzo-P-Dioxin (Tcdd)/2,3,7,8-Tetrachlorodibenzofuran (Tcdf), Polychlorinated Biphenyl Compounds (Pcbs), Hexachlorobenzene, Mercury, And Alkylated Lead. Emission Factor and Inventory Group (MD-14) Emissions MaADUSEPA.

U.S. EPA (2001) Residential wood combustion. Volume III, Chapter 2. 2001, Eastern Research Group, Inc.

U.S. EPA (2006) Evaluation report: EPA can improve emission factors development and management. 2006-P-00017. Washington: Office of Inspector general. Environmental protection agency. http://www.epa.gov/oig/reports/2006/20060322-2006-P-00017.pdf

U.S. EPA (2012) Report to congress on black carbon. Department of the interior, environment, and related agencies. EPA-450/R-12-001. http://www.epa.gov/blackcarbon/

Van Donkelaar, A., Martin, R.V., Brauer, M., Kahn, R., Levy, R., Verduzco, C., Villeneuve, P.J., (2010). Global estimates of ambient fine particulate matter concentrations from satellite-based aerosol optical depth: development and application. *Environmental Health Perspectives*, 118, 847–855.

Van Jaarsveld, J. A., Van Pul, W. A. J., & De Leeuw, F. A. A. M. (1997). Modelling transport and deposition of persistent organic pollutants in the European region. *Atmospheric Environment, 31*, 1011–1024.

Van Rooij, J. G. M., Veeger, M. M. S., & Bodelier-Bade, M. M. (1994). Smoking and dietary intake of polycyclic aromatic hydrocarbons as sources of interindividual variability in the baseline excretion of 1-hydroxypyrene in urine. *International Achieves of Occupational and Environmental Health, 66*, 55–65.

Vasconcelos, L. A., Macias, E. S., & White, W. H. (1994). Aerosol composition as a function of haze and humidity levels in the southwestern US. *Atmospheric Environment, 28*, 3679–3691.

Venier, M., & Hites, R. A. (2010). Regression model of partial pressures of PCBs, PAHs, and organochlorine pesticides in the Great Lakes' atmosphere. *Environmental Science and Technology, 44*, 618–623.

Venkataraman, C., & Friedlander, S. (1994). Size distributions of polycyclic aromatic hydrocarbons and elemental carbon. 2. ambient measurement and effects of atmospheric processes. *Environmental Science and Technology, 28*, 563–572.

Venkataraman, C., Lyons, J. M., & Friedlander, S.K. (1994). Size distribution of polycyclic aromatic hydrocarbons and elemental carbon. 1. sampling, measurement methods, and source characterization. *Environmental Science and Technology, 28*, 555–562.

Venkataraman, C., & Rao, G. U. M. (2001). Emission factors of carbon monoxide and size-resolved aerosols from biofuel combustion. *Environmental Science and Technology, 35*, 2100–2107.

Venkataraman, C., Negi, G., Sardar, S. B., & Rastogi, R. (2002). Size distributions of polycyclic aromatic hydrocarbons in aerosol emissions from biofuel combustion. *Journal of Aerosol Science, 33*, 503–518.

Veselovskii, I., Kolgotin, A., Griaznov, V., Müller, D., Franke, K., & Whiteman, D. N. (2004). Inversion of multiwavelength Raman lidar data for retrieval of bimodal aerosol size distribution. *Applied Optics, 43*, 1180–1195.

Vyskocil, A., Fiala, Z., & Fialova, D. (1997). Environmental exposure to polycyclic aromatic hydrocarbons in Czech Republic. *Human and Experimental Toxicology, 16*, 589–595.

Vyskocil, A., Fiala, Z., & Chenier, V. (2000). Assessment of multipathway exposure of small children to PAH. *Environmental Toxicology and Pharmacology, 8*, 111–118.

Vyskocil, A., Viau, C., & Camus, M. (2004). Risk assessment of lung cancer related to environmental PAH pollution sources. *Human and Experimental Toxicology, 23*, 115–127.

Walgraeve, C., Demesstere, K., Dewulf, J., Zimmermann, R., & van Langenhove, H. (2010). Oxygenated polycyclic aromatic hydrocarbons in atmospheric particulate matter: Molecular characterization and occurrence. *Atmospheric Environment, 44*, 1831–1846.

Wan, X. L., Chen, J. W., Tian, F. L., Sun, W. J., Yang, F. L., & Saiki, K. (2006). Source apportionment of PAHs in atmospheric particulates of Dalian: Factor analysis with nonnegative constraints and emission inventory analysis. *Atmospheric Environment, 40*, 6666–6675.

Wang, W. (2010). *Distribution and soil-air exchange of polycyclic aromatic hydrocarbons and derivatives in the Beijing-Tianjin region*. Ph.D thesis, Peking University (In Chinese).

Wen, J., Zheng, Y., Wang, W., Guang, L., & Ma, Q. (2006). Study on the chronic obstructive pulmonary disease in Shantou area. *Hainan Medical Journal, 17*, 122–123. (In Chinese).

Wang, G., & Kawamura, K. (2005). Molecular characteristics of Urban organic aerosols from Nanjing: A case study of a mega-city in China. *Environmental Science and Technology, 39*, 7430–7438.

Wang, G., Kawamura, K., Watanable, T., Lee, S., Ho, K., & Cao, J. (2006a). High loading s and source strengths of organic aerosols in China. *Geophysical Research Letter, 33*, L22801.

Wang, G., Kawamura, K., Lee, S., Ho, K., & Cao, J. (2006b). Molecular, seasonal and spatial distributions of organic aerosols from fourteen Chinese cities. *Environmental Science and Technology, 40*, 4619–4625.

Wang, G., Kawamura, K., Zhao, X., Li, Q., Niu, H., & Dai, Z. (2007a). Identification, abundance and seasonal variation of anthropogenic organic aerosols from a mega-city in China. *Atmospheric Environment, 41*, 407–416.

Wang, G., Kawamura, K., Hatakeyama, S., Takami, A., Li, H., & Wang, W. (2007b). Aircraft measurement of organic aerosols over China. *Environmental Science and Technology, 41*, 3115–3120.

Wang, W., Tao, S., Wang, W., Shen, G., Zhao, J., & Lam, K. (2009a). Airborne particulates and polycyclic aromatic hydrocarbons (PAHs) in ambient air in Donghe, Northern China. *Journal of Environmental Science and Health, Part A 44*, 854–860.

Wang, S., Wei, W., Du. Li., Li, G., & Hao, J. (2009b). Characteristics of gaseous pollutants from biofuel-stoves in rural China. *Atmospheric environment, 43*, 4148–4154.

Wang, R., Tao, S., Wang, B., Yang, Y., Lang, C., Zhang, Y., et al. (2010). Source and pathways of polycyclic aromatic hydrocarbons transported to Alert, the Canadian High Arctic. *Environmental Science and Technology, 44*, 1017–1022.

Wang, W. T., Staci, L. M. S., Basant G., Xue, M., Zhao, J. Y., Chen, S. J., Shen, H. Z., Shen, G. F., Wang, R., Cao, J., & Tao, S. (2011a). Spatial distribution and seasonal variation of atmospheric bulk deposition of polycyclic aromatic hydrocarbons in Beijing-Tianjin region, North China. *Environmental Pollution, 159*, 287–293.

Wang, W. T., Staci, L. M. S., Wang, W, Basant, G., Zhao, J. Y., Xue, M., Cao, J., Lu, X. X., Tao, S. (2011b). Atmospheric Polycyclic aromatic hydrocarbon concentrations and gas/particle partitioning at background, rural village and Urban sites in the North China Plain. *Atmospheric Research, 99*, 197–206.

Wang, R., Tao, S., Shen, H., Wang, X., Li, B., Shen, G., et al. (2012). Global emission of black carbon from motor vehicles from 1960–2006. *Environmental Science and Technology, 46*, 1278–1284.

Wania, F., & Mackay, D. (1993). Global fraction and cold condensation of low volatility organochloride compounds in polar-regions. *Ambio, 22*, 10–18.

Warneke, C., Bahreini, R., Brioude, J., Brock, C. A., deGouw, J. A., Fahey, D. W., et al. (2009). Biomass burning in Siberia and Kazakhstan as an important source for haze over the Alaskan Arctic in April 2008. *Geophysical Research Letter, 36*, L02813.

Warneke, C., Froyd, K. D., Brioude, J., Bahreini, R., Brock, C. A., Cozic, J., et al. (2010). An important contribution to springtime Arctic aerosol from biomass burning in Russia. *Geophysical Research Letters, 37*, L01801.

Wenborn, M., Coleman, P., Passant, N., Lymberidi, E., Sully, J., & Weir, R. (1999). Speciated PAH inventory for the UK. Department of the Environment TatR.

Whitby, K. T. (2007). The physical characteristics of sulfur aerosols. *Atmospheric Environment, 41*, 25–49.

Wilcke, W., & Amelung, W. (2000). Persistent organic pollutants in native grassland soils along a climosequence in North America A-1573-2008. *Soil Science Society of America Journal, 64*, 2140–2148.

Wild, S. R., & Jones, K. C. (1995). Polynuclear aromatic hydrocarbons in the United Kingdom environment: A preliminary source inventory and budget. *Environmental Pollution, 88*, 91–108.

Wilson, N. K., McCurdy, T. R., & Chuang, J. C. (1995). Concentrations and phase distributions of nitrated and oxygenated polycyclic aromatic hydrocarbons in ambient air. *Atmospheric Environment, 29*, 2575–2584.

World health organization (WHO) (2002a) The world health report: Reducing risks, promoting healthy life. Geneva, Switzerland.

World Health Organization(WHO) (2002b). The health effects of indoor air pollution exposure in developing countries, WHO/SDE/OEH/02.05.

World Health Organization (WHO) (2011) Database: Outdoor air pollution in cities. http://www.who.int/phe/health_topics/outdoorair/databases/en/index.html. Accessed April 2012.

Wu, S. P., Tao, S., Xu, F. L., Dawson, R., Lan, T., Li, B. G., et al. (2005). Polycyclic aromatic hydrocarbons in dustfall in Tianjin, China. *Science of the Total Environment, 345*, 115–126.

Wu, S. P., Tao, S., & Liu, W. X. (2006). Particle size distributions of polycyclic aromatic hydrocarbons in rural and Urban atmosphere of Tianjin, China. *Chemosphere, 62*, 357–367.

Wu, J., Hou, H. Y., & Ritz, B. (2010). Exposure to polycyclic aromatic hydrocarbons and missed abortion in early pregnancy in a Chinese population. *Science of the Total Environment, 408*, 2312–2318.

Xia, Z., Duan, X., Qiu, W., Liu, D., Wang, B., Tao, S., et al. (2010). Health risk assessment on dietary exposure to polycyclic aromatic hydrocarbons (PAHs) in Taiyuan, China. *Science of the Total Environment, 408*, 5331–5337.

Xie, M., Wang, G., Hu, S., Zhao, X., & Shen, G. (2008). Size distributions of particulate matter and PAHs from Nanjing in summer and autumn. *China Environmental Science, 28*, 867–871. (In Chinese).

Xu, X. (1984). Nitro-polycyclic aromatic hydrocarbons-direct-acting mutagens in environment. *Environmental Chemistry, 3*, 1–16. (In Chinese).

Xu, S. S., Liu, W. X., & Tao, S. (2006). Emission of polycyclic aromatic hydrocarbons in China. *Environmental Science and Technology, 40*, 702–708.

Xue, W. L., & Warshawsky, D. (2004). Metabolic activation of polycyclic and heterocyclic aromatic hydrocarbons and DNA damage: A review. *Toxicology and Applied Pharmacology, 206*, 73–93.

Youhei, K., Emiko, M., & Kazuhiko, S. (2004). Size distributions of mutagenic compounds and mutagenicity in atmospheric particulate matter collected with a low—pressure cascade impactor. *Atmospheric Environment, 38*, 2125–2132.

Yu, J., Yu, T., Yang, X., Shi, J., & Wang, X. (2004). Characteristics of organic carbon and elemental carbon in PM2.5 in Beijing in the winter. *Research of Environmental Sciences, 17*, 48–55. (In Chinese).

Yuan, Y., Liu, S., Castro, R., & Pan, X. (2012). PM2.5 monitoring and mitigation in the cities of China. *Environmental Science and Technology, 46*, 3627–3628.

Zhang, W., Zou, L., Li, Y., & Atkiston, S. (2003). The characterization of polycyclic aromatic hydrocarbons in particle emitted from different firewood burning in two conditions. *Environmental Monitoring in China, 19*, 23–27. (In Chinese).

Zhang, Y. (2010). *Emission, air transport and exposure risk of polycyclic aromatic hydrocarbons in China*. Ph.D thesis, Peking University (In Chinese).

Zhou, W., Fu, D., & Sun, C. (1990). A list of priority pollutants in water. *Environmental Monitoring in China, 6*, 1–4. (In Chinese).

Zhang, J., & Smith, K. R. (1999). Emissions of carbonyl compounds from various cookstoves in China. *Environmental Science and Technology, 33*, 2311–2320.

Zhang, J., Smith, K. R., Uma, R., Ma, Y., Kishore, V. V. N., Lata, K., et al. (1999). Carbon monoxide from cookstoves in developing countries: 1 emission factors. *Chemosphere, 1*, 353–366.

Zhang, J., Smith, K. R., Ma, Y., Ye, S., Jiang, F., Qi, W., et al. (2000). Greenhouse gases and other airborne pollutants from household stoves in China: A database for emission factors. *Atmospheric Environment, 34*, 4537–4549.

Zhang, J., & Smith, K. R. (2007). Household air pollution from coal and biomass fuels in China: Measurements, health impacts, and interventions. *Environmental Health Perspective, 115*, 848–855.

Zhang, Q., Streets, D. G., He, K., Klimont, Z. (2007a). Major components of China's anthropogenic primary particulate emissions. *Environmental Research Letters, 2*, 27–45.

Zhang, Y. X., Tao, S., Cao, J., & Coveney, R. M. (2007b). Emission of polycyclic aromatic hydrocarbons in China by county. *Environmental Science and Technology, 41*, 683–687.

Zhang, Y., Dou, H., Chang, B., Wei Z., Qiu, W., Liu, S., Liu, W., & Tao, S. (2008a). Emission of polycyclic aromatic hydrocarbons form indoor straw burning and emission inventory updating in China. *Annals of the New York Academy of Science, 1140*, 218–227.

Zhang, Y. X., Schauer, J. J., Zhang, Y., Zeng, L., Wei, Y., Liu, Y., & Shao, M. (2008b). Characteristics of particulate carbon emissions from real-world Chinese coal combustion. *Environmental Science and Technology, 42*, 5068–5073.

Zhang, H., Ye. X., Cheng, T., Chen, J., Yang, X., Wang, L., & Zhang, R. (2008c). A laboratory study of agricultural crop residue combustion in China. Emission factors and emission inventory. *Atmospheric Environment, 42*, 8432–8441.

Zhang, Y. X., & Tao, S. (2009). Global atmospheric emission inventory of polycyclic aromatic hydrocarbons (PAHs) for 2004. *Atmospheric Environment, 43*, 812–819.

Zhang, R. J., Han, Z. W., Cheng, T. T., & Tao, J. (2009a). Chemical properties and origin of dust aerosols in Beijing during springtime. *Particuology, 7*, 61–67.

Zhang, H., Wang, Z., Guo, P., & Wang, Z. (2009b). A modeling study of the effects of direct radiative forcing due to carbonaceous aerosol on the climate in East Asia. *Advances in Atmospheric sciences, 1*, 57–66.

Zhang, Y. X., Tao, S., Shen, H. Z., & Ma, J. M. (2009c). Inhalation exposure to ambient polycyclic aromatic hydrocarbons and lung cancer risk of Chinese population. *Proceedings of the National Academy of Sciences (PNAS), 106*, 21063–21067.

Zhang, H., Hu, D., Chen, J., Ye, X., Wang, S., Hao, J., et al. (2011). Particle size distribution and polycyclic aromatic hydrocarbons emissions from agricultural crop residue burning. *Environmental Science and Technology, 45*, 5477–5482.

Zhao, U., Duan, L., Xing, J., Larssen, T., Nielsen, C. P., & Hao, J. M. (2009). Soil acidification in China: Is controlling SO2 emissions enough. *Environmental Science and Technology, 43*, 8021–8026.

Zhi, G., Chen, Y., Feng, Y., Xiong, S., Li, J., Zhang, G., et al. (2008). Emission characteristics of carbonaceous particles from various residential coal-stoves in China. *Environmental Science and Technology, 42*, 3310–3315.

Zhi, G., Peng, C., Chen, Y., Liu, D., Sheng, G., & Fu, J. (2009). Deployment of coal briquettes and improved stoves: Possibly an option for both environment and climate. *Environmental Science and Technology, 43*, 5586–5591.

Zhou, J., Wang, T., Huang, Y., Mao, T., & Zhong, N. (2005). Seasonal variation and spatial distribution of polycyclic aromatic hydrocarbons in atmospheric PM10 of Beijing, People's Republic of China. *Bulletin of Environmental Contamination and Toxicology, 74*, 660–666.

Zhu, Y., Ioannidis, J. P. A., Li, H., Jone, K. C., & Martin, F. L. (2011). Understanding and harnessing the health effects of rapid urbanization in China. *Environmental Science and Technology, 45*, 5099–5104.

Chapter 3
Method

3.1 Fuel and Combustion Experiment

3.1.1 Simulated Kitchen

In the present study, the main objective is to measure emission factors of PM and PAHs from residential solid fuel combustion. To reflect the real fuel burning activity in rural household, and to investigate the influence of factors, we built a kitchen according to the real layout of rural family in Northern China. As shown in Fig. 3.1, the kitchen was connected to a bedroom, and a typical heating bed (known as Kang in Chinese) was used. The stove-Kang structure was very common in Northern China. It was estimated that there were about 66.85 million Kang in China, covering 43.64 million households and about 174.65 million populations by 2004 (Zhuang et al. 2009). The built kitchen was in the campus of Peking University. There was no other traffic or residential source nearby. A brick stove was used for the burning of biomass fuels and the exhaust from the stove chamber passed through the Kang and then emitted into the ambient air through the chimney.

3.1.2 Fuel Property

Traditional solid fuels like crop residue, wood and coal are main energy resources used in rural China, and also the major burning source of various incomplete pollutants (Bond et al. 2004; Cao et al. 2006, 2007; Lu et al. 2011; Streets et al. 2001, 2003; Tian et al. 2011; Xu et al. 2006; Zhang et al. 2009). Although the adoptions of cleaner fuels like biogas, liquid petroleum gas and biomass pellet have been advocated by the local government in the last several years (Chen et al. 2009; Kou et al. 2008; Liu and Zhou 2007; NDRC 2007; Wang et al. 2007), traditional solid fuels would be still the main fuel used in the years to come, affecting local/regional air quality and climate change significantly.

G. Shen, *Emission Factors of Carbonaceous Particulate Matter and Polycyclic Aromatic Hydrocarbons from Residential Solid Fuel Combustions*, Springer Theses, DOI: 10.1007/978-3-642-39762-2_3, © Springer-Verlag Berlin Heidelberg 2014

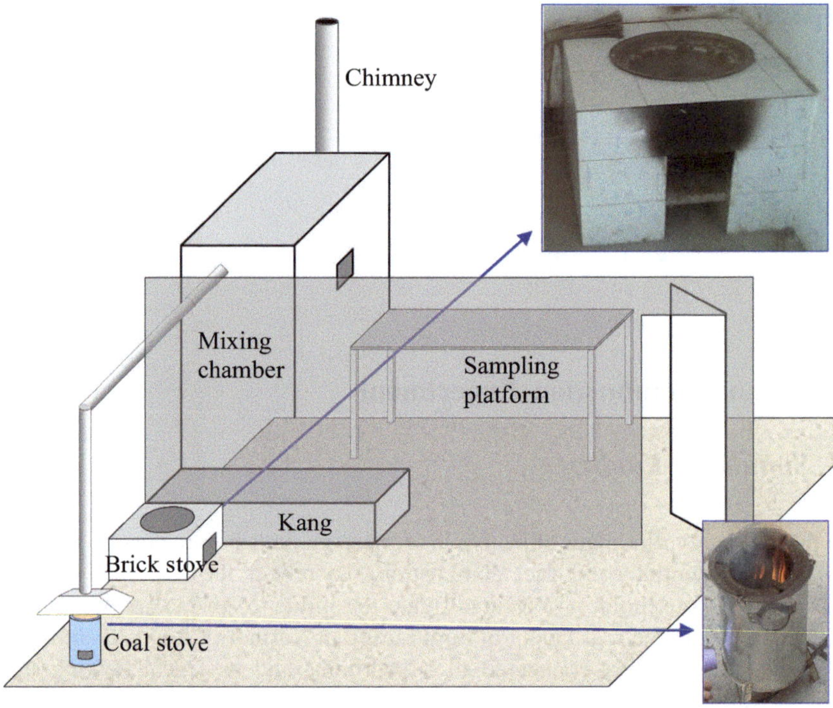

Fig. 3.1 The layout of built kitchen and pictures of a brick stove and a coal stove in this study. Reprinted from Shen et al. (2010) with permission of American Chemical Society

In this study, residential combustions of crop residue, wood and coal were investigated. The fuels were collected by the colleagues from different rural areas in China, like Heilongjiang, Beijing, Henan, Hubei, Hunan and Jiangsu Provinces. Coals were purchased from the local market of Beijing, Taiyuan and Yulin. Nine crop residues including the straws of rice, wheat, corn, soybean, horsebean, cotton, peanut, sesame and rape, contributed to more than 90 % of the total crop residue combusted in China. Contents of moisture, C, N, and H of crop residues and coal samples were measured by the Analytical Instrumentation Center, Peking University. The results are listed in Table 3.1.

A cast-iron stove purchased from the local market in suburban Beijing was used in coal combustion experiment. Five coals (two honeycomb briquettes from Beijing and Taiyuan and three chunk coals from Taiyuan and Yulin) were tested. The elemental and proximate analysis results for these five coals are listed in Table 3.2. The honeycomb briquette was 15 cm in diameter and 11 cm thick with 16 holes. The volatile matter contents of these two briquettes were 4 and 15 %, respectively. The three raw chunk coals (from Taiyuan and Yulin) were all medium volatile bituminous (MVB) with VM between 23 and 29 %.

Table 3.1 Properties of the crop residues tested (%)

Fuel		Moisture	C	H	N	C in ash
Horsebean	*Vicia faba*	2.84 ± 0.45	43.10 ± 0.18	7.39 ± 0.15	0.86 ± 0.08	10.65 ± 1.25
Peanut	*Arachis hypogaea*	3.25 ± 0.36	40.37 ± 0.07	6.75 ± 0.14	2.27 ± 0.17	5.11 ± 0.43
Soybean	*Cassia agnes*	2.79 ± 0.36	42.33 ± 0.16	7.55 ± 0.12	0.86 ± 0.28	41.68 ± 2.38
Cotton	*Anemone vitifolia*	3.34 ± 0.33	44.38 ± 0.08	7.35 ± 0.01	1.10 ± 0.11	18.10 ± 1.80
Rice	*Oryza sativa*	3.92 ± 0.40	38.36 ± 0.47	6.26 ± 0.16	0.87 ± 0.23	5.98 ± 0.53
Wheat	*Triticum aestivum*	1.41 ± 0.10	42.61 ± 0.04	6.67 ± 0.00	0.27 ± 0.12	11.86 ± 2.31
Rape	*Brassica napus*	2.22 ± 0.42	41.25 ± 0.63	6.99 ± 0.01	0.45 ± 0.01	7.69 ± 0.50
Sesame	*Sesamum indicum*	2.46 ± 0.31	43.20 ± 0.22	7.34 ± 0.08	1.73 ± 0.01	11.47 ± 3.02
Corn	*Zea mays*	2.53 ± 0.57	44.46 ± 0.44	7.08 ± 0.00	1.54 ± 0.15	9.11 ± 0.08

Reprinted from Shen et al. (2010) with permission of American Chemical Society

Table 3.2 Properties of the coals tested

Market type	Beijing honeycomb	Taiyuan honeycomb	Taiyuan Chunk	Yulin Chunk A	Yulin Chunk B
N, %	0.39 ± 0.04	0.36 ± 0.05	1.24 ± 0.13	0.95 ± 0.04	0.72 ± 0.03
C, %	65.5 ± 1.2	24.0 ± 2.0	81.2 ± 2.6	72.9 ± 0.5	59.9 ± 0.39
H, %	0.84 ± 0.01	1.73 ± 0.09	4.48 ± 0.04	4.33 ± 0.01	4.59 ± 0.09
Moisture, %	1.26 ± 0.10	2.18 ± 0.03	0.72 ± 0.07	3.73 ± 0.19	9.17 ± 0.68
Ash, %	27.4	61.7	7.78	15.3	4.45
Volatile Matter, %	3.99	14.7	22.8	28.0	28.5
Fixed Carbon, %	65.9	19.7	68.9	55.4	55.0
Heat value, MJ/kg	22.4	8.73	33.0	28.2	24.2
CRC	1	2	6	5	2

In wood combustion experiment, 27 fuels, including Chinese white poplar (*Populus tomentosa Carr.*), water Chinese fir (*Metasequoia glyptostroboides*), Chinese pine (*Pinus tabulaeformis Carr.*), cypress (*Cupressus funebris Endl.*), elm (*Ulmus pumila L.*), fir (*Cunninghamia lanceolata*), larch (*Larix gmelini (Rupr.) Rupr.*), maple (*Acer mono Maxim.*), oak (*Quercus mongolica*), paulowonia tomentosa (*P.tomentosa (Thunb.) Steud.*), toon (*Ailanthus altissima*), white birch (*Betula platyphylla Suk*), willow (*Salix babylonica*), locust (*Robinia pseudoacacia L.*), bamboo (*Phyllostachys heterocycla(Carr.)*), ribbed birch (*Betula dahurica Pall.*), paulownia elongata (*P. elongata S.Y. Hu*), black poplar (*Populus nigra L.*), aspen (*Populus adenopoda Maxim.*), chinaberry (*Melia azedarach*), the tree of jujube (*Ziziphus jujuba Mill.*), persimmon (*Diospyros kaki Thunb.*), mulberry (*Morus alba L.*), peach (*Prunus persica*), and 3 brushwood of lespedeza (*Leapedeza bicolor. Turcz*), holly (*Buxus megistophylla Lévl*) and buxus sinica shrubs (*Buxus sinica (Rehd. et Wils.) Cheng*), were burned in the brick stove. These wood fuels are the main tree species used for bio-energy in China (Gao et al. 1990). Measured fuel properties are listed in Tables 3.3 and 3.4. The average carbon content was 48 %, similar to that of the crop residue. Wood moisture ranged widely from 5.32 to 41.8 %. In comparison with coal, wood material had relatively higher volatile matter content with a mean of 81.0 ± 3.1 %.

3.1.3 Biomass Pellet

Two commercial biomass pellets (8 mm in diameter and 1.5–2.0 cm in length) made of pine wood and corn straw, respectively were burned in a modern pellet burner. Figure 3.2 shows the pictures of two pellet types and the specific burner for pellets. Fuel properties, including density, moisture, contents of C, H, N, O, volatile matter (VM), ash content, and lower heating value were measured and are listed in Table 3.5. The bulk densities of the pellets were much higher than those of the raw materials, especially for corn straw. The moisture of the two pellets

Table 3.3 Density (g/cm^3), volatile matter (%), ash (%), fixed carbon (%) content and HHV (MJ/kg) of tested fuels

Fuel		Density	Ash	VM	FC	HHV
Bamboo	*Phyllostachys heterocycla(Carr.)*	0.912	0.51	84.94	14.55	18.33
Chinese White Poplar	*Populus tomentosa Carr.*	0.463	0.90	81.69	17.41	18.35
Elm	*Ulmus pumila L.*	0.536	1.50	78.56	19.94	18.27
Yellow Locust	*Robinia pseudoacacia L.*	1.176	0.90	78.93	20.16	18.77
Maple	*Acer mono Maxim.*	0.949	1.80	86.12	12.08	19.03
Fir	*Cunninghamia lanceolata*	0.427	0.42	82.94	16.64	18.61
Larch	*Larix gmelini (Rupr.) Rupr.*	0.634	0.46	82.10	17.44	19.22
Water Chinese fir	*Metasequoia glyptostroboides*	0.410	0.71	79.91	19.38	19.49
Cypress	*Cupressus funebris Endl.*	0.667	1.47	76.60	21.93	20.02
Oak	*Quercus mongolica*	1.114	1.85	78.33	19.82	19.00
Chinese Pine	*Pinus tabulaeformis Carr.*	0.443	0.25	84.77	14.98	18.51
Willow	*Salix babylonica*	0.551	1.41	82.84	15.75	16.99
Paulownia tomentosa	*P.tomentosa (Thunb.) Steud.*	0.284	0.30	85.03	14.67	16.00
Toon	*Ailanthus altissima*	0.734	1.49	82.94	15.57	17.28
White Birch	*Betula platyphylla Suk*	0.798	0.39	87.45	12.17	20.22
Ribbed Birch	*Betula dahurica Pall.*	1.059	5.05	84.08	10.87	18.14
Paulownia elongata	*P. elongata S. Y. Hu*	0.763	0.34	81.68	17.98	19.33
Black Poplar	*Populus nigra L.*	1.080	1.53	77.11	21.36	18.93
China Aspen	*Populus adenopoda Maxim.*	0.867	1.45	81.89	16.67	19.33
Chinaberry	*Melia azedarach*	0.654	2.07	76.95	20.98	17.67
Jujube tree	*Ziziphus jujuba Mill.*	0.805	0.88	79.81	19.30	18.61
Persimmon tree	*Diospyros kaki Thunb.*	0.710	1.71	79.56	18.72	18.05
Mulberry tree	*Morus alba L.*	0.909	1.37	77.13	21.50	18.73
Peach tree	*Prunus persica*	0.770	3.81	77.79	18.40	19.35
Lespedeza	*Leapedeza bicolor. Turcz*		4.60	81.08	14.32	
Buxus sinica	*Buxus sinica (Rehd. et Wils.) Cheng*		5.84	78.36	15.80	
Holly	*Buxus megistophylla Lévl*		8.56	77.58	13.86	

Proximate analysis was conducted by the Analytical Center of Chinese Academy of Agricultural Engineering. Reprinted from Shen et al. (2012) with permission of American Chemical Society

were 5.63 and 5.80 %, respectively, and generally lower than those of raw pine wood (9.10 %) and corn straw (7.02 %). Biomass pellets had lower VM and higher ash contents than the corresponding raw fuels. Ash content of corn straw pellet was higher than that of pine wood pellet. Such difference was previously reported in the literature (Yao et al. 2010; Johansson et al. 2004; Schimidl et al. 2011). Pre-weighed pellets were added into the burner and fired. After ignition, the burner was set up under a stainless steel hood. The pellets were combusted in the burner in two modes, without (mode I) and with (mode II) secondary side air supply.

It was noted that ash contents of the pellet fuels were much higher than those made and used in Europe and United States (Bäfver et al. 2011; Boman et al. 2011; Chen et al. 2010; García-Maraver et al. 2011; Johansson et al. 2004; Schimidl et al. 2011). A comparison study on properties of biomass pellets found that ash contents

Table 3.4 Moisture and elemental analysis of C, H, N, O content (%, dry basis) in tested wood fuels

Fuel		Moisture	C	N	H	O
Bamboo	*Phyllostachys heterocycla(Carr.)*	8.18	48.75	0.26	5.98	45.02
Chinese White Poplar	*Populus tomentosa Carr.*	5.32	47.75	0.08	6.15	46.03
Elm	*Ulmus pumila L.*	6.52	46.89	0.84	5.89	46.40
Yellow Locust	*Robinia pseudoacacia L.*	33.33	45.37	1.37	6.05	47.22
Maple	*Acer mono Maxim.*	31.41	47.73	0.18	6.10	46.00
Fir	*Cunninghamia lanceolata*	9.12	49.83	0.13	6.18	43.87
Larch	*Larix gmelini (Rupr.) Rupr.*	12.77	48.23	0.14	6.20	45.44
Water Chinese fir	*Metasequoia glyptostroboides*	12.83	49.42	0.29	6.07	44.22
Cypress	*Cupressus funebris Endl.*	12.71	50.13	0.36	6.02	43.49
Oak	*Quercus mongolica*	29.41	47.42	0.30	6.15	46.14
Chinese Pine	*Pinus tabulaeformis Carr.*	9.10	49.10	0.18	6.32	44.41
Willow	*Salix babylonica*	9.92	47.39	0.23	6.14	46.25
Paulownia tomentosa	*P.tomentosa (Thunb.) Steud.*	8.69	48.76	0.13	6.11	45.02
Toon	*Ailanthus altissima*	7.17	47.71	0.18	6.13	45.99
White Birch	*Betula platyphylla Suk*	32.21	48.33	0.21	6.20	45.28
Ribbed Birch	*Betula dahurica Pall.*	32.33	48.41	0.19	6.22	45.19
Paulownia elongata	*P. elongata S. Y. Hu*	41.78	48.85	0.08	6.09	44.98
Black Poplar	*Populus nigra L.*	39.45	47.45	0.32	6.06	46.18
China Aspen	*Populus adenopoda Maxim.*	38.27	47.42	0.38	6.08	46.14
Chinaberry	*Melia azedarach*	18.45	48.72	0.46	5.93	44.89
Jujube tree	*Ziziphus jujuba Mill.*	15.22	47.19	0.59	6.03	46.20
Persimmon tree	*Diospyros kaki Thunb.*	18.69	46.52	0.30	5.94	47.25
Mulberry tree	*Morus alba L.*	17.79	48.27	0.20	6.01	45.52
Peach tree	*Prunus persica*	27.16	48.10	0.26	6.01	45.65
Lespedeza	*Leapedeza bicolor. Turcz*	6.04	48.59			
Buxus sinica	*Buxus sinica (Rehd. et Wils.) Cheng*	6.96	48.19			
Holly	*Buxus megistophylla Lévl*	6.76	45.85			

C, H, and N contents were analyzed by the Analytical Instrumentation Center, Peking University (*Elementar Vario MICRO CUBE, German*) and O content is calculated by the difference. Reprinted from Shen et al. (2012) with permission of American Chemical Society

Fig. 3.2 Pictures of the pellet burner and two types of biomass pellet fuels in this study. Reprinted from Shen et al. (2012) with permission of American Chemical Society

Table 3.5 Fuel properties (dry basis) of raw and pelletized corn straw (*Zea mays*) and pine wood (*Pinus tabulaeformis Carr.*)

	Raw pine wood	Pelletized pine wood	Raw corn straw	Pelletized corn straw
Moisture, %	9.10	5.63	7.02	5.80
Density, g/cm^3	0.44	1.30	0.069	1.41
Proximate Analysis, %				
Ash content	0.25	3.75	4.01	10.84
Volatile matter	84.77	76.77	77.94	70.62
Fixed carbon	14.98	19.48	18.05	18.54
LHV, MJ/kg	16.72	16.48	15.44	13.89
Elemental analysis, %				
N	0.18	0.16	0.96	1.23
C	49.10	46.78	43.47	43.14
H	6.32	5.84	6.02	5.48
O (*by difference*)	44.41	47.23	49.56	50.16

Reprinted from Shen et al. (2012) with permission of American Chemical Society

of the pellets from China (7.71–21.7 % and 1.01–9.25 % for crop straw pellets and wood pellets, respectively) were significantly higher than those of 4.70–7.9 % and 0.30–3.40 % from Sweden (Yao et al. 2010). It was believed that these high ash content pellets usually had lower softening temperature and were much easier to form slag. The use of anti-slagging additives to reduce slag formation was adopted (Xiong et al. 2008, 2010; Yuan et al. 2009a, b). It was reported that the slagging rates in the burning of corn straw pellets with 3 % of $MgCO_3$, $CaCO_3$, Al_2O_3, and kaolin added were 1.12, 5.07, 25.8, and 43.9 %, respectively, while a slagging rate of 46.5 % was found for pellet fuels without additives (Yuan et al. 2009b). The use of additive restrains the slagging formation, but increases fuel ash content.

3.2 Sampling

3.2.1 A Mixing and Sampling Chamber

The exited smoke from the wok stove (passed through a "Kang") and cast-iron stove entered a mixing chamber (4.5 m^3) with a build-in mixing fan. The mixing chamber was used to deposit large PM; to cool the exhaust temperature to avoid the adverse impacts of high temperature on sampling efficiencies; and to mix the exhaust (Hildemann et al. 1989; Purvis et al. 2000), even though the emission factor calculation was based on carbon mass balance method and the site of sampling probe was flexible (Zhang et al. 2000).

3.2.2 *Total and Size Segregated PM*

Low-volume pumps (XQC-15E, Tianyue, China) with quartz fiber filters were used to collect PM (as total suspended particles) in the mixing chamber at a flow rate of 1.5 L/min. A nine stage cascade impactor (FA-3, Kangjie, China) with glass fiber filters was used to collect PM_{10} samples with different aerodynamic diameters (D_a) (<0.4, 0.4–0.7, 0.7–1.1, 1.1–2.1, 2.1–3.3, 3.3–4.7, 4.7–5.8, 5.8–9.0, and 9.0–10.0 μm) at a flow rate of 28.3 L/min. The filters were baked at 450 °C for 6 h and stored in a desiccator for 24 h prior to weighing and sampling. After sampling, particle-loaded filters were packed with aluminum foil and stored in a desiccator before further analysis.

3.2.3 *PAHs and the Derivatives*

Gaseous and particulate phase PAHs, and their derivatives, were collected on polyurethane foam plugs (PUF, 22 mm diameter × 7.6 cm, 0.024 g/cm^3) and quartz fiber filters (QFFs, 22 mm in diameter) respectively, using active samplers at a flow rate of 1.5 L/min. All filters were baked at 450 °C for 6 h and equilibrated in a desiccator. PUF were pre-extracted with acetone and dichloromethane, followed by hexane for 8 h each. After sampling, the PUFs and QFFs were packed in aluminum foil.

3.3 Field Measurement

3.3.1 *Site*

Field measurements on residential solid fuel combustion emission were conducted in selected households in rural Shanxi and Jiangsu Provinces. In rural Shanxi, coal is the main fuel used by residents. Corn stalk was occasionally used by the residents as cooking fuel or materials to feed livestock. Woody materials are also frequently used. In addition to the briquette or chunk coal purchased from the local market, rural residents often make coal cake by themselves at home by mixing the coal and clay. Home-made coal cake is much cheaper in comparison to the purchased coals, and thus, is another widely used solid fuel in rural households. Figure 3.3 shows the pictures of fuel-stove combinations measured in rural Shanxi. Wood was burned in a simple movable metal stove, honeycomb briquette was burned in an improved metallic stove with a chimney, and coal cake was burned in a brick stove with an outdoor flue pipe. These three combinations are most widely found in the studied area.

Fig. 3.3 Pictures of three fuel/stove combinations investigated in this study. For *left* to the *right* wood, honeycomb briquette, and coal cake burned in a simple metal stove, an improved metallic stove with a chimney and a brick stove with a outdoor flue pipe, respectively. Reprinted from Shen et al. (2013) with permission of American Chemical Society

In the field study in Jiangsu, a large fraction of these crop residues were used as residential fuels for cooking. Rice, wheat, cotton, and rape are main crops in this area. An improved two-pot brick stove with a chimney is commonly used in this area by most households. A relatively new stove of 1-year old and an old stove of 15-year old were selected in this study so as to compare the emissions from stoves of different ages.

3.3.2 Sampling

Samples were collected during regular cooking time and the residents were asked to conduct fuel burning activities as normal. For sample collection using a commonly adopted method, the probes were placed near the chimney center or 1.0 m above the stoves without a chimney. The exhausted smoke was naturally diluted by the ambient air during the initial few seconds and then it was sampled directly (Fig. 3.4).

PM was collected on pre-baked quartz fiber filters (QFFs, 22 mm in diameter) using an active sampler at a flow rate of 1.5 L/min (XQC-15E, Tianyue, China). The flows were calibrated using a primary flow calibrator (Bios. Defender 510, USA). Polyurethane foam plugs (PUFs, 22 mm diameter × 7.6 cm) were used to collect gaseous PAHs and derivatives.

Fig. 3.4 Pictures of two brick stoves tested in this study. The structures of these two stoves were similar with two pots in the middle and an outdoor chimney. These two stoves had different usages. The stove in the *left* was a new built one (about 1 year ago), and the *right* one was an old stove used for about 15 years. Reprinted from Environmental Pollution 184, Wei et al., Field measurement on the emissions of PM, OC, EC and PAHs from indoor crop straw burning in rural China, 18–24, with permission from Elsevier

3.4 Chemical Analysis

3.4.1 CO and CO_2

The EFs in the present study were calculated using the carbon mass balance method under the assumption that the carbon was released in the forms of CO_2, CO, total gaseous hydrocarbons (THC), and carbonaceous carbon in PM. EFs of targets (CPM and PAHs) can be calculated from the EFs of CO_2 and their mass emission ratios. In this study, CO_2 and CO were measured every 2 s using an on-line non-dispersive infrared sensor. The instrument (GXH-3051, Technical Institute, China) was calibrated using a span gas before each combustion experiment (CO 1.00 %; CO_2 5.00 %). The THC was not analzyed. Since most gaseous carbon was in the forms of CO and CO_2, the omission of total hydrocarbon in gas would cause a very small error which was estimated in 4 % (Roden et al. 2006).

3.4.2 PM and EC/OC Mass

Gravimetric measurements were conducted using a high precision (0.00001 g) digital balance (Mettler Toledo XS105). EC and organic carbon (OC) were analyzed using Sunset EC/OC analyzer (Sunset Lab, USA). The filter was heated in a pure helium at 600, 840, 550 °C for OC detection, and then at 550, 650, 870 °C in an oxygen/helium atmosphere to determine EC. The carbon results were calculated using methane at the end of each analysis cycle, and pryolyzed OC, produced in inert helium when temperature increased, was subtracted from EC results according to the initial laser value.

3.4.3 PAHs and Derivatives

The PUFs were Soxhlet extracted using 150 ml of dichloromethane for 8 h. A microwave accelerated system (CEM Mars Xpress, USA) was used to extract the PAHs from the PM collected on QFFs using 25 ml of a hexane/acetone mixture (1:1, v/v) at 1200 W (100 %). The temperature was increased to 110 °C over 10 min and held for an additional 10 min. Both PUF and QFF extracts were concentrated to 1 ml and then transferred to a silica/alumina gel column for cleanup (12 cm silica gel, 12 cm alumina and 1 cm anhydrous sodium sulfate from bottom to top). The column was eluted with 20 ml hexane, followed by 70 ml hexane/dichloromethane (1:1, v/v). The hexane/dichloromethane eluate was concentrated to 1 ml and spiked with deuterated internal standards (J&W Chemical Ltd., USA).

Parent PAHs were measured using a gas chromatograph (GC, Agilent 6890) connected to a mass spectrometer (MS, Agilent 5973 in electron ionization mode. A HP-5MS capillary column (30 m × 0.25 mm × 0.25 μm) was used, and the oven temperature was held at 50 °C for 1 min, increased to 150 °C at a rate of 10 °C/min, to 240 °C at 3 °C/min, and then to 280 °C for another 20 min. Helium was used as the carrier gas. PAHs were identified based on retention time and qualifying ions of standards in selected ion monitoring mode. In the beginning, 16 U.S. EPA priority PAHs were analyzed. Later for samples from wood combustion, we added 12 other non-priority but of increasing concern PAHs into the analytical list. So, a total of twenty-eight parent PAHs were measured, including naphthalene (NAP), acenaphthylene (ACY), acenaphthene (ACE), fluorene (FLO), phenanthrene (PHE), anthracene (ANT), fluoranthene (FLA), pyrene (PYR), retene (RET), benzo[c]phenanthrene (BcP), cyclopenta[c,d]pyrene(CPP), benzo(a)anthracene (BaA), chrysene (CHR), benzo(b)fluoranthene (BbF), benzo(k)fluoranthene (BkF), benzo(a)pyrene (BaP), benzo(e)pyrene (BeP), perylene (PER), dibenz(a,h)anthracene (DahA), indeno(l,2,3-cd)pyrene (IcdP), benzo(g,h,i)perylene (BghiP), dibenzo[a,c]pyrene (DacP), dibenzo[a,l]pyrene (DalP), dibenzo[a,e]flluoranthene (DaeF),Coronene(Corn), dibenzo[a,e]pyrene (DaeP), dibenzo[a,i]pyrene (DaiP), dibenzo[a,h]pyrene (DahP).

Nitro- and oxygenated PAH analyses were performed using a gas chromatography (GC, Agilent 6890) connected to a mass spectrometer (MS, Agilent 5975) equipped with a HP-5MS capillary column (30 m × 0.25 mm × 0.25 μm) in a negative chemical ionization mode. The oven temperature was programmed at 60 °C, increased to 150 °C at a rate of 15 °C/min, and then to 300 °C at 5 °C/min held for 15 min. High-purity helium at the flow rate of 1.0 L/min and methane at the rate of 2.5 L/min were used as the carrier and reagent gas, respectively. nPAHs and oPAHs were identified and quantified based on the retention times and selected ions of the standards (J&W Chemical, USA). Twelve nPAHs and 4 oPAHs were quantified, including 1-nitro-naphthalene (1 N-NAP), 2-nitro-naphthalene (2 N-NAP), 5-nitro-acenaohthene (5 N-ACE), 2-nitro-fluorene (2 N-FLO), 9-nitro-anthracene (9 N-ANT), 9-nitro-phenanthrene (9 N-PHE), 3-nitro-phenanthrene (3 N-PHE), 3-nitro-fluoranthene (3 N-FLA), 1-nitro-pyrene (1 N-PYR), 7-nitro-benzo[a]anthracene (7 N-BaA), 6-nitro-chrysene (6 N-CHR), 6-nitro-benzo[a]pyrene (6 N-BaP), and 9-fluorenone (9FO), anthracene-9,10-dione (ATQ), benzanthrone (BZO), and benzo[a]anthracene-7,12-dione (BaAQ).

3.4.4 Quality Control

All solvents were from Beijing Reagent, China and re-distillated and checked for PAHs blank before use. The silica gel and alumina were baked at 450 °C for 6 h, activated at 300 °C for 12 h, and deactivated with deionized water (3 %, w/w) prior to use. The anhydrous sodium sulfate was baked at 450 °C for 8 h. All glassware was cleaned in an ultrasonic cleaner and baked at 500 °C for at least 10 h.

Instrumental detection limits (IDLs) for the pPAHs ranged from 0.13 ng (ACY) to 0.92 ng (BghiP). Laboratory analysis method detection limits (MDLs) ranged from 0.23 ng/mL (NAP) to 1.42 ng/mL (BghiP) for gaseous phase PAHs and from 0.53 ng/mL (PHE) to 1.32 ng/mL (BghiP) for particulate phase PAHs. Recoveries of the spiked standard PAHs ranged from 70 to 121 % for gaseous and 68–120 % for particulate phase compounds. IDLs ranged from 0.12 to 0.49 ng for nPAHs and 0.06 to 0.24 ng for oPAHs. MDLs for gaseous and particulate phase nPAH ranged from 0.18 to 0.58 and 0.12 to 0.53 ng/mL, respectively. Recoveries of the spiked standards were 78 ± 6 to 92 ± 10 % and 93 ± 19 to 124 ± 30 % for gaseous and particulate phase nPAH, respectively. MDLs for oPAHs were 0.32 to 0.60 ng/mL and 0.11 to 0.44 ng/mL for gaseous and particulate phase PAHs, respectively. Recoveries of the spiked oPAH standards were 72 ± 4 to 96 ± 14 % for gaseous phase and 82 ± 26 to 125 ± 10 % for particulate phase.

3.5 Carbon Mass Balance Method

Pollutant emission factors were calculated based on carbon mass balance method (Zhang et al. 2000). It follows the assumption that total carbon released into the gaseous form of CO_2, CO, total hydrocarbon (THC) and particle phase carbon.

$$m_f - m_a = m_{C-CO_2} + m_{C-CO} + m_{C-THC} + m_{C-PM},$$

where m_f and m_a are carbon mass in fuel and ash, respectively. $m_{C\text{-}CO2}$, $m_{C\text{-}CO}$, $m_{C\text{-}THC}$, and $m_{C\text{-}PM}$ are the carbon released in CO_2, CO, THC and PM, respectively.

By defining a parameter "K":

$$k = \frac{m_{C-CO} + m_{C-THC} + m_{C-PM}}{m_{C-CO_2}}.$$

Emission factor of carbon in CO_2 from per mass of burnt fuel (M) can be calculated from the equation:

$$EF_{C-CO_2} = \frac{m_f - m_a}{(1+K)M}.$$

With a factor (f_{CO_2}) for conversing carbon mass in CO_2 to CO_2 mass, EF of CO_2 is:

$$EF_{CO_2} = EF_{C-CO_2} f_{CO_2} = \frac{(m_f - m_a) f_{CO_2}}{(1+K)M}.$$

And emission factors of other compounds (EF_X) can be calculated from emission factor of CO_2 and their mass concentration ratios to CO_2:

$$EF_X = EF_{CO_2} \frac{C_X}{C_{CO_2}},$$

where C_X and C_{CO_2} is the mass concentration of X and CO_2, respectively.

In most cases, carbon mass in ash was much smaller compared to that in fuel ($m_a \ll m_f$), so:

$$EF_{CO_2} = \frac{(m_f - m_a) f_{CO_2}}{(1+K)M} = \frac{f_{c-fuel} f_{CO_2}}{(1+K)}$$

Where $f_{C\text{-}fuel}$ is the carbon content of fuel.

Modified combustion efficiency (MCE), defined as $CO_2/(CO + CO_2)$ (molar basis) is calculated to characterize the combustion. Since most carbon emitted as gaseous CO and CO_2, the difference between MCE and combustion efficiency, which is defined as carbon in CO_2 divided by the total carbon released, is very small (Dhammapala et al. 2006; Janhall et al. 2010; McMeeking et al. 2009), and the former is widely used.

References

Bäfver, L. S., Leckner, B., Tullin, C., & Berntsen, M. (2011). Particle emissions from pellets stoves and modern and old-type wood stoves. *Biomass and Bioenergy, 35*, 3648–3655.

Boman, C., Pettersson, E., Westerhol, R., Boström, D., & Nordin, A. (2011). Stove performance and emission characteristics in residential wood log and pellet combustion, Part 1: pellet stoves. *Energy & Fuels, 25*, 307–314.

Bond, T. C., Streets, D. G., Yarber, K. F., Nelson, S. M., Woo, J., & Klimont, Z. (2004). A technology-based global inventory of black and organic carbon emissions from combustion. *Journal of Geophysical Reaserch, 109*, D14203. doi:10.1029/2003JD003697

Cao, G., Zhang, X., & Zheng, F. (2006). Inventory of black carbon and organic carbon emissions from China. *Atmospheric Environment, 40*, 6516–6527.

Cao, G., Zhang, X., Wang, Y., & Zheng, F. (2007). Emission inventory of air pollutants from open burning of crop straw in China. *Chinese Science Bulletin, 52*, 1826–1831. (In Chinese).

Chen, L., Xing, L., & Han, L. (2009). Renewable energy from agro-residues in China: solid biofuels and biomass briquetting technology. *Renewable and Suatinable Energy Review, 13*, 2689–2695.

Chen, Q., Zhang, X., Bradford, D., Sharifi, V., & Swithenbank, J. (2010). Comparison of emission characteristics of small-scale heating systems using biomass instead of coal. *Energy & Fuels, 24*, 4255–4265.

Dhammapala, R., Claiborn, C., Corkill, J., & Gullett, B. (2006). Particulate emissions from wheat and Kentucky bluegrass stubble burning in eastern Washington and northern Idaho. *Atmospheric Environment, 40*, 1007–1015.

Gao, S. W., & Ma, W. Y. (1990) Tree species as main bio-energy resources in China. China Forestry Press, Beijing, P. R. China, (In Chinese).

García-Maraver, A., Popov, V., & Zamorano, M. (2011). A review of European standards for pellet quality. *Renewable Energy, 36*, 3537–3540.

Hildemann, L. M., Cass, C. R., & Markowski, G. R. (1989). A dilution stack sampler for collection of organic aerosol emissions: Design, characterization and field tests. *Aerosol Science and Technology, 10*, 193–204.

Janhall, S., Andreae, M. O., & Pöschl, U. (2010). Biomass burning aerosol emission from vegation fires: Particle numver and mass emission factors and size distributions. *Atmospheric Chemistry and Physics, 10*, 1427–1439.

Johansson, L. S., Leckner, B., Gustavsson, L., Cooper, D., Tullin, C., & Potter, A. (2004). Emission characteristics of modern and old-type residential boilers fired with wood logs and wood pellets. *Atmospheric Environment, 38*, 4183–4195.

Kou, J., Zhao, L., Hao, X., & Tian, Y. (2008). Development status and trend of rural renewable energy in China. *Renewable Energy Resources, 26*, 1–6. (In Chinese).

Liu, R., & Zhou, X. (2007). Development of biomass energy in China's rural areas and countermeasures. *Chinese Agricultural Science Bulletin, 23*, 434–436. (In Chinese).

Lu, B., Kong, S., Han, B., Wang, X., & Bai, Z. (2011). Inventory of atmospheric pollutants discharged from biomass burning in China continent in 2007. *China Environmental Science, 31*, 186–194. (In Chinese).

McMeeking, G. R., Kreidenweis, S. M., Baker, S., Carrico, C. M., Chow, J. C., & Collett J. L., Jr., et al. (2009). Emission of trace gases and aerosols during the open combustion of biomass in the laboratory. *Journal of Geophysical Research, 114*, D19210, doi:10.1029/2009JD011836

National Development and Reformation Committee (NDRC). (2007). National Medium and Long Term Strategy Plan for Renewable Energy Development. Beijing, China (In Chinese).

Purvis, C. R., McCrillis, R. C., & Kariher, P. H. (2000). Fine particulate matter (PM) and organic speciation of fireplace emissions. *Environmental Science and Technology, 34*, 1653–1658.

Roden, C. A., Bond, T. C., Conway, S., & Pinel, A. B. O. (2006). Emission factors and real-time optical properties of particles emitted from traditional wood burning cookstoves. *Environmental Science and Technology, 40*, 6750–6757.

Schimidl, C., Luisser, M., Padouvas, E., Lasselsberger, L., Rzaca, M., Curz, C. R., et al. (2011). Particulate and gaseous emissions form manually and automatically fired small scale combustion systems. *Atmospheric Environment, 45*, 7443–7454.

Shen, G. F., Yang, Y., Wang, W., Tao, S., Zhu, C., Min, Y., Xue, N., Ding, J., Wang, B., Wang, R., Shen, H., Li, W., Wang, X., Russell, A.. (2010). Emission factors of particulate matter and elemental carbon for crop residues and coals burned in typical household stoves in China. *Environmental Science & Technology*, *44*, 7157–7162.

Shen, G. F., Tao, S., Wei, S., Zhang, Y., Wang, R., Wang, B., Li, W., Shen, H. Z., Huang, Y., Yang, Y., Wang, W., Wei, W., Wang, X., Liu, W., Wang, X., Simonich, S. (2012). Reductions in emissions of carbonaceous particulate matter and polycyclic aromatic hydrocarbons from combustion of biomass pellets in comparison with raw fuel burning. *Environmental Science & Technology, 46*, 6409–6416.

Shen, G. F., Tao, S., Wei, S., Chen, Y., Zhang, Y., Shen, H., Huang, Y., Zhu, D., Yuan, C., Wang, H., Wang, Y., Pei, L., Lioa, Y., Duan, Y., Wang, B., Wang, R., Lv, Y., Li, W., Wang, X., Zheng, X. (2013). Field measurement of emission factors of PM, EC, OC, parent, nitro- and oxy-polycyclic aromatic hydrocarbons for residential briquette, coal cake and wood in rural Shanxi, China. *Environmental Science & Technology, 47*, 2998–3005.

Streets, D. G., Bond, T. C., Carmichael, G. R., Fernandes, S.D., Fu, Q., & He, D., et al. (2003). An inventory of gaseous and primary aerosol emissions in Asia in the year 2000. *Journal of Geophysical Research, 108*(D21,8809), 1–23.

Streets, D. G., Gupta, S., Waldhof, S. T., Wang, M. Q., Bond, T. C., & Yi, B. (2001). Black carbon emissions in China. *Atmospheric Environment, 35*(428), 1–4296.

Tian, H., Zhao, D., & Wang, Y. (2011). Emission inventories of atmospheric pollutants discharged from biomass burning in China. *Acta Scientiae Circumstantiae, 31*, 349–357. (In Chinese).

Wang, J., Dai, L., Tian, Y., & Qing, S. (2007). Analysis of the development status and trends of biomass energy industry in China. *Transactions of the Chinese Society of Agricultural Engineering, 23*, 276–282. (In Chinese).

Xiong, S., Burvall, J., Örberg, H., Kalen, G., Thyrel, M., Öhman, M., Boström, D. (2008) Slagging characteristics during combustion of corn stovers with and without Kaolin and Calcite. *Energy & Fuels, 22*, 3465–3470.

Xiong, S., Öhman, M., Zhang, Y., Lestander, T. (2010) Corn stalk ash composition and its melting (slagging) behavior during combustion. *Energy & Fuels, 24*, 4866–4871.

Xu, S. S., Liu, W. X., & Tao, S. (2006). Emission of polycyclic aromatic hydrocarbons in China. *Environmental Science and Technology, 40*, 702–708.

Yao, Z., Zhao, L., Ronnback, M., Meng, H., Luo, J., & Tian, Y. (2010). Comparison on characterization effect of biomass pellet fuels on combustion behavior. *Transactions of the Chinese Society for Agricultural Machinery, 41*, 97–102. (In Chinese).

Yuan, Y., Lin, C., Zhao, L., Tian, Y., & Meng, H. (2009a). The research progress of anti-slagging for biomass pellet fuel. *Renewable Energy, 27*, 48–51. (In Chinese).

Yuan, Y., Zhao, L., Meng, H., Lin, C., & Tian, Y. (2009b). Effects comparison on anti-slagging additives of corn straw biomass pellet fuel. *Transactions of the Chinese Society of Agricultural Engineering, 26*, 251–255. (In Chinese).

Zhang, J., Smith, K. R., Ma, Y., Ye, S., Jiang, F., Qi, W., et al. (2000). Greenhouse gases and other airborne pollutants from household stoves in China: A database for emission factors. *Atmospheric Environment, 34*, 4537–4549.

Zhang, Y. X., Tao, S., Shen, H. Z., & Ma, J. M. (2009). Inhalation exposure to ambient polycyclic aromatic hydrocarbons and lung cancer risk of Chinese population. *Proceedings of the National Academy of Sciences (PNAS), 106*, 21063–21067.

Zhuang, Z., Li, Y., Chen, B., & Guo, J. (2009). Chinese kang as a domestic heating system in rural northern China- a review. *Energy Building, 41*, 111–119.

Chapter 4
Carbonaceous Particulate Matter

Ambient particulate matter affects local/regional air quality, human health and climate change significantly. The carbonaceous fractions like organic carbon (OC) and elemental carbon (EC, sometimes known as black carbon) are of high interest for scientists all over the world. In this section, measured EFs of PM, OC and EC for various solid fuels, and the influences of fuel properties and burning conditions were reported. The size distribution of PM, EC/OC ratio in PM and total carbon mass percent in PM were discussed and compared among three fuel types.

4.1 Coal Combustion

4.1.1 EFs and Influencing Factor

Measured EFs of PM, OC, and EC (EF_{PM}, EF_{OC} and EF_{EC}, respectively) from the coal combustions are listed in Table 4.1. EF_{PM} for coal combustion varied widely from 0.065 ± 0.002 for anthracite (honeycomb coals from Beijing) to 10.8 ± 0.55 g/kg for bituminous coals (raw chunk from Yulin) with a mean and a standard deviation of 3.17 ± 4.67 g/kg, depending on origin and type of the coals. Zhang et al. (2000) reported comparable EF_{PM} of 1.30 (0.026–10.0) g/kg for coals burned in stoves. EF_{PM} for anthracite and bituminous coals were measured at 0.62–2.20 and 4.05–37.8 g/kg, respectively (Chen et al. 2004, 2005, 2006; Zhi et al. 2008, 2009). The differences in EF_{PM} between anthracite and bituminous coals and between raw chunk and honeycomb coals are well recognized. In addition to the coal property and form, the stove design, which may cause different air supply rate and fuel-air mixing ratio that are directly associated with the burning status in stove chamber also affects the emission significantly.

EF_{EC} varied from 0.006 for the anthracite coal (Beijing, honeycomb) to 0.83 ± 0.34 g/kg for the MVB (Median Volatile Bituminous) chunk coals from Taiyuan. The mean and standard deviation of EF_{EC} for coal was 0.23 ± 0.36 g/kg. Similarly, EF_{OC} ranged from 0.007 to 1.00 g/kg for these coals. It has been previously reported by Chen et al. (2006) that EF_{EC} and EF_{OC} for raw chunk anthracite

G. Shen, *Emission Factors of Carbonaceous Particulate Matter and Polycyclic Aromatic Hydrocarbons from Residential Solid Fuel Combustions*, Springer Theses,
DOI: 10.1007/978-3-642-39762-2_4, © Springer-Verlag Berlin Heidelberg 2014

Table 4.1 EFs of CO, CO_2, OC, BC, and PM (g/kg) from the burning of different types of coals

	CO_2	CO	OC	EC	PM
Honeycomb, Beijing	291 ± 43	35.0 ± 8.7	0.007 ± 0.003	0.006 ± 0.001	0.065 ± 0.023
Honeycomb, Taiyuan	639 ± 94	54.6 ± 3.5	0.021 ± 0.002	0.004 ± 0.002	0.173 ± 0.010
Chunk, Taiyuan	2290 ± 106	241 ± 136	1.00 ± 0.48	0.825 ± 0.343	4.55 ± 1.66
Chunk A, Yulin	1810 ± 0	288 ± 0	0.660 ± 0.022	0.310 ± 0.020	10.8 ± 0.55
Chunk B, Yulin	2050 ± 9	112 ± 23	0.104 ± 0.015	0.006 ± 0.002	0.258 ± 0.070

were 0.030–0.051 and 0.002–0.007 g/kg, respectively, and for raw bituminous coals were 2.66–17.0 and 0.20–12.7 g/kg, respectively. In their another study, EFs of CPM for various honeycomb briquette coals were measured, and EF_{EC} and EF_{OC} were only 0.017 and 0.004 g/kg for anthracite briquette, but for the bituminous briquette, they were 3.58–13.8 and 0.064–0.675 g/kg, respectively (Chen et al. 2005). If the bituminous coal was further classified into three categories of low volatile bituminous, median volatile bituminous and high volatile bituminous (LVB, MVB, and HVB, respectively), it was found that MVB coals with VM content at 20–35 % had relatively higher pollutant emissions than the LVB and HVB (Chen et al. 2006, 2009). In Chen et al. (2009)'s study, EF_{OC} values for LVB, MVB and HVB in the form of briquette were 2.14, 8.44, and 2.89 g/kg, and in the form of raw chunk were 4.68, 10.6 and 3.36 g/kg, respectively. For EF_{EC}, they were 0.043, 0.25 and 0.060 g/kg for briquette and 2.15, 13.2 and 0.99 for raw chunk, respectively. Again, distinct difference between the raw chunk and briquette was shown. It was estimated that if all coals were burned in the form of briquette, the annual emissions of PM, OC and EC would reduce by about 64, 61 and 98 % in comparison with those from the burning of raw chunk (Zhi et al. 2009).

Coal properties such as volatile matter content and heating values significantly influence the formation and emission of incomplete pollutant. In general, high VM coals are difficult to achieve complete burning and hence produce relatively high pollutants in comparison with low VM coals (Bond et al. 2004; Chen et al. 2005). It is thought that the VM content is closely related to the tar and hydrocarbon fractions in coal, and the later usually acts as the nuclei in the formation of light absorption carbon (Bond et al. 2002; Mitra et al. 1987; Ledesma et al. 2000).

To address the factors affecting CPM emission from coal combustion, a number of parameters including moisture, ash content, VM, heat value, and MCE were tested using a stepwise regression. The result showed that MCE ($p = 7.0 \times 10^{-7}$) and VM ($p = 0.0003$) were two most significant factors affecting EF_{PM}, and about 92 % of the variation could be explained. The predicted EFs were comparable with the measured results (Fig. 4.1). For EF_{EC}, in addition to VM and MCE, ash content and heating value also affect the EC emission significantly ($p < 0.05$), and these four factors accounted approximately 95 % of the total variation of EF_{EC}.

$$EF_{PM}(g/kg) = 66 - 73 \times MCE + 0.17 \times VM(\%) \quad R^2 = 0.918$$

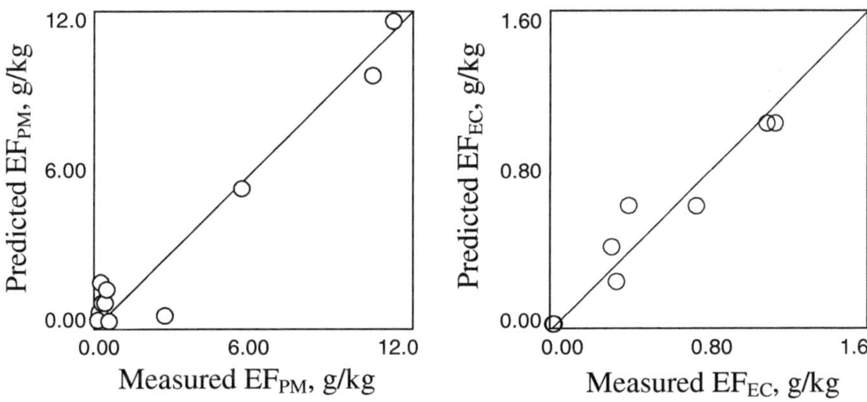

Fig. 4.1 Comparison between the measured and calculated EF_{PM} and EF_{EC} for coals. The calculation was based on a stepwise regression model for predicting EF_{PM} based on MCE and VM, and for EF_{EC} based on MCE, VM, ash content and heating value. Reprinted from Shen et al. (2010) with permission of American Chemical Society

4.1.2 EC/OC Ratio

The ratio of EC/OC and carbon mass percent in PM (TC/PM) are two important factors in the analysis of source and impact on climate change of carbonaceous particulate matter (Chow et al. 2011; Zhi et al. 2009). Calculated EC/OC varied obviously among five coals in the present study, ranging from 0.06 to 0.86. It was reported than because the reduction in EC emission was much larger than that in OC reduction when replacing raw chunk coal with briquette ones, it could be expected that the EC/OC ratio would decrease as well after the deployment of coal briquette (Zhi et al. 2009). Due to limited data in our present study, no significant dependence of EC/OC ratio on coal properties like moisture and VM content was found. Overall, the average EC/OC for coals in this study was 0.48 ± 0.38.

Total carbon mass percentage in PM ranged from 9 to 46 % (Table 4.2). With all data included, there was no significant correlation found between TC/PM ratio and coal property. If the lowest 9 % from the combustion of bituminous coal from Yulin was excluded, it appears that the TC/PM for anthracite was generally lower than that for bituminous, and the percent for briquette was lower than that for raw chunk.

4.1.3 Size Distribution

Fine particles dominate the PM emission from coal combustion. On average, $PM_{2.5}$ fractions were more than 77 ± 5 % of the total. Size distributions of PM for five coals fell into two distinguished categories with dominant size ranges of 0.7–2.1 μm or <0.7 μm (Fig. 4.2). For chunk coal from Taiyuan and chunk coal A

Table 4.2 Calculated EC/OC and TC/PM ratios in emissions from coal combustion

Coal type	Briquette		Raw chunk			Average
Location	Beijing	Taiyuan	Taiyuan	Yulin-A	Yulin-B	
Type	Anthracite	Bituminous	Bituminous	Bituminous	Bituminous	
BC/OC	0.86 ± 0.32	0.20 ± 0.12	0.84 ± 0.14	0.47 ± 0.05	0.06 ± 0.01	0.48 ± 0.38
TC/PM	0.17 ± 0.08	0.15 ± 0.01	0.45 ± 0.09	0.09 ± 0.00	0.46 ± 0.15	0.32 ± 0.19

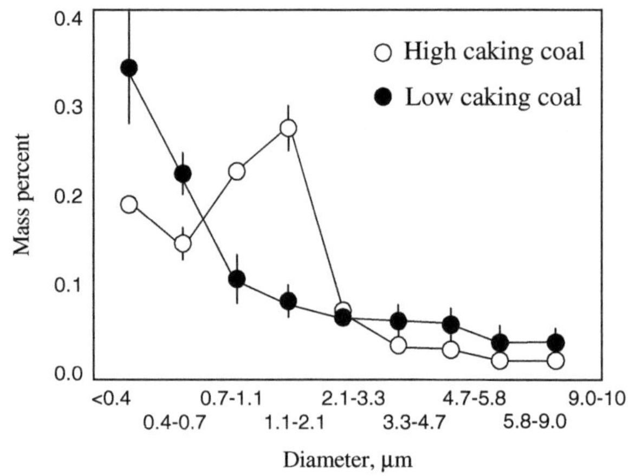

Fig. 4.2 Size distribution of PM from residential coal combustion. The size distributions of PM_{10} from five coals can be divided into two categories according to the coal caking property. Reprinted from Shen et al. (2010) with permission of American Chemical Society

from Yulin, size fraction between 0.7 and 2.1 μm contributed 49 ± 11 % of the total mass of PM_{10}, while the dominant fraction of two honeycomb coals and chunk coal B from Yulin was those with D_a less than 0.7 μm, accounting for 52 ± 18 to 60 ± 1 % of the total. Of all coal properties and combustion status determined in the study, the only one which distinguished the two categories was Char Residue Characteristics (CRC, an index describing caking property of combusted coal residue, the higher the CRC the tighter the combusted residue). The CRCs of the 2 coals emitted PM_{10} with dominant D_a range of 0.7–2.1 μm were 5 and 6, while CRCs of the 3 coals emitted PM_{10} with dominant D_a range of <0.7 μm was 1 or 2. It appeared that the coals with higher CRC had stronger caking potential and tended to emit larger particles during the combustion, primarily due to decrease in particle surface area and increase in contact time between the volatiles and char.

Mass median aerodynamic diameter (MMAD, the diameter at which 50 % of the mass are larger and the other 50 % are smaller) is one frequently used parameter in the analysis of particle size and particle-bound pollutants. It was believed that the cumulative mass percentage (y) was linearly correlated with log-transformed up diameter (*Dp*) of each size fraction (Francois et al. 1989; Parterson and Gillette 1977; Whitby 2007; Zhang et al. 1994).

$$y = a\lg(Dp) + b$$

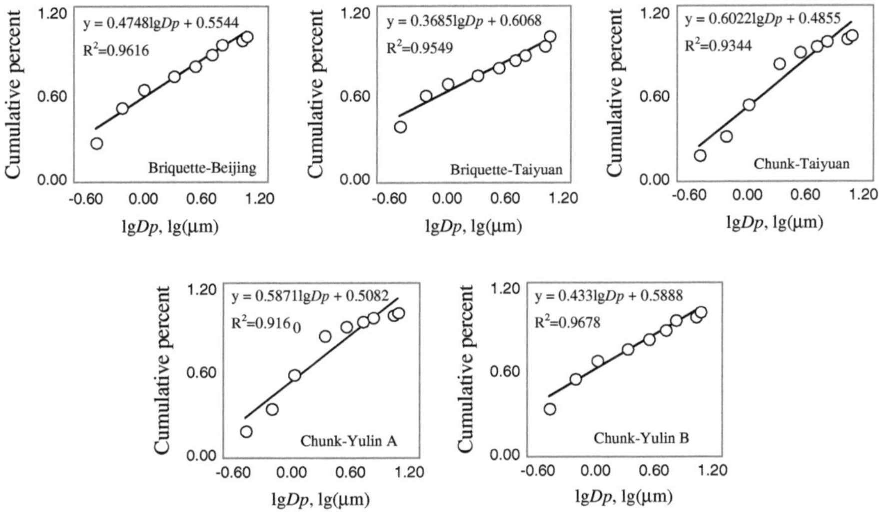

Fig. 4.3 Relationship between the cumulative mass percent and the upper diameter for size segregated PM from coal combustion

Table 4.3 Calculated MMAD of PM for five coals

Coal type	Briquette		Raw chunk		
Location	Beijing	Taiyuan	Location	Beijing	Taiyuan
R^2	0.9616	0.9549	0.9344	0.9160	0.9678
MMAD	0.77	0.51	1.06	0.97	0.62
δ_g	5.23	8.44	3.69	3.81	6.14

The slope coefficient (a) and intercept (b) can be used to calculated MMAD:

$$MMAD = 10^{\frac{50-b}{a}}.$$

For five coals in the present study, the cumulative mass percentage was significantly linear correlated with the up diameter, as shown in Fig. 4.3. The regression coefficients were between 0.92–0.97. Calculated MMAD for each coal is shown in Table 4.3. For two briquette coals from Beijing and Taiyuan and one chunk from Yulin, all of which are low caking coals, the MMAD values were 0.77, 0.51 and 0.62 μm, while for the other two high caking chunk coals, MMAD were 0.97 and 1.06 μm, respectively.

4.1.4 Correlation Among Co-emitted Pollutants

During the fuel combustion, a series of pollutants can be formed and emitted simultaneously. It is often interesting to look into the correlation among these

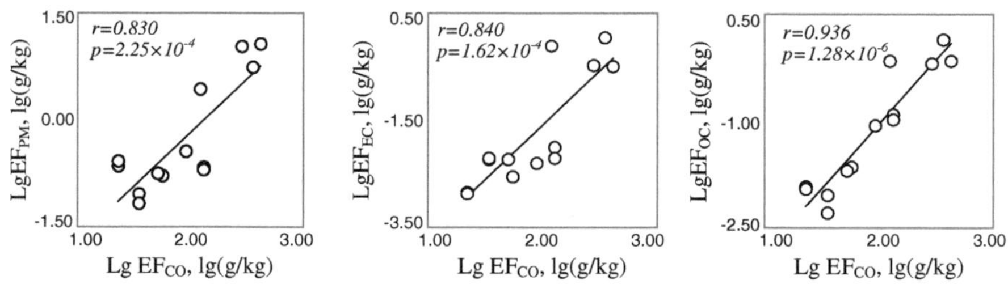

Fig. 4.4 Correlation between EF_{CO} and EFs of PM, OC and EC for coal

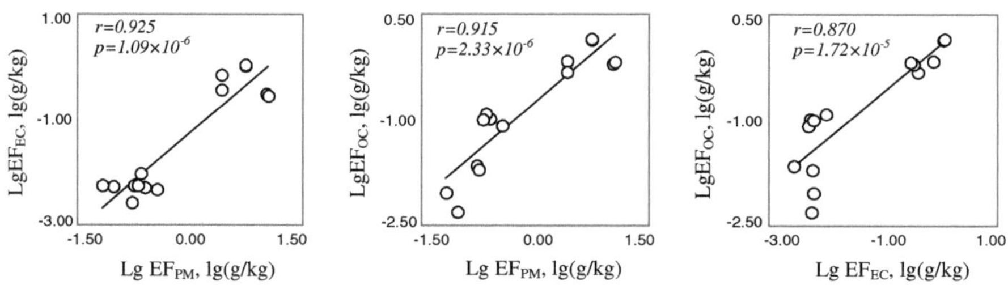

Fig. 4.5 Correlation between EF_{PM}, EF_{OC} and EF_{EC} for coal

co-emitted incomplete pollutants. In Fig. 4.4, the EF of CO was found positively correlated with EFs of PM, OC and EC. In fact, the EFs of PM, OC and EC were also correlated with each other (Fig. 4.5).

4.2 Indoor Crop Straw Burning

4.2.1 EFs and Influencing Factor

Emissions of PM and EC vary over the whole burning period of crop residues, which can be at least divided into flaming (with obvious fire) and smoldering (without observed fire) phases. Both CO and CO_2 increased in the flaming phase and decreased during the smoldering phase. The difference in PM and EC emissions between the two phases was expected. Therefore, in addition to the whole burning cycle experiment, the two phases were tested individually in duplicates for all crop residues.

Measured EF_{PM}, EF_{EC} and EF_{OC} for crop residue are listed in Table 4.4. EF_{PM} for crop residues varied from 3.41 ± 0.11 (cotton) to 16.8 ± 4.81 g/kg (rape) with a mean and a standard deviation of 8.19 ± 4.27 g/kg. It was previously reported that for crop residue burned in residential cooking stoves, the EF_{PM} was 8.05 (1.12–29.0 as range) g/kg (Zhang et al. 2000). Andreae and Merlet (2001)

Table 4.4 EFs of CO, CO_2, OC, BC, and PM (g/kg) for different crop straws from the whole burning cycle

Fuel type	CO_2	CO	OC	EC	PM
Horsebean straw	1430 ± 3	123 ± 8	1.20 ± 0.17	1.28 ± 0.10	7.45 ± 0.21
Peanut straw	1280	170	1.08	0.493	7.2
Soybean straw	1380 ± 13	72.3 ± 0.5	1.17 ± 0.18	1.37 ± 0.04	5.46 ± 0.22
Cotton straw	1520 ± 6	78.7 ± 1.2	0.354 ± 0.041	1.34 ± 0.58	3.41 ± 0.11
Rice straw	1140 ± 2	215 ± 0.2	1.50 ± 0.16	0.749 ± 0.427	6.39 ± 0.51
Wheat straw	1370 ± 9	126 ± 5	2.27 ± 1.30	2.64 ± 1.01	13.7 ± 5.1
Rape straw	1330 ± 16	154 ± 23	1.75 ± 0.63	2.34 ± 0.92	16.8 ± 4.8
Sesame straw	1440 ± 59	92.7 ± 60.5	2.34 ± 1.12	1.07 ± 0.03	7.41 ± 1.90
Corn straw	1490 ± 25	104 ± 19	1.36 ± 0.29	1.11 ± 0.08	5.84 ± 0.63

Reprinted from Shen et al. (2010) with permission of American Chemical Society

Table 4.5 EFs of CO, CO_2, OC, BC, and PM (g/kg) for different crop straws from the flaming and smoldering phases

	CO_2	CO	OC	EC	PM
Flaming phase					
Horsebean straw	1430 ± 34	135 ± 26	1.65 ± 0.64	2.24 ± 1.41	10.6 ± 3.4
Peanut straw	1290	184	1.16	0.643	9.69
Soybean straw	1470 ± 1	74.0 ± 3	1.09 ± 0.29	1.66 ± 0.29	6.88 ± 1.78
Cotton straw	1560 ± 2	62.2 ± 3	0.277 ± 0.207	0.951 ± 1.345	5.05 ± 0.27
Rice straw	1190 ± 65	205 ± 62	1.82 ± 0.36	0.988 ± 0.49	8.44 ± 0.84
Wheat straw	1390 ± 6	146 ± 1	2.99 ± 0.24	3.23 ± 0.24	15.7 ± 4.7
Rape straw	1300 ± 6	181 ± 31	1.58 ± 0.25	2.15 ± 0.25	14.2 ± 1.7
Sesame straw	1470 ± 62	98.8 ± 52	2.12 ± 0.07	1.47 ± 0.07	9.03 ± 0.40
Corn straw	1520 ± 20	100 ± 19	1.91 ± 0.50	1.76 ± 0.08	8.52 ± 0.36
Smoldering phase					
Horsebean straw	1450 ± 1	123 ± 3	1.14 ± 0.37	1.10 ± 0.17	9.05 ± 3.92
Peanut straw	1310	168	1.09	0.46	6.13
Soybean straw	1470 ± 2	74.5 ± 2.5	1.33 ± 0.22	1.37 ± 0.07	5.35 ± 0.30
Cotton straw	1540 ± 2	85.1 ± 1.1	0.393 ± 0.017	1.53 ± 0.32	3.44 ± 0.47
Rice straw	1170 ± 26	226 ± 27	1.33 ± 0.11	0.604 ± 0.324	4.59 ± 1.45
Wheat straw	1420 ± 1	122 ± 9	2.06 ± 1.20	2.61 ± 1.44	14.8 ± 5.6
Rape straw	1350 ± 14	150 ± 18	1.38 ± 0.28	1.96 ± 0.65	13.5 ± 2.3
Sesame straw	1480 ± 68	92.8 ± 65.6	1.30 ± 0.04	0.844 ± 0.040	5.68 ± 1.61
Corn straw	1520 ± 20	107 ± 19	1.04 ± 0.26	0.674 ± 0.074	4.63 ± 1.40

Reprinted from Shen et al. (2010) with permission of American Chemical Society

reviewed published EF_{PM} data and came up with a similar value of 9.4 ± 6.0 g/kg for domestic biomass combustion. For the two phases of flaming and smoldering burning (Table 4.5), EF_{PM} of the flaming phase (9.51 ± 3.02 g/kg) were significantly higher than those of the smoldering one (7.09 ± 3.87 g/kg) ($p < 0.05$). In fact, smoke observed in the smoldering phase was less thick than that in the flaming phase.

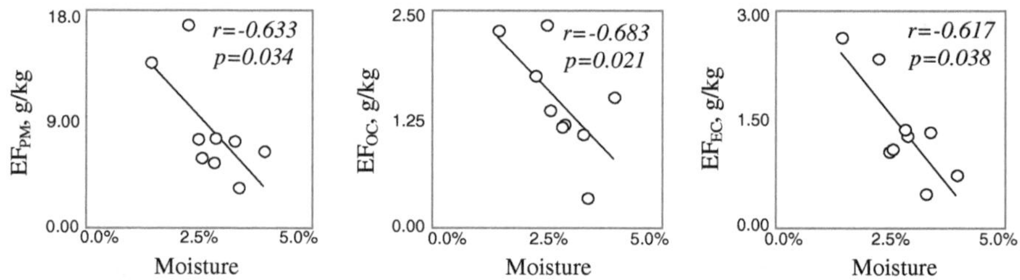

Fig. 4.6 Dependence of EFs of PM, OC and EC on fuel moisture for crop residue. Adapted from Shen et al. (2010) with permission of American Chemical Society

EF_{EC} and EF_{OC} for crop residue were 1.38 ± 0.70 and 1.45 ± 0.62 g/kg, respectively. Due to differences in crop residue property and fuel burning conditions, relatively large variations could be found in measured EFs. In general, our results are more or less similar to those measured using cooking stoves in the literature (Li et al. 2009; Venkataraman et al. 2005). For example, Li et al. (2009) reported that EF_{EC} and EF_{OC} for crop residues in residential stoves were 0.09–0.94 and 0.85–3.21 g/kg, respectively.

It is noted that the EF_{EC} measured for residential stoves (both our study and those reported in the literature) were often higher than those measured in laboratory chambers or open field. It was reported that EF_{EC} and EF_{OC} were 0.08 and 6.2 g/kg for rice residue (McMeeking et al. 2009) and 0.35 and 1.9 g/kg for wheat (Jimenez et al. 2007) burned in laboratory chambers. EF_{EC} and EF_{OC} of open fire burning for wheat were 0.16–0.17 and 0.29–2.81 g/kg, respectively (Jimenez et al. 2007; Sahai et al. 2007). EF_{EC} and EF_{OC} for rice straw were 0.749 and 1.50 g/kg, respectively, and 2.64 and 2.27 g/kg for wheat straw, respectively. The difference is likely due to the restricted air supply and poor mixing in residential stoves compared with those in chambers and open field, resulting in relatively lower combustion efficiency and higher combustion temperature, which is favorable for EC formation.

For the emissions emitted from crop straw burning, it could be found that EFs were negatively correlated with fuel moisture significantly (Fig. 4.6). The stepwise regression analysis using moisture, C, H, and N of the fuels, and the calculated MCE as independent variables showed that fuel moisture and MCE were two most significant influencing factors accounting for 63–83 % of total variation in EF_{PM} (Fig. 4.7). For EF_{EC}, in addition to fuel moisture, fuel N content was also significant and these two factors can explain 63–72 % of the total variations in EF_{EC} (Fig. 4.8). The effect of N on EC emission is unclear at this stage (Fig. 4.9), which required more detailed studies in future.

4.2.2 EC/OC Ratio

The EC/OC in emissions from crop residue burning was 1.25 ± 1.04, and total carbon mass percent in PM was 38 %. The EC/OC ratios were comparable among

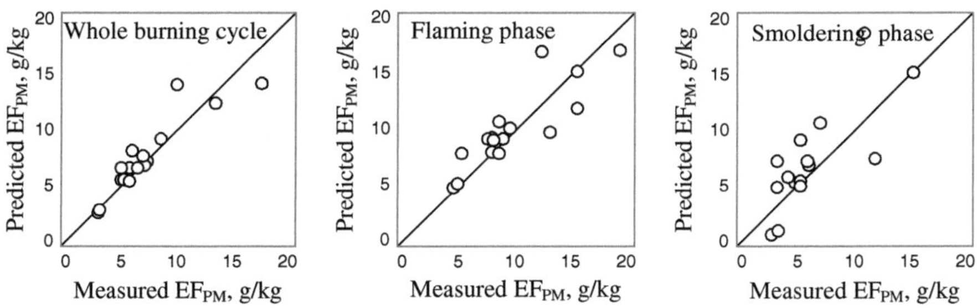

Fig. 4.7 Comparison of measured EF_{PM} and predicted EF_{PM} from crop residue burning. Modified from Shen et al. (2010) with permission of American Chemical Society

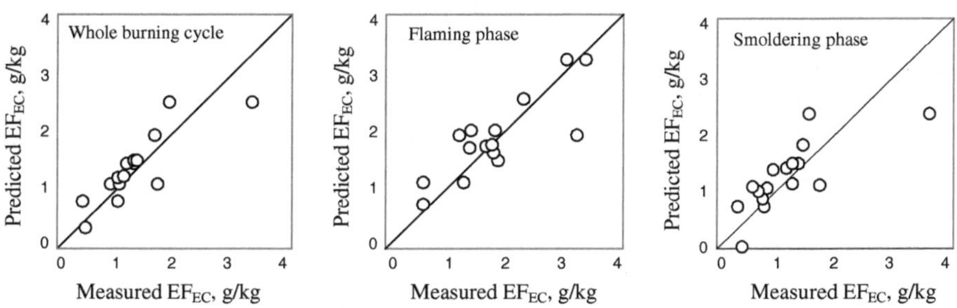

Fig. 4.8 Comparison of measured EF_{EC} and predicted EF_{EC} from crop residue burning. The prediction was based on two independent variables of moisture and N

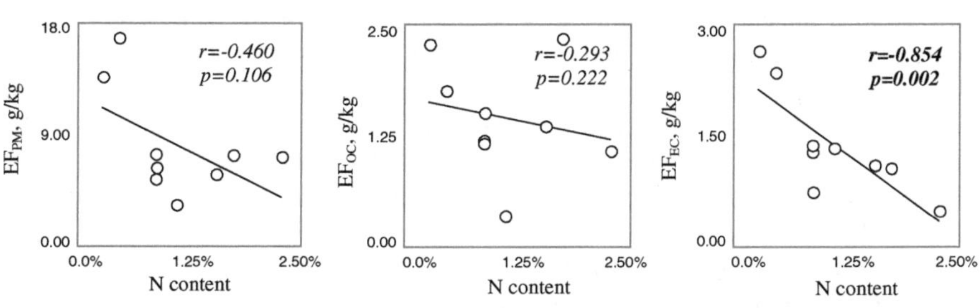

Fig. 4.9 Dependence of EF_{PM}, EF_{OC}, and EF_{EC} on fuel N content for crop residue

different crop straw types, and insignificantly different between the flaming and smoldering phases. In comparison with literature reported ratios (Li et al. 2009; Chow et al. 2011), it appears that the ratio in our present study was higher. For example, Li et al. (2009) reported ratios of 0.08–0.10 and 0.21–0.25 in emissions from the burnings of rice and corn stalk in Chongqing household, at 0.09 ± 0.01 and 0.10 ± 0.08 for rice and corn straws in a Shandong household, and at 0.22 ± 0.05, 0.14–0.16 and 0.18–0.19 in measurements in Henan area, respectively, but a high ratio of 1.08 ± 0.22 was also found by them in a study on bean straw burning. The MCE for crop residue burning the present study was 92.2 % larger than 90 %, and the ratio of CO/CO_2 averaged at 8.98 % lower than 10 %, both of which indicated that the high temperature fire flaming phase dominated the whole burning cycle, and hence preferable for the formation of light absorption carbon (Bonsang et al. 1995; Hurst et al. 1994; Li et al. 2009; Novakov et al. 2000).

4.2.3 Size Distribution

For all crop residues tested in whole burning cycle, the distributions were similar and unimodal with the peak between 0.7–1.1 and 1.1–2.1 μm (Fig. 4.10). On average, over 81 % of the total mass of PM_{10} from crop residues was $PM_{2.5}$ and approximately 12 % were finer particles with diameter less than 0.4 μm ($PM_{0.4}$). Calculated MMAD of PM for each crop straw type is provided in Table 4.6. The values are in the range of 0.91–1.34 μm. PM emitted during the smoldering phase was generally finer in comparison with that produced from the flaming phase (Fig. 4.11). Fine $PM_{2.5}$ contributed over 78 ± 7 and 83 ± 4 % of total PM mass during the two burning phases, respectively.

PM size distribution can be affected by a variety of factors such as temperature, moisture, and excess air ratio (Chang et al. 2004; Hays et al. 2003; Lighty et al. 2000; Maguhn et al. 2003; Purvis et al. 2000; Venkataraman and Rao 2001). Although the size distributions of PM from burning of various crop residues were similar to one another, the minor difference could be found among different crop straws, and it appears to be moisture dependent. Of the 9 size stages, correlation coefficients between moisture and relative fractions of 6 stages with D_a larger than 1.1 μm were positive (5 out of the 6 were significant at $p < 0.05$), while correlation coefficients between moisture and relative fractions of the remained 3 stages with D_a less than 1.1 μm were negative ($p = 0.098$, 0.041, and 0.054 for 0.7–1.1, 0.4–0.7, and <0.4 μm, respectively). Such a relationship was aggregately characterized by the significantly negative correlation between moisture and fine/coarse ratio (mass of PM smaller than 1.1 μm divided by that larger than 1.1 μm) ($r = -0.651$, $p = 0.002$), as shown in Fig. 4.12.

It is believed that the increase of moisture can reduce combustion temperature and efficiency (Chomanee et al. 2009; Hays et al. 2003; Purvis et al. 2000; Venkataraman et al. 2002). Since larger particles are produced under lower combustion temperature, higher moisture is favorable for emission of larger particles. Higher temperature may also shift mass distribution of particles to smaller diameter by limiting partitioning of organics on particles. In addition, fuel moisture may also affect relative humidity of flue gas, subsequently particle condensation in flue gas and the size of new emitted particles.

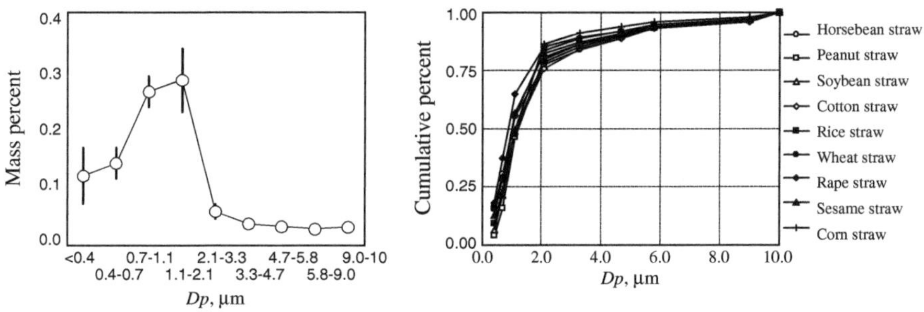

Fig. 4.10 Size distribution of PM from indoor crop straw burning

Table 4.6 Calculated MMAD of PM for different crop straw types

Fuel	Horsebean	Peanut	Soybean	Cotton	Rice	Wheat	Rape	Sesame	Corn
R^2	0.9447	0.9172	0.9267	0.9539	0.9327	0.9129	0.8942	0.9184	0.8978
MMAD	1.15	1.44	1.35	1.22	1.34	1.09	0.91	1.13	1.22
δ_g	3.71	3.04	3.18	3.59	3.25	3.63	4.12	3.57	3.23

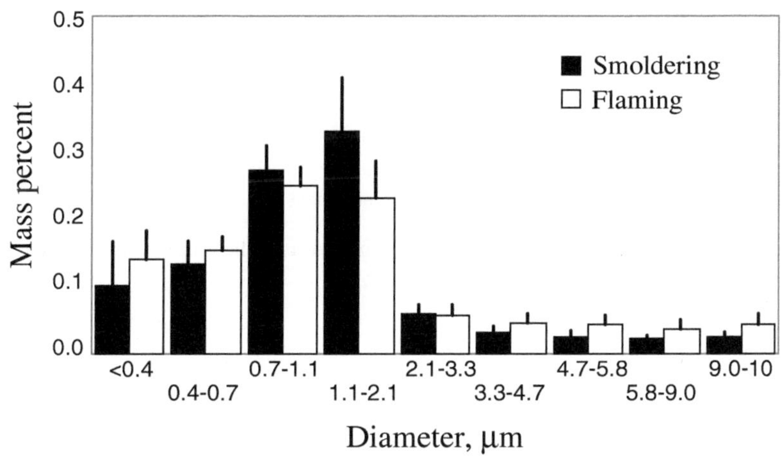

Fig. 4.11 Comparison of PM size distribution in emissions from the flaming and smoldering phases of indoor crop residue burning. Modified from Shen et al. (2010) with permission of American Chemical Society

4.2.4 Correlation Among Co-emitted Pollutants

PM, OC and EC were positively correlated with one another (Fig. 4.13). Figure 4.14 shows the relationship between EFs of PM, OC and EC and EF of CO from the crop straw burning. Only PM was found to be positively correlated with

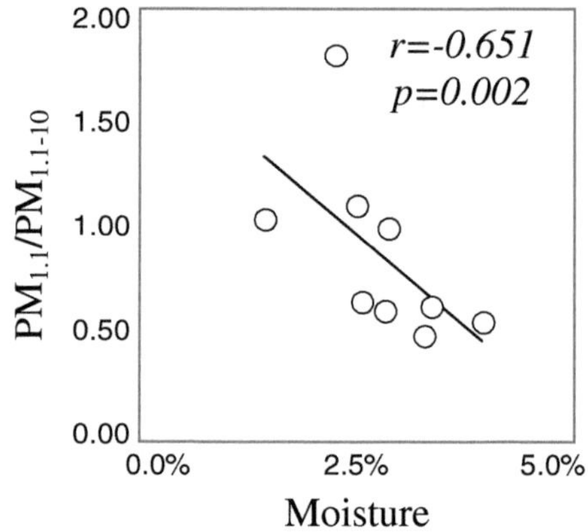

Fig. 4.12 Dependence of $PM_{1.1}/PM_{1.1-10}$ on fuel moisture in emissions from indoor crop residue burning. Adapted from Shen et al. (2010) with permission of American Chemical Society

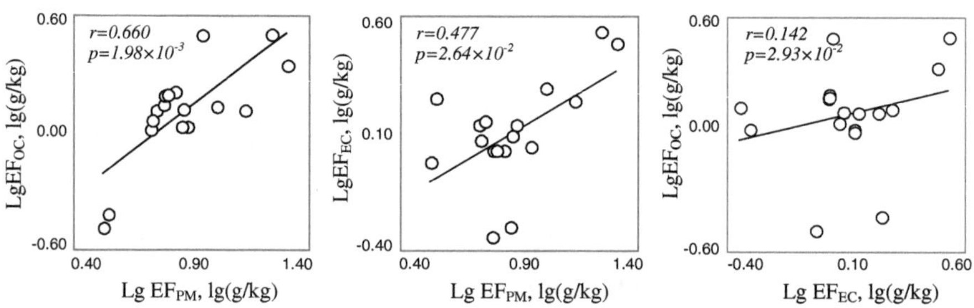

Fig. 4.13 Relationship among PM, OC and EC emitted from the crop residue burning

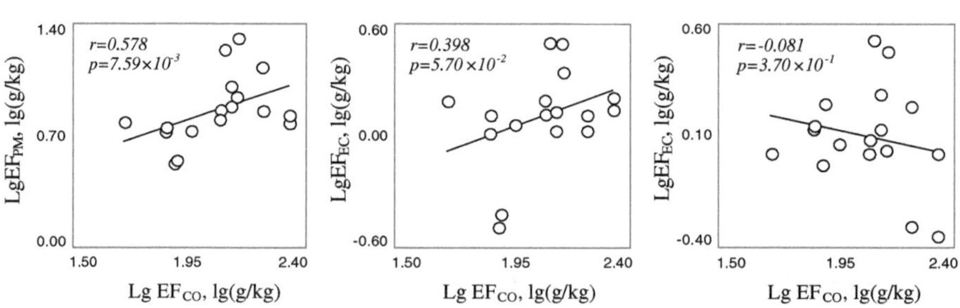

Fig. 4.14 Relationship between EF_{CO} and EFs of PM, OC and EC for crop residue

CO, while the correlation between OC and CO and between EC and CO was statistically insignificant ($p > 0.05$). This may indicate that although CO is the main incomplete product from the fuel combustion, the relationship between CO and other co-emitted incomplete pollutants may vary dramatically depending on the combustion conditions and experimental circumstance (Li et al. 2007; Venkataraman et al. 2004). So, the use of CO as a marker for other pollutants may be in caution before a sound conclusion reached based on sufficient data and detailed information on factors like fuel property, burning conditions, as well as pollutant formation and emission mechanisms.

4.3 Residential Wood Combustion

4.3.1 EFs and Influencing Factor

Table 4.7 lists measured EFs of PM, OC and EC for various wood materials in the present study. For 23 wood logs, EF_{PM}, EF_{OC} and EF_{EC} were in the range of 0.71–6.23, 0.11–3.81 and 0.06–1.19 g/kg, with means of 2.04 ± 1.38, 0.80 ± 0.85 and 0.50 ± 0.36 g/kg, respectively. For 3 brushwood fuels, they were 3.10–4.63, 0.21–1.48 and 0.48–2.49 g/kg with means of 3.74 ± 0.80, 0.81 ± 0.64 and 1.53 ± 1.01 g/kg, respectively. The EFs for brushwood were generally higher than that for fuel wood log, as shown in Fig. 4.15.

Since there was no significant difference in chemical composition (C, H, N, and O contents) between these two wood types, and the burning rates of the fuel wood log were significantly lower than those of brushwood fuels ($p < 0.05$), the relatively higher EFs of shrubby biomass likely resulted from the faster burning which may cause severe oxygen shortage in a stove hearth with limited volume (Rogge et al. 1998; Simoneit 2002). Relatively high pollutant emissions for brushwood or wood branches had been previously mentioned in the literature (Edwards et al. 2004; Zhang et al. 2000). In practice, brushwood and/or wood branches are widely used for daily cooking, thus, relatively higher emissions of incomplete pollutant would lead to high emissions of these pollutants from residential wood combustion which would subsequently affect the air quality and human health outcome significantly.

PM emissions from wood burning in China were reported at 1.17–5.87 g/kg in the instant combustion, and 1.51–8.73 g/kg in the ultimate combustion (Zhang et al. 2000). In a field measurement on residential wood combustion in rural area, OC, EC, and $PM_{2.5}$ EFs were 1.14 ± 0.40, 1.49 ± 0.69, and 3.08 ± 0.82 g/kg, respectively (Li et al. 2009). It can be seen that these EFs were all highly variable depending on fuel types and combustion conditions. In many developed countries and regions, firewood was often combusted in fireplace in addition to woodstove. It was reported that for the same fuel, EFs of fine PM from cooking stove combustion were significantly lower than those in fireplaces (Fine et al. 2004b). Based on the previously published EFs in literature and those measured in this study, EFs

Table 4.7 Emission factors of CO, OC, EC and PM from residential wood combustion

		CO	OC	EC	PM
1.	White Poplar	49.83 ± 5.45	0.66 ± 0.32	0.88 ± 0.49	1.75 ± 0.25
2.	Elm	44.34 ± 2.24	0.79 ± 0.15	1.19 ± 0.32	2.16 ± 0.22
3.	Locust	42.27 ± 2.53	0.19 ± 0.13	0.52 ± 0.41	0.99 ± 0.39
4.	Maple	53.02 ± 1.96	0.39 ± 0.12	0.94 ± 0.51	1.89 ± 0.21
5.	Fir	38.4 ± 11.8	0.60 ± 0.35	0.94 ± 0.40	1.59 ± 0.32
6.	Larch	36.91 ± 2.43	0.97 ± 0.88	0.95 ± 0.20	1.68 ± 0.70
7.	Water Chinese fir	44.21 ± 1.06	0.21 ± 0.13	0.41 ± 0.39	0.87 ± 0.29
8.	Cypress	56.06 ± 6.52	0.82 ± 0.45	0.71 ± 0.39	2.02 ± 0.59
9.	Oak	32.75 ± 3.20	0.14 ± 0.11	0.35 ± 0.34	0.71 ± 0.36
10.	Chinese Pine	35.67 ± 0.16	0.36 ± 0.17	0.85 ± 0.45	1.79 ± 0.19
11.	Willow	63.90 ± 0.45	0.50 ± 0.11	0.97 ± 0.28	2.30 ± 0.74
12.	Paulownia tomentosa	62.72 ± 2.02	1.21 ± 0.63	0.28 ± 0.08	2.31 ± 0.13
13.	Toon	82.52 ± 18.4	2.01 ± 1.53	0.21 ± 0.20	3.43 ± 1.97
14.	White Birch	57.48 ± 2.76	0.14 ± 0.03	0.10 ± 0.06	1.35 ± 0.10
15.	Ribbed Birch	44.53 ± 0.11	0.26 ± 0.25	0.10 ± 0.06	1.29 ± 0.22
16.	Paulownia elongata	45.31 ± 5.40	0.54 ± 0.63	0.13 ± 0.13	1.68 ± 1.00
17.	Black Poplar	37.51 ± 3.11	0.11 ± 0.01	0.06 ± 0.00	0.79 ± 0.42
18.	China Aspen	49.74 ± 3.01	0.69 ± 0.32	0.67 ± 0.61	1.90 ± 0.42
19.	Chinaberry	35.73 ± 0.91	0.16 ± 0.17	0.12 ± 0.14	0.74 ± 0.13
20.	Jujube	57.30 ± 0.62	1.98 ± 1.50	0.21 ± 0.15	1.65 ± 0.65
21.	Persimmon	54.48 ± 8.02	0.51 ± 0.61	0.13 ± 0.17	2.15 ± 0.58
22.	Mulberry	66.78 ± 0.27	1.25 ± 1.41	0.39 ± 0.45	5.65 ± 5.31
23.	Peach	140.0 ± 13.5	3.81 ± 1.31	0.32 ± 0.04	6.23 ± 2.08
24.	Lespedeza	62.44 ± 3.71	0.21 ± 0.17	0.48 ± 0.40	3.48 ± 2.45
25.	Buxus sinica	45.70 ± 6.85	1.48 ± 0.57	2.49 ± 1.23	4.63 ± 0.51
26.	Holly	34.67 ± 3.12	0.73 ± 0.28	1.63 ± 0.37	3.10 ± 0.14
27.	Bamboo	44.12 ± 11.8	0.13 ± 0.09	1.19 ± 0.54	2.13 ± 0.46

Data shown are average and standard derivations of triplicate combustion experiments. Modified from Shen et al. (2012) with permission of American Chemical Society

of PM, OC, and EC were 2.95 ± 3.04 (0.31–16.3, as range, $n = 85$), 1.10 ± 1.56 (0.02–8.09, $n = 109$), and 0.68 ± 0.64 (0.04–3.77, $n = 85$) g/kg for woodstoves and 7.96 ± 4.32 (1.6–20.2, n = 61), 4.80 ± 2.06 (1.09–9.17, n = 36), and 0.66 ± 1.23 (0.04–6.58, n = 36) for fireplaces, respectively. The differences between them were indeed significant ($p < 0.05$).

Figure 4.16 shows the dependence of EFs of PM, OC and EC on MCE. PM and OC emissions decreased with the increase of MCE, however, the relationship between EC and MCE was statically insignificant ($p > 0.05$). Individually, most of fuel property factors were not significantly correlated with the measured EFs, as shown in Fig. 4.17 using fuel moisture as an example. This could be partly explained by the fact that the impacts of these factors were often interacted and relatively greater variation of these factors and measured EFs may prevent from seeing the effect of an individual factor (Roden et al. 2006).

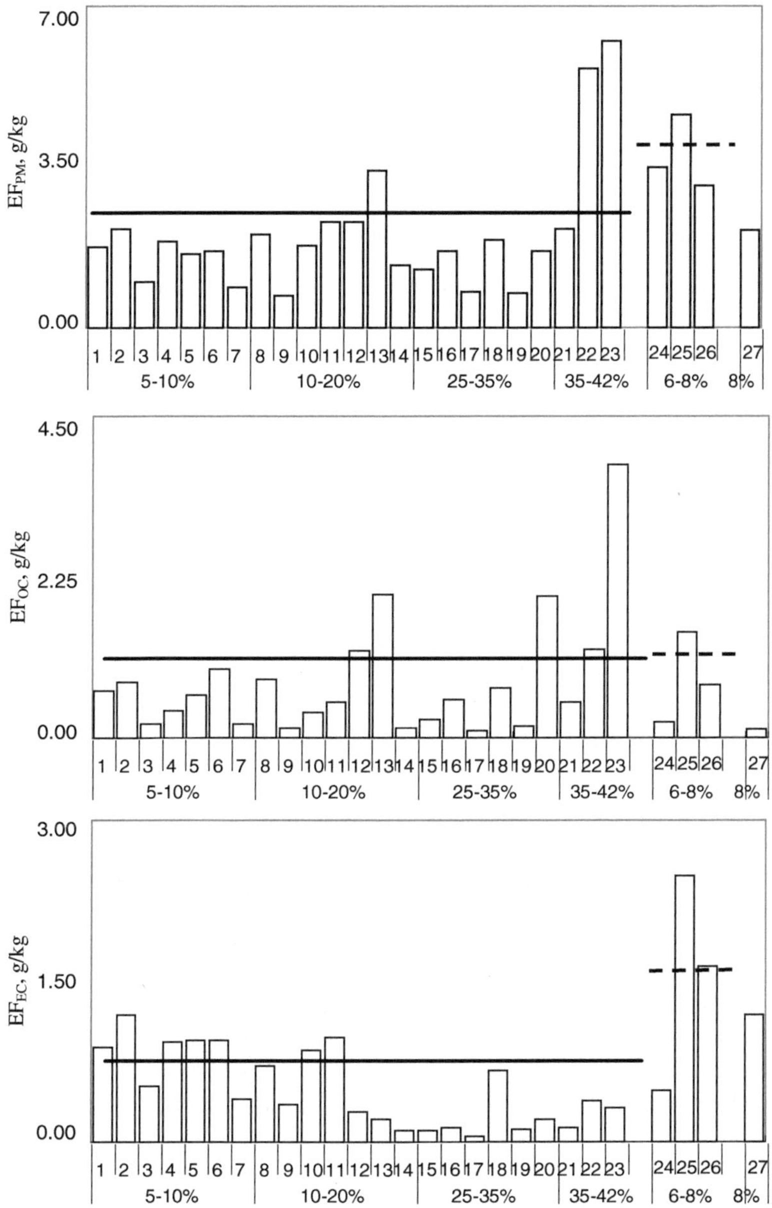

Fig. 4.15 Comparison of EF_{PM}, EF_{EC} and EF_{OC} for various wood materials. Fuel woods were classified into groups with moisture of 5–10, 10–20, 25–35, and >35 %. The *lines* shown are overall averages for fuel wood log and brushwood, respectively

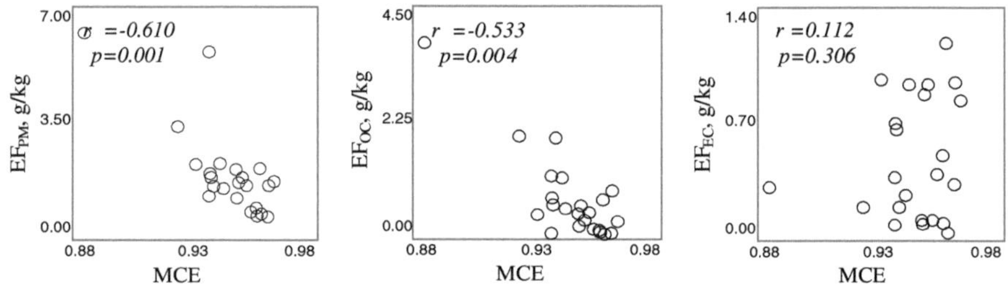

Fig. 4.16 Dependence of measured EFs of PM, OC and EC on MCE for wood combustion

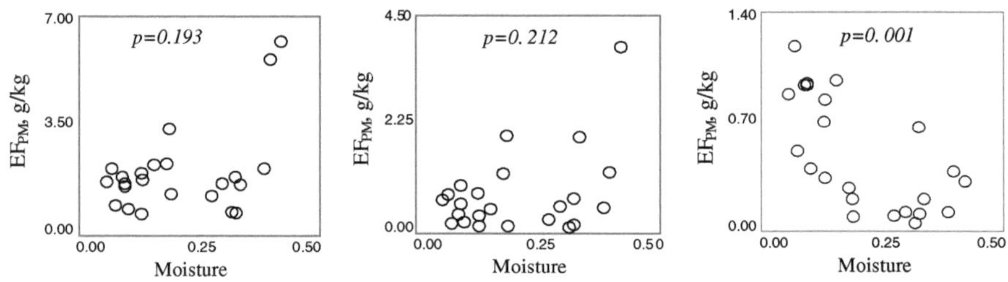

Fig. 4.17 Dependence of measured EFs of PM, OC and EC on fuel moisture in residential wood combustion

4.3.2 EC/OC Ratio

The ratio of EC/OC in emissions from residential wood combustion was 1.19 ± 0.88. The result was very close to 1.41 ± 0.57 reported by Li et al. (2009) for wood combustion in rural households in China. Both were found to be higher than those reported in the literature (Chow et al. 2011; Schauer et al. 2001; Fine et al. 2001, 2002, 2004a, b). CO/CO_2 ratios measured in the present wood combustion were lower than 10 %, indicating the hot flaming domination in the tested wood combustion. As such, more EC would be emitted from the relatively high temperature combustion in the flaming phase (Bonsang et al. 1995; Li et al. 2009; Novakov et al. 2000). Total carbon mass percent in PM emitted from fuel wood and brushwood combustions were 64 and 61 %, respectively.

4.3.3 Size Distribution

Freshly emitted PM from residential wood combustion is very small with the domination of fine $PM_{0.4}$. On average, over 80 % of total PM was fine $PM_{2.5}$. In brushwood combustion, there was a high fraction of coarse PM in comparison with PM from fuel wood combustion (Fig. 4.18). PM was dominated by those between 2.1 and 3.3 μm (21.0 ± 9.4 %), followed by PM at diameter between 1.1 and

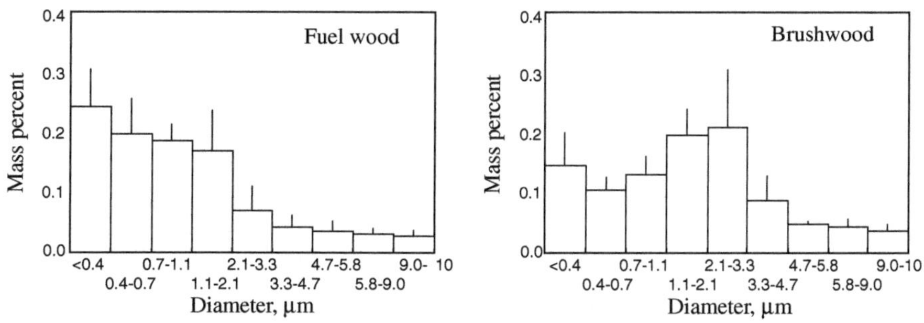

Fig. 4.18 Size distribution of PM emitted from fuel wood log and brushwood combustions. Modified from Shen et al. (2012) with permission of American Chemical Society

2.1 μm (19.6 ± 4.4 %). Overall, $PM_{0.4}$ and $PM_{2.1}$ fractions were 14.6 ± 5.4 % and 57.7 ± 10.8 %, respectively. Distinct size distribution of PM between the fuel wood log and brushwood combustions should be taken into consideration in the health exposure study in future since PM with different size had different health outcome.

MCE was found to be negatively correlated with 3 PM fractions with diameters less than 1.1 μm and positively correlated with all other PM fractions ($p < 0.05$), which means that under higher MCE, the mass median diameters of emitted PM would be larger. Meanwhile, moisture affects the size distribution in an opposite direction to MCE. It is negatively correlated with coarse PM and positively correlated with fine PM ($p < 0.05$), indicating that the median mass diameters of PM from lower moisture wood combustions were larger. Such a relationship was aggregately characterized by the significantly positive correlation between moisture and fine/coarse ratio ($PM_{1.1}/PM_{1.1-10}$) and negative correlation between MCE and the $PM_{1.1}/PM_{1.1-10}$ ratio (Fig. 4.19).

The result was opposite to that found in crop residue burning. The influence of fuel moisture is believed to be complicated. Lower moisture may result in a higher temperature which is favorable for formation of small particles. But fuels with too low moisture may burn too fast to result in an oxygen limited atmosphere in the stove with relatively small chamber and only natural air ventilation. These, in turn, produce a large number of large particles since lower oxygen levels are expected to promote less intense smoldering conditions producing large particles due to agglomeration and condensation processes (Rogge et al. 1998; Hays et al. 2003). It is also noted that the wood moisture was generally higher than that of crop residue, and moreover, wood moisture ranged widely in comparison with relatively small moisture range in crop residue.

4.3.4 Correlation Among Co-emitted Pollutants

It appears that the correlation between PM and OC was significant, but the correlation between EC and OC, and between EC and PM were statistically

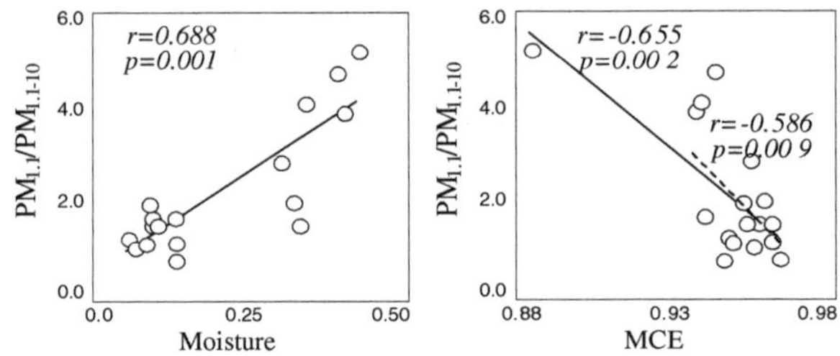

Fig. 4.19 Dependence of $PM_{1.1}/PM_{1.1-10}$ ratio on fuel moisture and MCE in residential wood combustion

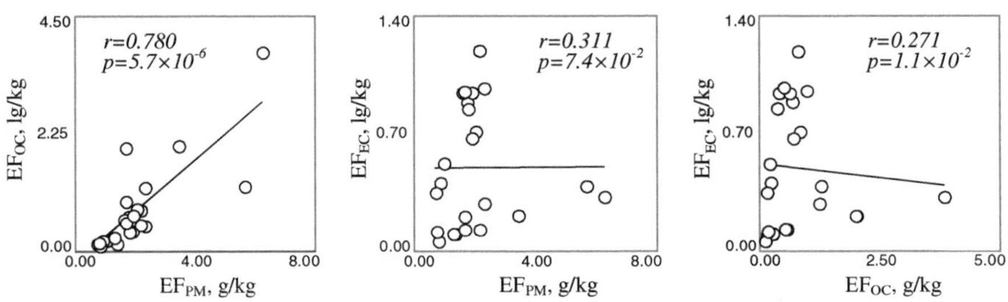

Fig. 4.20 Correlations among EFs of PM, EC, and OC for wood combusted in the residential brick stove

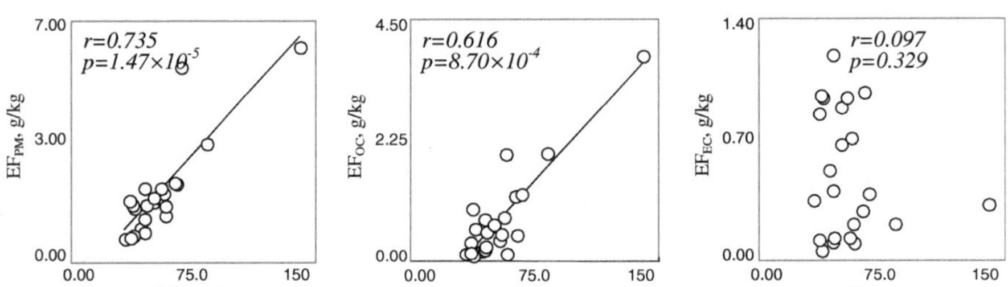

Fig. 4.21 Correlations between EF_{CO} and EFs of PM, EC, and OC for wood

insignificant (Fig. 4.20). PM and OC were positively correlated with CO, but no significant correlation was found between OC and CO (Fig. 4.21). Although CO is also an incomplete combustion product and is occasionally used as a surrogate for the emissions of PM and other pollutants (Bignal et al. 2008; Venkataraman and Rao 2001), variable dependence of other pollutants on CO had been reported in the literature. In some circumstances, CO may even be negatively correlated with other incomplete pollutants (Li et al. 2007; Venkataraman et al. 2004).

4.4 Fuel Comparison

4.4.1 EFs and EC/OC Ratio

In Table 4.8, we summarized measured EFs of PM, OC and EC, as well as EC/OC and TC/PM ratios for different fuels. The EFs for coal have much larger variations in comparison with biomass fuels, and highly depend on the coal property (anthracite/bituminous) and form (raw chunk/briquette). For biomass fuel, although varying obviously among different fuel types, the EFs are generally in the same order of magnitude. Brushwood combustion produced much higher pollutant emissions than the burning of fuel wood log.

The ratios of EC/OC for various biomass fuels were comparable, and appeared to be higher than that for coal. The ratio varied not only among different fuel types, but obviously within the fuel type. This may suggest that in the use of it for source analysis or environmental impact analysis, a single specific value would cause considerable bias. The total carbon mass percents in PM emitted from coal and crop residue combustions were 32 and 38 %, respectively. In emissions from residential wood combustion, carbonaceous carbon comprised up to 64 % of the total PM mass, which was significantly higher than those for crop residue and coal.

4.4.2 Size Distribution

Size distributions of PM for different fuels are compared in Fig. 4.22. The distribution of PM from coal combustion is strongly associated with coal property, such as caking property. High caking coal would produce much more coarse particles. The size distribution of PM from biomass burning was similar with the fuel type, but different between fuel wood log and brushwood, and the later was comparable to that for crop residue. In emissions from brushwood and crop residue burning, the dominated fraction was PM with diameter between 1.1–2.1 μm, whereas in emissions from fuel wood log burning, fine $PM_{0.4}$ was the most abundant. Overall, freshly emitted PM from residential solid fuel combustion is very small with over 80 % was fine $PM_{2.5}$.

4.5 Summary

EF_{PM}, EF_{OC} and EF_{EC} for crop residue burned in residential brick stove were 8.19 ± 4.27 (3.41–16.8 as range), 1.38 ± 0.70 (0.493–2.64) and 1.45 ± 0.62 (0.354–2.34) g/kg, respectively. In the stove burning, the fast fuel combustion is usually easy to form oxygenated deficient atmosphere in the stove chamber with

Table 4.8 Comparison of EF_{PM}, EF_{OC}, EF_{EC} and EC/OC ratio among crop residue, coal, fuel wood and brushwood

	EF_{PM}, g/kg	EF_{OC}, g/kg	EF_{EC}, g/kg	EC/OC	TC/PM
Coal	2.23 ± 3.44	0.42 ± 0.49	0.26 ± 0.40	0.48 ± 0.38	0.32 ± 0.19
	(0.065–10.8)	(0.007–1.00)	(0.006–0.83)	(0.06–0.86)	(0.09–0.46)
Crop residue	8.19 ± 4.27	1.38 ± 0.70	1.45 ± 0.62	1.25 ± 1.04	0.38 ± 0.10
	(3.41–16.8)	(0.493–2.64)	(0.354–2.34)	(0.32–4.57)	(0.22–0.61)
Fuel wood	2.04 ± 1.38	0.80 ± 0.85	0.50 ± 0.36	1.08 ± 0.87	0.64 ± 0.29
	(0.71–6.23)	(0.11–3.81)	(0.06–1.19)	(0.09–2.67)	(0.18–1.32)
Brushwood	3.74 ± 0.80	0.81 ± 0.64	1.53 ± 1.01	2.06 ± 0.33	0.61 ± 0.36
	(3.10–4.63)	(0.21–1.48)	(0.48–2.49)	(1.68–2.26)	(0.20–0.86)

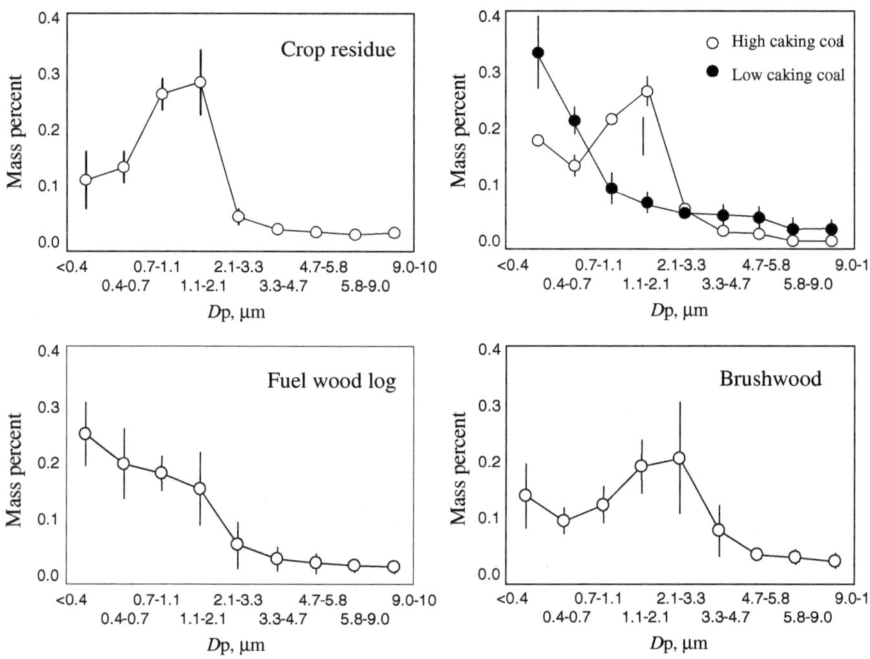

Fig. 4.22 Comparison of PM size distribution in emissions from residential crop residue, wood and coal combustions

small volume and limited air supply, and hence produces relatively higher emissions of incomplete pollutant.

In residential wood combustion, the EF_{PM}, EF_{OC} and EF_{EC} for brushwood were 3.74 ± 0.80 (3.10–4.63), 0.81 ± 0.64 (0.21–1.48) and 1.53 ± 1.01 (0.48–2.49) g/kg, respectively, significantly higher than those of 2.04 ± 1.38 (0.71–6.23), 0.80 ± 0.85 (0.11–3.81) and 0.50 ± 0.36 (0.06–1.19) g/kg for fuel wood log. The carbonaceous carbon mass percent in PM emitted from residential wood combustion was 61 %, which was higher ($p < 0.05$) than that of 38 % in emissions from indoor crop residue burning.

EF_{PM}, EF_{OC} and EF_{EC} for coal were in the range of 0.065–10.8, 0.007–1.00 and 0.006–0.825 g/kg, respectively. The pollutant emission factors for anthracite were smaller compared to bituminous, and pollutant emissions for two studied briquette coals were significantly lower than those for the raw chunk.

In residential solid fuel combustion, the domination of fine PM was observed in freshly emitted PM. The mass percents of $PM_{2.5}$ in total PM emitted were 77.5, 81.0 and 79.4 % in emissions from coal, crop residue and wood combustions. The size distributions of PM for different crop straw types were similar in general. Significant difference was found in emissions between fuel wood log and brushwood. PM emitted from the fuel wood log burning was dominated by very fine PM, such as $PM_{0.4}$ in the present study while PM with diameter between 0.7–2.1 μm was the most abundant fraction and a relatively higher percents of coarse particles could be found in the emission from brushwood burning.

References

Andreae, M. O., & Merlet, P. (2001). Emission of trace gases and aerosols from biomass burning. *Global Biogeochemistry Cycle, 15*, 955–966.

Bignal, K. L., Langridge, S., & Zhou, J. L. (2008). Release of polycyclic aromatic hydrocarbons, carbon monoxide and particulate matter form biomass combustion in a wood-fired boiler under varying boiler conditions. *Atmospheric Environment, 42*, 8863–8871.

Bond, T. C., Covert, D. S., Kramlich, J. C., Larson, T. V., & Charlson, R. J. (2002). Primary particle emissions from residential coal burning: Optical properties and size distributions. *Journal of Geophysical Research-Atmospheres, 107*(D21), 8347. doi: 10.1029/2001JDO00571.

Bond, T. C., Streets, D. G., Yarber, K. F., Nelson, S. M., Woo, J., & Klimont, Z. (2004). A technology-based global inventory of black and organic carbon emissions from combustion. *Journal of Geophysical Research, 109*, D14203. doi:10.1029/2003JD003697.

Bonsang, B., Boissard, C., Lecloarec, M. F., Rudolph, J., & Lacaux, J. P. (1995). Methane, carbon monoxide and light non-methane hydrocarbon emissions from African savanna burning during the FOS/DECAFE experiment. *Journal of Atmospheric Chemistry, 22*, 149–162.

Chang, M. C., Chow, J. C., Watson, J. G., Hopke, P. K., Yi, S. M., & England, G. C. (2004). Measurement of ultrafine particle size distributions from coal-, oil-, and gas-fired stationary combustion sources. *Journal of the Air and Waste Management Association, 54*, 1494–1505.

Chen, Y. J., Bi, X. H., Mai, B. X., Sheng, G. Y., & Fu, J. M. (2004). Emission characterization of particulate/gaseous phases and size association for polycyclic aromatic hydrocarbons from residential coal combustion. *Fuel, 83*, 781–790.

Chen, Y. J., Sheng, G. Y., Bi, X. H., Feng, Y. L., Mai, B. X., & Fu, J. M. (2005). Emission factors for carbonaceous particles and polycyclic aromatic hydrocarbons from residential coal combustion in China. *Environmental Science and Technology, 39*, 1861–1867.

Chen, Y. J., Zhi, G., Feng, Y., Fu, J., Feng, J., Sheng, G., et al. (2006). Measurement of emission factors for primary carbonaceous particles from residential raw-coal combustion in China. *Geophysical Research Letter, 33*, L20815. doi:10.1029/2006GL026966.

Chen, Y., Zhi, G., Feng, Y., Liu, D., Zhang, G., Li, J., et al. (2009). Measurements of black and organic carbon emission factors for household coal combustion in China: Implication for emission reduction. *Environmental Science and Technology, 43*, 9495–9500.

Chomanee, C., Tekasakul, S., Tekasakul, P., Furuuchi, M., & Otani, Y. (2009). Effects of moisture content and burning period on concentration of smoke particles and particle-bound polycyclic aromatic hydrocarbons from rubber wood combustion. *Aerosol and Air Quality Research, 9*, 404–411.

Chow, J.C., Watson, J.G., Lowenthal, D. H., Antony Chen, L.-W., & Motallebi, N. (2011). PM2.5 source profiles for black and organic carbon emission inventories. *Atmospheric Environment, 45*, 5407–5414.

Edwards, R. D., Smith, K. R., Zhang, J., & Ma, Y. (2004). Implications of changes in household stoves and fuel use in China. *Energy Policy, 32*, 395–411.

Fine, P. M., Cass, G. R., & Simoneit, B. R. T. (2001). Chemical characterization of fine particle emissions from fireplace combustion of woods grown in the Northeastern United States. *Environmental Science and Technology, 35*, 2665–2675.

Fine, P. M., Cass, G. R., & Simoneit, B. R. T. (2002). Chemical characterization of fine particle emissions from fireplace combustion of woods grown in the Southern United States. *Environmental Science and Technology, 36*, 1442–1451.

Fine, P. M., Cass, G. R., & Simoneit, B. R. T. (2004a). Chemical characterization of fine particle emissions from fireplace combustion of wood stove combustion of prevalent United States tree species. *Environmental Engineering and Science, 21*, 705–721.

Fine, P. M., Cass, G. R., & Simoneit, B. R. T. (2004b). Chemical characterization of fine particle emissions from fireplace combustion of wood types grown in the Midwestern and Western United States. *Environmental Engineering and Science, 21*, 387–409.

Francois, D., Patrick, B. M., Ulith, E., Sami, M., & Gilles, B. (1989). Atmospheric input of trace metals to the western Mediterranean: Uncertainties in modeling dry deposition from cascade impactor data. *Tellus, 41*, 362–378.

Hays, M. D., Smith, N. D., Kinsey, J., Dong, Y., & Kariher, P. (2003). Polycyclic aromatic hydrocarbon size distributions in aerosols from appliances of residential wood combustion as determined by direct thermal desorption-GC/MS. *Journal of Aerosol Science, 34*, 1061–1084.

Hurst, D. F., Griffith, D. W. T., Carras, J. N., Williams, D. J., & Fraser, P. J. (1994). Measurements of trace gases emitted by Australian savanna fires during the 1990 dry season. *Journal of Atmospheric Chemistry, 18*, 33–56.

Jimenez, J. R., Claiborn, C. S., Dhammapala, R. S., & Simpson, C. D. (2007). Methoxyphenols and levoglucosan ratios in PM2.5 from wheat and Kentucky bluegrass stubble burning in Eastern Washington and Northern Idaho. *Environmental Science and Technology, 41*, 7824–7829.

Ledesma, E. B., Kalish, M. A., Nelson, P. F., Wornat, M. J., & Mackie, J. C. (2000). Formation and fate of PAH during the pyrolusis and fuel rich combustion of coal primary tar. *Fuel, 79*, 1801–1814.

Li, X., Duan, L., Wang, S., Duan, J., Guo, X., Yi, H., et al. (2007). Emission characteristics of particulate matter from rural household biofuel combustion in China. *Energy & Fuels, 21*, 845–851.

Li, X., Wang, S., Duan, L., Hao, J., & Nie, Y. (2009). Carbonaceous aerosol emissions from household biofuel combustion in China. *Environmental Science and Technology, 43*, 6076–6081.

Lighty, J. S., Veranth, J. M., & Sarofim, A. F. (2000). Combustion aerosols: Factors governing their size and composition and implications to human health. *Journal of the Air and Waste Management Association, 50*, 1565–1618.

Maguhn, J., Karg, E., Kettrup, A., & Zimmermann, R. (2003). On-line analysis of the size distribution of fine and ultrafine aerosol particles in flue and stack gas of a municipal waste incineration plant: Effects of dynamic process control measures and emission reduction devices. *Environmental Science and Technology, 37*, 4761–4770.

McMeeking, G. R., Kreidenweis, S. M., Baker, S., Carrico, C. M., Chow, J. C., Collett, J. L, Jr, et al. (2009). Emission of trace gases and aerosols during the open combustion of biomass in the laboratory. *Journal of Geophysical Research, 114*, D19210. doi:10.1029/2009JD011836.

Mitra, A., Sarofim, A., & Bar-Ziv, E. (1987). The influence of coal type on the evolution of polycyclic aromatic hydrocarbons during coal devolatilization. *Aerosol Science and Technology, 6*, 261–271.

Novakov, T., Andreae, M. O., Gabriel, R., Kirchstetter, T., Mayol-Bracero, O., & Ramanathan, V. (2000). Origin of carbonaceous aerosols over the tropical Indian Ocean: Biomass burning or fossil fuels. *Geophysical Research Letter, 27*, 4061–4064.

Parterson, C. C., & Gillette, D. A. (1977). Commonalities in measured size distributions for aerosols having a soil-derived component. *Journal of Geophysics Research, 82*, 2074–2082.

Purvis, C. R., McCrillis, R. C., & Kariher, P. H. (2000). Fine particulate matter (PM) and organic speciation of fireplace emissions. *Environmental Science and Technology, 34*, 1653–1658.

Roden, C. A., Bond, T. C., Conway, S., & Pinel, A. B. O. (2006). Emission factors and real-time optical properties of particles emitted from traditional wood burning cookstoves. *Environmental Science and Technology, 40*, 6750–6757.

Rogge, W. F., Hildemann, L. M., Mazurek, M. A., Cass, G. R., & Simoneit, B. R. T. (1998). Sources of fine organic aerosol. 9. pine, oak and synthetic log combustion in residential fireplaces. *Environmental Science and Technology, 32*, 13–22.

Sahai, S., Sharma, C., Singh, D. P., Dixit, C. K., Singh, N., Sharma, P., et al. (2007). A study for development of emission factors for trace gases and carbonaceous particulate species from in situ burning of wheat straw in agricultural fields in india. *Atmospheric Environment, 41*, 9173–9186.

Schauer, J. J., Kleeman, M. J., Cass, G. R., & Simoneit, B. R. T. (2001). Measurement of emissions from air pollution sources 3. C1 through C29 organic compounds from fireplace combustion of wood. *Environmental Science and Technology, 33*, 1716–1728.

Shen, G. F., Yang, Y., Wang, W., Tao, S., Zhu, C., Min, Y., Xue, N., Ding, J., Wang, B., Wang, R., Shen, H., Li, W., Wang, X., Russell, A. (2010) Emission factors of particulate matter and elemental carbon for crop residues and coals burned in typical household stoves in China. *Environmental Science & Technology, 44*, 7157–7162.

Shen, G. F., Wei, S., Wei, W., Zhang, Y., Min, Y., Wang, B., Wang, R., Li, W., Shen, H., Huang, Y., Yang, Y., Wang, W., Wang, X., Wang, X., Tao, S. (2012) Emission Factors, Size Distributions and Emission Inventories of Carbonaceous Particulate Matter from Residential Wood Combustion in Rural China. *Environmental Science & Technology*, *46*, 4207–4214.

Simoneit, B. R. T. (2002). Biomass burning—A review of organic tracers for smoke from incomplete combustion. *Applied Geochemistry, 17*, 129–162.

Venkataraman, C., & Rao, G. U. M. (2001). Emission factors of carbon monoxide and size-resolved aerosols from biofuel combustion. *Environmental Science and Technology, 35*, 2100–2107.

Venkataraman, C., Negi, G., Sardar, S. B., & Rastogi, R. (2002). Size distributions of polycyclic aromatic hydrocarbons in aerosol emissions from biofuel combustion. *Journal of Aerosol Science, 33*, 503–518.

Venkataraman, C., Joshi, P., Sethi, V., Kohli, S., & Ravi, M. R. (2004). Aerosol and carbon monoxide emissions from low temperature combustion in a sawdust packed-bed stove. *Aerosol Science and Technology, 38*, 50–61.

Venkataraman, C., Habib, G., Eiguren-Fernandez, A., Miguel, A. H., & Friedlander, S. K. (2005). Residential biofuel in South Asia: Carbonaceous aerosol emissions and climate impacts. *Science, 307*, 1454–1456.

Whitby, K. T. (2007). The physical characteristics of sulfur aerosols. *Atmospheric Environment, 41*, 25–49.

Zhang, X., An, Z., Zhang, G., Chen, T., Liu, D., Arimoto, R., et al. (1994). Mass transport, deposition and climate impacts of atmospheric particulate matter in mainland China. *Science in China (Series B), 24*(11), 1206–1215. (In Chinese).

Zhang, J., Smith, K. R., Ma, Y., Ye, S., Jiang, F., Qi, W., et al. (2000). Greenhouse gases and other airborne pollutants from household stoves in China: A database for emission factors. *Atmospheric Environment, 34*, 4537–4549.

Zhi, G., Chen, Y., Feng, Y., Xiong, S., Li, J., Zhang, G., et al. (2008). Emission characteristics of carbonaceous particles from various residential coal-stoves in China. *Environmental Science and Technology, 42*, 3310–3315.

Zhi, G., Peng, C., Chen, Y., Liu, D., Sheng, G., & Fu, J. (2009). Deployment of coal briquettes and improved stoves: Possibly an option for both environment and climate. *Environmental Science and Technology, 43*, 5586–5591.

Chapter 5
Parent Polycyclic Aromatic Hydrocarbons

Polycyclic aromatic hydrocarbons (PAHs) are a group of pollutants mainly produced from the incomplete burning process. Anthropogenic sources such as residential solid fuel combustion, vehicle emission, and industrial coal combustion are the main sources of PAHs in environment while the contributions of natural sources are relatively small. PAHs can undergo long range transport after emission and cause detrimental impacts on human health. In this section, emission factors of PAHs (EF_{PAHs}) were reported for coal, crop residue and wood burned in residential stoves. The influence of fuel property and burning condition was discussed. In emissions from coal and crop residue burning, we mainly analyzed 16 priority PAHs listed by U.S. EPA, and in the emissions from residential wood combustion, some non-priority PAHs which had relatively high molecular weight and strong carcinogenic toxic potentials were also analyzed.

5.1 Coal Combustion

5.1.1 EFs and Influencing Factor

EFs of 16 EPA priority PAHs from residential coal combustion ranged from 4.78 mg/kg (anthracite) to 373 mg/kg (bituminous), averaged at 82.9 mg/kg. Gaseous PAH EFs were in the range of 3.83–174 mg/kg with most abundance of low molecular weight PAHs (NAP-PHE), while average PAH EFs in particulate phase ranged from 0.851 to 214 mg/kg dominated by high molecular weight PAHs (from BaA to BghiP). Means and standard deviations of PAH EFs for each coal are listed in Table 5.1. The EF_{PAHs} for two honeycomb briquettes were 6.25 ± 1.16 mg/kg (Beijing) and 14.3 ± 0.3 mg/kg (Taiyuan), respectively, significantly lower than those at 140 ± 35, 253 ± 170 and 78.7 ± 93.4 mg/kg for three chunk coals.

It has been widely documented that PAH EFs among different coal types (honeycomb/chunk, and anthracite/bituminous) varied largely (Chen et al. 2005;

G. Shen, *Emission Factors of Carbonaceous Particulate Matter and Polycyclic Aromatic Hydrocarbons from Residential Solid Fuel Combustions*, Springer Theses,
DOI: 10.1007/978-3-642-39762-2_5,

Table 5.1 Emission factors (mg/kg) of PAHs for five tested coals

PAH	Phase	Beijing Honeycomb	Taiyuan Honeycomb	Taiyuan Chunk	Yulin Chunk-A	Yulin Chunk-B
NAP	Gaseous	$1.6 \pm 1.3 \times 10^{0}$	$5.8 \pm 0.6 \times 10^{0}$	$1.4 \pm 0.6 \times 10^{1}$	$1.3 \pm 1.6 \times 10^{1}$	$1.1 \pm 1.9 \times 10^{1}$
	Particulate	$4.2 \pm 2.2 \times 10^{-2}$	$3.1 \pm 1.9 \times 10^{-2}$	$1.4 \pm 0.2 \times 10^{-1}$	$1.8 \pm 0.6 \times 10^{-1}$	$1.0 \pm 1.1 \times 10^{-1}$
ACY	Gaseous	$8.3 \pm 3.7 \times 10^{-1}$	$1.9 \pm 0.5 \times 10^{0}$	$2.0 \pm 0.7 \times 10^{1}$	$5.7 \pm 3.7 \times 10^{1}$	$8.9 \pm 9.8 \times 10^{0}$
	Particulate	$2.8 \pm 0.6 \times 10^{-3}$	$2.0 \pm 1.1 \times 10^{-3}$	$1.0 \pm 1.3 \times 10^{0}$	$3.4 \pm 4.5 \times 10^{0}$	$8.2 \pm 8.4 \times 10^{-3}$
ACE	Gaseous	$7.0 \pm 3.1 \times 10^{-2}$	$1.4 \pm 0.1 \times 10^{-1}$	$1.3 \pm 0.6 \times 10^{0}$	$1.7 \pm 1.0 \times 10^{0}$	$9.5 \pm 9.5 \times 10^{-1}$
	Particulate	$7.7 \pm 3.8 \times 10^{-3}$	$4.1 \pm 1.8 \times 10^{-3}$	$1.3 \pm 1.4 \times 10^{-1}$	$1.2 \pm 1.2 \times 10^{-1}$	$1.9 \pm 2.1 \times 10^{-2}$
FLO	Gaseous	$4.8 \pm 1.7 \times 10^{-1}$	$1.1 \pm 0.0 \times 10^{0}$	$7.1 \pm 1.1 \times 10^{0}$	$1.3 \pm 0.6 \times 10^{1}$	$1.1 \pm 1.3 \times 10^{1}$
	Particulate	$4.9 \pm 2.4 \times 10^{-2}$	$2.4 \pm 1.0 \times 10^{-2}$	$2.0 \pm 2.1 \times 10^{0}$	$3.7 \pm 4.2 \times 10^{0}$	$1.1 \pm 1.1 \times 10^{-1}$
PHE	Gaseous	$1.1 \pm 0.6 \times 10^{0}$	$2.7 \pm 0.0 \times 10^{0}$	$7.7 \pm 0.7 \times 10^{0}$	$1.9 \pm 0.4 \times 10^{1}$	$3.0 \pm 3.6 \times 10^{1}$
	Particulate	$3.1 \pm 0.9 \times 10^{-1}$	$1.5 \pm 0.7 \times 10^{-1}$	$1.3 \pm 1.0 \times 10^{1}$	$2.3 \pm 2.0 \times 10^{1}$	$9.0 \pm 8.0 \times 10^{-1}$
ANT	Gaseous	$1.4 \pm 1.1 \times 10^{-1}$	$2.5 \pm 0.7 \times 10^{-1}$	$1.9 \pm 0.3 \times 10^{0}$	$5.8 \pm 2.7 \times 10^{0}$	$3.9 \pm 3.8 \times 10^{0}$
	Particulate	$4.6 \pm 1.4 \times 10^{-2}$	$2.6 \pm 1.5 \times 10^{-2}$	$4.5 \pm 3.4 \times 10^{0}$	$9.1 \pm 8.7 \times 10^{0}$	$1.5 \pm 1.5 \times 10^{-1}$
FLA	Gaseous	$1.5 \pm 0.7 \times 10^{-1}$	$3.8 \pm 0.8 \times 10^{-1}$	$1.2 \pm 0.6 \times 10^{0}$	$2.9 \pm 2.3 \times 10^{0}$	$3.4 \pm 3.5 \times 10^{0}$
	Particulate	$2.5 \pm 1.7 \times 10^{-1}$	$7.8 \pm 4.7 \times 10^{-2}$	$1.1 \pm 0.4 \times 10^{1}$	$2.1 \pm 1.4 \times 10^{1}$	$1.0 \pm 0.9 \times 10^{0}$
PYR	Gaseous	$1.2 \pm 0.5 \times 10^{-1}$	$2.8 \pm 0.7 \times 10^{-1}$	$9.0 \pm 4.8 \times 10^{-1}$	$2.0 \pm 1.7 \times 10^{0}$	$2.4 \pm 2.3 \times 10^{0}$
	Particulate	$2.4 \pm 1.9 \times 10^{-1}$	$6.3 \pm 4.2 \times 10^{-2}$	$8.5 \pm 3.1 \times 10^{0}$	$1.6 \pm 1.1 \times 10^{1}$	$1.0 \pm 0.9 \times 10^{0}$
BaA	Gaseous	$7.8 \pm 5.1 \times 10^{-3}$	$7.3 \pm 9.1 \times 10^{-2}$	$1.9 \pm 1.7 \times 10^{-1}$	$2.3 \pm 2.4 \times 10^{-1}$	$4.2 \pm 6.2 \times 10^{-2}$
	Particulate	$1.2 \pm 0.9 \times 10^{-1}$	$8.5 \pm 4.3 \times 10^{-2}$	$5.5 \pm 1.5 \times 10^{0}$	$9.8 \pm 7.0 \times 10^{0}$	$6.4 \pm 3.7 \times 10^{-1}$
CHR	Gaseous	$7.2 \pm 5.1 \times 10^{-3}$	$1.5 \pm 1.8 \times 10^{-1}$	$1.4 \pm 1.1 \times 10^{-1}$	$1.7 \pm 1.7 \times 10^{-1}$	$3.6 \pm 5.3 \times 10^{-2}$
	Particulate	$1.1 \pm 0.7 \times 10^{-1}$	$2.2 \pm 0.4 \times 10^{-1}$	$4.3 \pm 1.5 \times 10^{0}$	$6.4 \pm 4.0 \times 10^{0}$	$5.7 \pm 3.0 \times 10^{-1}$
BbF	Gaseous	$4.2 \pm 2.9 \times 10^{-3}$	$8.0 \pm 0.1 \times 10^{-2}$	$1.7 \pm 1.5 \times 10^{-1}$	$1.1 \pm 1.1 \times 10^{-1}$	$1.9 \pm 0.4 \times 10^{-2}$
	Particulate	$9.5 \pm 7.3 \times 10^{-2}$	$1.3 \pm 0.5 \times 10^{-1}$	$6.4 \pm 2.3 \times 10^{0}$	$8.3 \pm 6.1 \times 10^{0}$	$3.9 \pm 2.1 \times 10^{-1}$
BkF	Gaseous	$5.2 \pm 3.7 \times 10^{-3}$	$6.1 \pm 0.8 \times 10^{-2}$	$1.3 \pm 1.2 \times 10^{-1}$	$1.3 \pm 1.3 \times 10^{-1}$	$2.3 \pm 0.4 \times 10^{-2}$
	Particulate	$9.0 \pm 6.0 \times 10^{-2}$	$9.7 \pm 2.9 \times 10^{-2}$	$3.8 \pm 1.5 \times 10^{0}$	$6.3 \pm 3.8 \times 10^{0}$	$4.4 \pm 3.0 \times 10^{-1}$
BaP	Gaseous	$2.0 \pm 3.4 \times 10^{-2}$	$5.5 \pm 0.8 \times 10^{-2}$	$1.7 \pm 1.4 \times 10^{-1}$	$1.8 \pm 1.1 \times 10^{-1}$	$8.1 \pm 1.4 \times 10^{-2}$

(continued)

Table 5.1 (continued)

PAH	Phase	Beijing Honeycomb	Taiyuan Honeycomb	Taiyuan Chunk	Yulin Chunk-A	Yulin Chunk-B
	Particulate	$1.3 \pm 1.1 \times 10^{-1}$	$8.0 \pm 3.1 \times 10^{-2}$	$6.1 \pm 1.9 \times 10^{0}$	$9.4 \pm 7.0 \times 10^{0}$	$4.4 \pm 2.8 \times 10^{-1}$
IcdP	Gaseous	$3.6 \pm 2.6 \times 10^{-3}$	$4.5 \pm 0.6 \times 10^{-2}$	$1.9 \pm 2.1 \times 10^{-1}$	$1.0 \pm 1.1 \times 10^{-1}$	$1.5 \pm 0.3 \times 10^{-2}$
	Particulate	$1.3 \pm 1.4 \times 10^{-1}$	$7.0 \pm 5.3 \times 10^{-2}$	$9.5 \pm 2.8 \times 10^{0}$	$1.4 \pm 1.2 \times 10^{1}$	$4.3 \pm 52.6 \times 10^{-1}$
DahA	Gaseous	$1.6 \pm 0.3 \times 10^{-3}$	$3.0 \pm 4.2 \times 10^{-2}$	$5.2 \pm 4.7 \times 10^{-2}$	$1.7 \pm 1.6 \times 10^{-2}$	$2.7 \pm 0.4 \times 10^{-3}$
	Particulate	$1.1 \pm 1.0 \times 10^{-2}$	$3.1 \pm 0.9 \times 10^{-2}$	$1.8 \pm 0.4 \times 10^{0}$	$1.8 \pm 1.7 \times 10^{0}$	$3.0 \pm 1.8 \times 10^{-2}$
BghiP	Gaseous	$2.8 \pm 2.3 \times 10^{-3}$	$8.8 \pm 1.2 \times 10^{-3}$	$2.4 \pm 2.6 \times 10^{-1}$	$5.2 \pm 5.4 \times 10^{-2}$	$8.1 \pm 1.3 \times 10^{-3}$
	Particulate	$6.4 \pm 5.1 \times 10^{-2}$	$1.4 \pm 0.8 \times 10^{-1}$	$7.1 \pm 0.3 \times 10^{0}$	$5.2 \pm 4.4 \times 10^{0}$	$2.0 \pm 1.2 \times 10^{-1}$
Total	Gaseous	$4.6 \pm 0.7 \times 10^{0}$	$1.3 \pm 0.3 \times 10^{1}$	$5.6 \pm 0.4 \times 10^{1}$	$1.2 \pm 0.6 \times 10^{2}$	$7.2 \pm 8.9 \times 10^{1}$
	Particulate	$1.7 \pm 0.8 \times 10^{0}$	$1.2 \pm 0.5 \times 10^{0}$	$8.5 \pm 3.6 \times 10^{1}$	$1.4 \pm 1.1 \times 10^{2}$	$6.5 \pm 4.8 \times 10^{0}$

Data are listed as means and standard deviations of gaseous and particulate phase EFs from two duplicated combustion experiments. Reprinted from Atmospheric Environment 44, Shen et al., Emission factors and particulate matter size distribution of polycyclic aromatic hydrocarbons from residential coal combustion in rural Northern China, 5237–5243, with permission from Elsevier

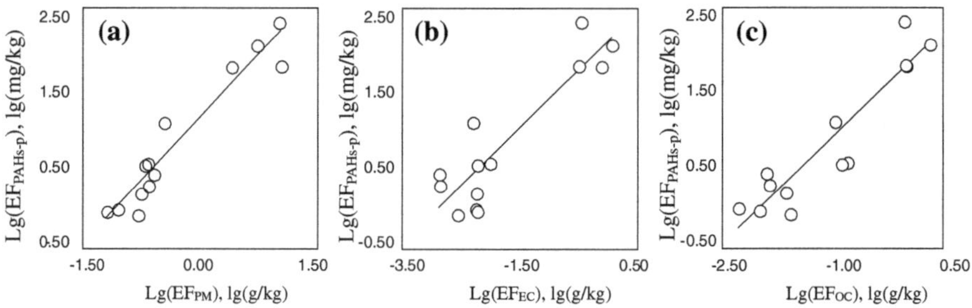

Fig. 5.1 Relationship between particulate phase PAHs and co-emitted PM, and EC and OC fraction in PM from residential coal combustion

Kim Oanh et al. 1999; Levendis et al. 1996; Liu et al. 2009; Mastral and Calleän 2000; Mitra et al. 1987; Zhang et al. 2008). It is believed that chunk coals generally produce more PAHs than honeycomb ones, and bituminous coals emit more PAHs than anthracite. Chen et al. (2004, 2005) measured PAH EFs for residential coals in laboratory equipped with combustion hood and dilution sampling system, and reported EFs of thirteen parent PAHs (without NAP, ACY and ACE) for honeycomb anthracite with 7 % VM were 0.12 mg/kg, which was lower than those (3.72 ± 1.80 mg/kg) from our honeycomb anthracite coals with VM content at 4 %. They also reported bituminous coals emitted thirteen PAHs at 66–151 mg/kg. In another laboratory study, Liu et al. (2009) measured 15 PAH EFs (except NAP) from Beijing and Shanxi coals ranging in 53–405 mg/kg and 78–1435 mg/kg in high heat (mainly used for cooking and heating in daytime) and low heat (mainly in night with decreased oxygen supply) modes, respectively.

PAHs were significantly correlated with other co-emitted pollutants, including CO, PM, black carbon (BC) and organic carbon (OC) ($p < 0.05$). Figure 5.1 shows the positive correlation between particulate phase PAHs and PM. It is realized that the formation mechanism(s) of these pollutants are not the same, so even a positive linear correlation was found, the use of such a correlation may need further evaluation.

A variety of factors affect pollutant emissions from coal combustion. Generally, coals with higher VM are often more difficult to achieve complete combustion, and hence produce more PAHs. Significantly positive correlation between VM content and PAH EFs was found ($p = 0.001$). Coals from Beijing, which had the lowest VM (4 %) and were the only anthracite coal tested, emitted the least PAHs when comparing to the others. Heat values of coals were also positively correlated to EFs of PAHs ($p = 0.004$), which might be related to the suitable PAHs formation temperature under residential conditions (Mitra et al. 1987; Mastral and Calleän 2000).

Similar to the approach in analysis of factors affecting PM emission, stepwise regression models were applied to address key factors influencing PAH emission from coal combustion. The results suggested that coal moisture (M) and VM content were two most significant factors, explaining about 51 % of the variation.

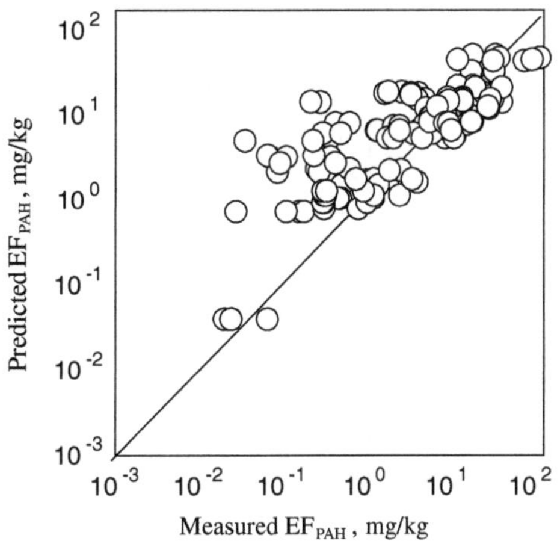

Fig. 5.2 Comparison of predicted EF_{PAHs} and measured results for coal. The prediction was calculated from fuel moisture and volatile matter content. Reprinted from Atmospheric Environment 44, Shen et al., Emission factors and particulate matter size distribution of polycyclic aromatic hydrocarbons from residential coal combustion in rural Northern China, 5237–5243. Copyright 2010, with permission from Elsevier

The predicted EFs based on the following equation agreed with the measured results (Fig. 5.2).

$$EF_{PAHs}(\text{mg/kg}) = 9.71 \times VM(\%, p = 0.010) - 16.7 \times M(\%, p = 0.119) - 31.9$$

5.1.2 Composition Profile and Isomer Ratio

Parent PAH isomer ratios are often used in receptor modeling for source apportionment by comparing PAH ratios between sources and receptors (Watson 1984; Yunker et al. 2002; Zhang et al. 2005). Several commonly used isomer ratios includes ANT/(ANT + PHE), FLA/(FLA + PYR), BaA/(BaA + CHR), IcdP/(IcdP + BghiP), BbF/(BbF + BkF), and BaP/(BaP + BghiP). For example, it is suggested that the FLA/(FLA + PYR) being larger than 0.5 indicates from coal or biomass burning and petroleum combustion if it is smaller than 0.5 (Yunker et al. 2002). Means and standard deviations of all five coals were 0.17 ± 0.07, 0.56 ± 0.02, 0.48 ± 0.13, 0.58 ± 0.16, 0.54 ± 0.06, and 0.58 ± 0.16 for ANT/(ANT + PHE), FLA/(FLA + PYR), BaA/(BaA + CHR), IcdP/(IcdP + BghiP), BbF/(BbF + BkF), and BaP/(BaP + BghiP), respectively. These values were generally comparable to those reported in literature for coal combustion (Chen et al. 2004, 2005; Yunker et al. 2002; Zhang et al. 2008).

The calculated isomer ratios for each coal are listed in Table 5.2. Except FLA/(FLA + PYR), the difference in the ratios among the five coals were significant.

Table 5.2 Parent PAH isomer ratios for tested coals measured in this study

	ANT/ANT + PHE	FLA/FLA + PYR	BaA/BaA + CHR	IcdP/IcdP + BghiP	BbF/BbF + BkF	BaP/BaP + BghiP
Beijing, Honeycomb	0.119 ± 0.004	0.554 ± 0.018	0.455 ± 0.071	0.582 ± 0.060	0.473 ± 0.002	0.730 ± 0.155
Taiyuan, Honeycomb	0.087 ± 0.007	0.574 ± 0.017	0.272 ± 0.070	0.316 ± 0.050	0.566 ± 0.014	0.390 ± 0.033
Taiyuan, Chunk	0.237 ± 0.010	0.564 ± 0.002	0.565 ± 0.021	0.565 ± 0.083	0.624 ± 0.005	0.454 ± 0.087
Yulin, Chunk-A	0.245 ± 0.048	0.583 ± 0.008	0.599 ± 0.024	0.730 ± 0.001	0.551 ± 0.047	0.664 ± 0.039
Yulin, Chunk-B	0.154 ± 0.001	0.543 ± 0.003	0.513 ± 0.009	0.687 ± 0.010	0.502 ± 0.003	0.686 ± 0.001

Data listed are means with standard deviations from duplicated combustion experiments

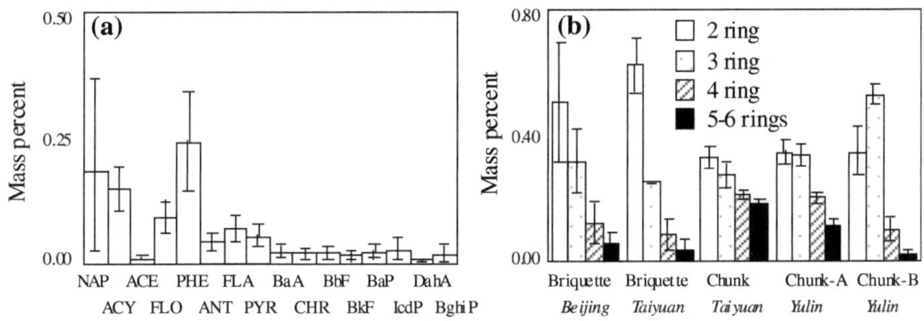

Fig. 5.3 Composition profile of PAHs from residential coal combustion (**a**) and relative distribution of 4 groups with different PAHs rings for each coal (**b**). Reprinted from Atmospheric Environment 44, Shen et al., Emission factors and particulate matter size distribution of polycyclic aromatic hydrocarbons from residential coal combustion in rural Northern China, 5237–5243, with permission from Elsevier

The coefficients of variation of these ratios were high (26–41 %), except 3 % in FLA/(FLA + PYR) and 11 % in BbF/(BbF + BkF). The ratio of IcdP/(IcdP + BghiP) larger than 0.50 is often thought to indicate solid fuel combustion sources while petroleum combustion source has the ratio lower than 0.5 (Yunker et al. 2002). However, in our present study, a low IcdP/(IcdP + BghiP) ratio of 0.316 was found in emissions from briquette combustion from Taiyuan, and the IcdP/(IcdP + BghiP) ratio for the other four coals were 0.582–0.730. In a study by Zhang et al. (2008), the IcdP/(IcdP + BghiP) ratio was reported at 0.50 and 0.57 for coal burned in industrial and residential stove, but a low value of 0.35 was also found in one briquette burned in residential stove. Chen et al. (2005) also reported a low IcdP/(IcdP + BghiP) value of 0.33 in emissions from residential briquette combustion. Thus, the use of isomer ratios in PAH source apportionment should be in caution since they might be similar among different source, moreover for the same source type, the ratios could vary dramatically depending on not only fuel properties, but also burning conditions. In addition, once emitted into the environment, the ratio would change causing the difficulty and uncertainties in their use (Wu et al. 2006; Zhang et al. 2005).

Normalized composition profile of PAHs is shown in Fig. 5.3a. PHE and NAP were two most abundant compounds, contributing about 24 and 19 %, respectively. In comparison to lower molecular weight PAHs, compounds with higher molecular weight often show considerable tendency to accumulate on PM. Taking coals with higher VM emit more PM into consideration, it is thought that the mass percent of PAHs with different molecular weight might be correlated with coal VM content. 16 PAHs were classified into four groups with different rings. As shown in Fig. 5.3b, the fractions of high molecular weight PAHs were higher in emissions from the burning of coals with relatively high volatile matter content. The percents of 2-, 3-, 4-, and 5-6 ring PAHs in emissions from the burning of chunk coal from Taiyuan were 33 ± 4, 27 ± 4, 21 ± 1 and 18 ± 2 %, respectively, while in emissions from the briquette coal from Taiyuan, they were 63 ± 9,

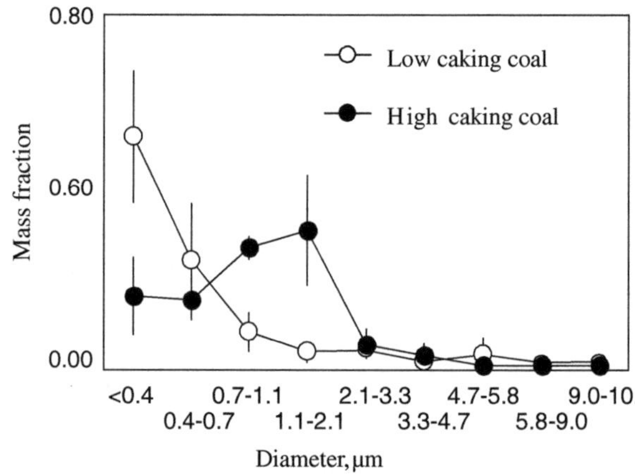

Fig. 5.4 Size distribution of particulate phase PAHs from residential coal combustion. The distributions were classified into two categories of low and high caking coals. Reprinted from Atmospheric Environment 44, Shen et al., Emission factors and particulate matter size distribution of polycyclic aromatic hydrocarbons from residential coal combustion in rural Northern China, 5237–5243. Copyright 2010, with permission from Elsevier

25 ± 0, 9 ± 5 and 4 ± 3 %. Since high volatile matter coal usually produces more PAHs, relatively higher fractions of high molecular weight organics suggest that the emission smoke would be much more toxic and cause detrimental impacts on human health for these coals.

5.1.3 Size Distribution of Particulate Phase PAHs

More than 89 ± 6 % particulate phase PAHs were found in $PM_{2.5}$ and similar to the distribution of PM, the size distributions of particle-bound PAHs fell into two categories (Fig. 5.4). For low caking coals (two honeycomb coals and one chunk coal from Yulin), there was about 52 ± 15 % of particulate phase PAHs were fine $PM_{0.4}$, followed by 24 ± 13 % in $PM_{0.4–0.7}$ fraction. For the other two coals, most (58 ± 13 %) were found in PM with size of 0.7–2.1 μm and only 16 % in fine $PM_{0.4}$. The calculated MMAD for these five coals were 0.13, 0.13, 0.98, 0.95 and 0.11 μm, respectively.

Figure 5.5 compares the partitioning of PAH individuals in fine $PM_{2.1}$ and coarse $PM_{2.1–10}$. As expected, with the increase of molecular weight, the mass percent of fine $PM_{2.1}$-bound organics increased as well. For example, 55, 78 and 73 % of particulate phase NAP, ACY and PHE were in fine $PM_{2.1}$, while there were 93, 96 and 95 % of BaA, BaP and BghiP present in fine PM fraction. The increase of mass percent in fine PM increased with the increase of MW was also found in ambient atmosphere (Wang et al. 2009; Wu et al. 2006). This might be explained by that: (1) the difference in diffusivity that is correlated to molecular

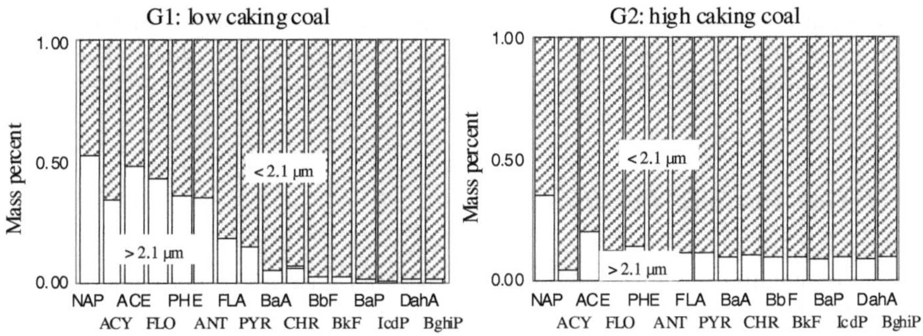

Fig. 5.5 Distribution of particulate phase PAH individuals between fine PM2.1 and coarse PM2.1-10 from coal combustion. Reprinted from Atmospheric Environment 44, Shen et al., Emission factors and particulate matter size distribution of polycyclic aromatic hydrocarbons from residential coal combustion in rural Northern China, 5237–5243, with permission from Elsevier

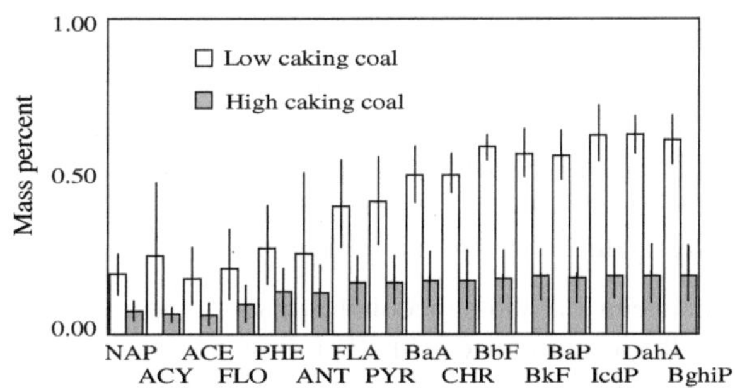

Fig. 5.6 Mass percent of fine PM0.4 bound PAHs to the total particulate phase mass for each individual. Reprinted from Atmospheric Environment 44, Shen et al., Emission factors and particulate matter size distribution of polycyclic aromatic hydrocarbons from residential coal combustion in rural Northern China, 5237–5243, with permission from Elsevier

weight; (2) enhanced vapor pressures of lower molecular weight compounds which having higher volatilization rates from small particles due to their curved surfaces; (3) the difference in particle surface area and organic matter content affecting absorption/absorption and (4) varied chemical composition and concentration gradients that govern the gas to particle transfer (Allen et al. 1996; Hays et al. 2003; Venkataraman et al. 1999).

In addition to the distribution between fine and coarse PM, it is more interesting to look into the fraction in finest $PM_{0.4}$ in detail. The mass percent of fine $PM_{0.4}$-bound fraction to the total particulate phase mass was shown in Fig. 5.6 for two coal groups (low and high caking coals). As discussed above, PM from the combustion of low caking coal was smaller than that from other coals. These coals might also produce more PAHs in fine PM, especially in $PM_{0.4}$. For each individual PAH from coals in category 1-low caking coals, there was about 18–63 %

of the total in $PM_{0.4}$, while for those in category 2, PAHs bound in $PM_{0.4}$ only made up to 6–19 % of the total, varying among different individuals. The difference was significant for 4- to 6-ring PAHs from PYR to BghiP ($p < 0.05$).

5.1.4 Gas-Particle Partitioning

Like most volatile organics, PAHs can be in either gaseous or particulate phases. Generally, low molecular weight PAHs are mainly in gaseous phase, while high molecular weight PAHs are more preferable to be associated with PM. The gas-particle partitioning of PAHs can be described by a partition coefficient ($K_P = F/(A \times PM)$), where F and A are concentrations in particulate (ng/m^3) and gaseous phases (ng/m^3), respectively and PM is particulate matter concentration (μg/m^3) in air (Pankow 1987):

$$K_p = F/(A \times PM)$$

It is speculated that the partitioning can be controlled by absorption, or adsorption, or both (Hays et al. 2003; Pankow 1987; Venkataraman et al. 1999). The adsorption is related to factors like surface area of PM, contents of adsorbs such as EC, while absorption is mainly associated to the present of absorbs like OC (Goss and Schwarzenbach 1998; Lohmann and Lammel 2004). Several empirical approaches to investigate the partitioning mechanism(s) are:

1. Relationship between *Kp* and subcooled liquid–vapor pressure (P_L^0)
The *Kp* is usually linear correlated with P_L^0 following:

$$\lg K_p = m_r \lg P_L^0 + b_r$$

It is suggested that a steeper slope than −1 indicates adsorption dominance, while a shallower slope than −0.6 is responsible for absorption governance, and the slope between −0.6 and −1.0 indicates the governance of both absorption and adsorption (Goss and Schwarzenbach 1998).
2. Relationship between *Kp* and octanol-air partition coefficient (K_{OA})
In the absorption controlled partitioning, the partitioning coefficient usually positively correlated with the K_{OA} (Lohmann and Lammel 2004).
3. Dependence of PAHs/PM in each size fraction on PM size
It is believed that in the adsorption controlled partitioning, the surface area is a critical factor. Since the particle surface area is directly related to PM size, the mass ratios of particle-bound PAHs in PM would be size dependent if adsorption controlled the partitioning (Hays et al. 2003; Venkataraman et al. 1999).
4. Relationship between PAHs and OC and EC fractions
As mentioned, EC and OC play an important role in the partitioning of volatile organics, acting as the adsorb or absorb for the organics. Thus, the relationship between PAHs and OC and between PAHs and EC can be also another piece of evidence for the absorption or adsorption dominance.

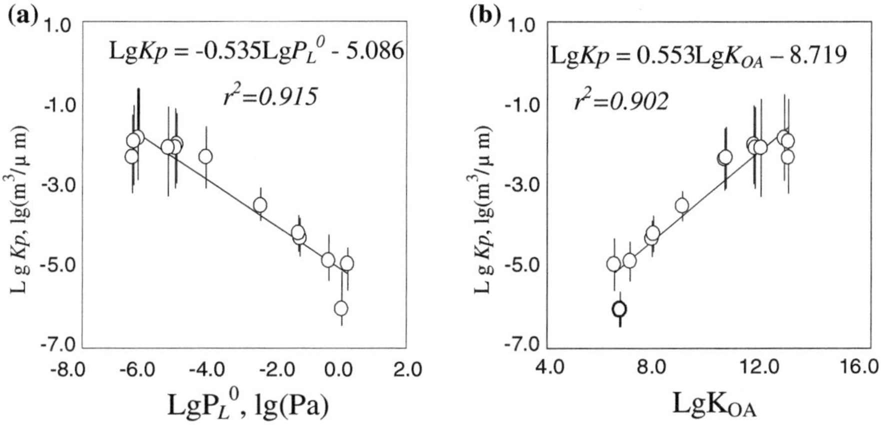

Fig. 5.7 Dependence of $\log(K_p)$ on $\log(P_L^0)$ (**a**) and $\log(K_{OA})$ (**b**) for PAHs from coal combustion. P_L^0 and K_{OA} were calculated based on the measured temperatures and equations established by Odabasi et al. (2006). Reprinted from Atmospheric Environment 44, Shen et al., Emission factors and particulate matter size distribution of polycyclic aromatic hydrocarbons from residential coal combustion in rural Northern China, 5237–5243. Copyright 2010, with permission from Elsevier

The mass percents of particulate phase PAHs were 26 ± 10, 9 ± 4, 59 ± 10, 52 ± 8 and 12 ± 5 % of the total for Beijing honeycomb, Taiyuan honeycomb, Taiyuan chunk and two Yulin chunk coals, respectively. Though two chunk coals produced more particulate-bound PAHs than the others, there was no significant difference in *Kp* of PAHs from different coals, which implied that partitioning of freshly emitted PAHs from residential coal combustion might be controlled by the similar mechanism(s).

The slope in the regression linear relationship between *Kp* and P_L^0, was −0.54 ($p = 1.3 \times 10^{-6}$), larger than −0.6 (Fig. 5.7), and the *Kp* significantly positively correlated ($p = 7.3 \times 10^{-6}$) to K_{OA}. The correlation between PAHs and OC ($r = 0.830$, $p = 2.3 \times 10^{-4}$) were also more significant than that between PAHs and EC ($r = 0.657$, $p = 7.4 \times 10^{-3}$). The mass ratios of particle-bound PAHs were not particle size (D_a) dependent (Fig. 5.8). All these indicated that absorption rather than adsorption governed the partitioning of freshly emitted PAHs from residential coal combustions.

5.2 Indoor Crop Residue Burning

5.2.1 EFs and Influencing Factor

EF_{PAHs} for crop residue ranged from 27 mg/kg (cotton stalk) to 142 mg/kg (wheat straw), with an overall mean of 62 ± 35 mg/kg. These results were higher than

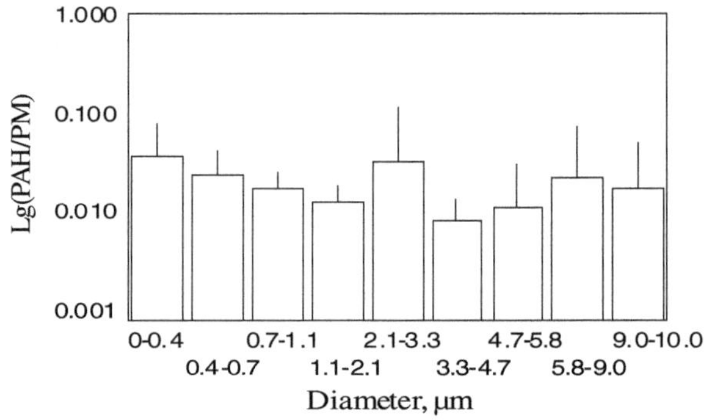

Fig. 5.8 Mass ratios of total PAHs in particulate matter over different size fractions

most data reported in literature (4–76 mg/kg) (Dhammapala et al. 2007a, b; Hays et al. 2005; Jenkins et al. 1996a, b; Lu et al. 2009). This might be mainly explained by that in real cooking stoves, air flow is more limited, oxygen deficiency is more severe, and mixing is less complete, resulting in lower combustion efficiency and higher PAHs emissions (Chen et al. 2005; Zhang et al. 2008). In addition, higher combustion temperature (measured as 550–750 °C in this study) due to limited heat loss in the stove is also favorable for PAHs formation. It was reported that PAHs emissions increased dramatically when combustion temperature increased from 200 to 700 °C (Lu et al. 2009). EFs of total PAHs in the flaming and smoldering phases were 99 ± 60 and 58 ± 32 mg/kg, respectively (Tables 5.3, 5.4 and 5.5), and the former was significantly higher than the later ($p < 0.05$). One possible explanation for the higher emission in flaming phase is that combustion temperatures are more optimal for PAHs formation, while the lower temperature during smoldering leads to slower formation.

EF_{PAHs} were negatively correlated with MCE significantly. The results of a stepwise regression showed that moisture and MCE were the two most significant factors affecting EF_{PAHs} ($p < 0.05$) for crop residue. PAH EFs can be then predicted from these two factors using the following equations:

$$\text{Whole cycle:}\quad EF_{P16} = -21 \times moisture(p = 0.0085) - 492 \times MCE(p = 0.0043) + 545(p = 0.0015)$$

$$\text{Flaming phase:}\quad EF_{P16} = -33 \times moisture(p = 0.0336) - 881 \times MCE(p = 0.0101) + 960(p = 0.0045)$$

$$\text{Smoldering phase:}\quad EF_{P16} = -19 \times moisture(p = 0.0102) - 360 \times MCE(p = 0.0202) + 413(p = 0.0073)$$

The calculated results were compared with measured results in Fig. 5.9. Approximately 60 % of total variation can be explained. Since PM was also affected by the moisture and MCE, the correlation between PAHs and PM could be expected in emissions from indoor crop residue burning, as shown in Fig. 5.10.

5.2.2 Composition Profile and Isomer Ratio

2–3 ring PAHs (from NAP to ANT) accounted for about 78 % of total PAH emissions (Fig. 5.11). NAP was the most abundant compound found in gaseous phase

Table 5.3 EFs of PAHs in gaseous (G) and particulate-bound (P) phases for crop residues in the whole burning cycle

PAH		Horsebean	Peanut	Soybean	Cotton	Rice	Wheat	Rape	Sesame	Corn
NAP	G	$1.4 \pm 0.2 \times 10^{1}$	1.8×10^{1}	$1.9 \pm 0.5 \times 10^{1}$	$1.2 \pm 0.4 \times 10^{1}$	$3.7 \pm 0.9 \times 10^{1}$	$3.6 \pm 0.6 \times 10^{1}$	$2.8 \pm 1.2 \times 10^{1}$	$5.7 \pm 2.5 \times 10^{0}$	$1.4 \pm 0.3 \times 10^{1}$
	P	$2.9 \pm 1.0 \times 10^{-1}$	1.4×10^{-1}	$1.1 \pm 0.3 \times 10^{-1}$	$5.7 \pm 0.7 \times 10^{-2}$	$4.2 \pm 1.8 \times 10^{-1}$	$4.2 \pm 4.6 \times 10^{-1}$	$2.4 \pm 1.2 \times 10^{-1}$	$6.9 \pm 8.4 \times 10^{-1}$	$2.3 \pm 2.0 \times 10^{-1}$
ACY	G	$2.9 \pm 0.7 \times 10^{0}$	6.4×10^{0}	$4.2 \pm 1.1 \times 10^{0}$	$2.9 \pm 1.1 \times 10^{0}$	$4.9 \pm 2.4 \times 10^{0}$	$6.5 \pm 1.7 \times 10^{0}$	$5.5 \pm 2.6 \times 10^{0}$	$8.6 \pm 4.9 \times 10^{-1}$	$2.1 \pm 0.5 \times 10^{0}$
	P	$2.1 \pm 1.0 \times 10^{-1}$	1.3×10^{-1}	$9.7 \pm 5.2 \times 10^{-2}$	$5.4 \pm 0.3 \times 10^{-2}$	$2.6 \pm 0.3 \times 10^{-1}$	$5.0 \pm 4.8 \times 10^{-1}$	$6.1 \pm 3.4 \times 10^{-1}$	$1.8 \pm 0.7 \times 10^{-1}$	$1.3 \pm 0.6 \times 10^{-1}$
ACE	G	$1.5 \pm 0.4 \times 10^{-1}$	3.2×10^{-1}	$1.9 \pm 0.4 \times 10^{-1}$	$9.2 \pm 4.7 \times 10^{-2}$	$3.7 \pm 0.8 \times 10^{-1}$	$3.4 \pm 1.3 \times 10^{-1}$	$2.7 \pm 0.9 \times 10^{-1}$	$1.5 \pm 1.1 \times 10^{-1}$	$1.8 \pm 0.7 \times 10^{-1}$
	P	$3.7 \pm 2.1 \times 10^{0}$	4.6×10^{0}	$2.4 \pm 1.4 \times 10^{0}$	$8.0 \pm 1.2 \times 10^{-1}$	$4.9 \pm 3.3 \times 10^{0}$	$7.8 \pm 8.2 \times 10^{0}$	$1.4 \pm 0.8 \times 10^{1}$	$2.0 \pm 1.3 \times 10^{0}$	$2.4 \pm 2.3 \times 10^{0}$
FLO	G	$4.3 \pm 1.8 \times 10^{-1}$	8.6×10^{-1}	$6.0 \pm 1.3 \times 10^{-1}$	$2.7 \pm 1.2 \times 10^{-1}$	$7.4 \pm 1.2 \times 10^{-1}$	$1.0 \pm 0.5 \times 10^{0}$	$7.2 \pm 3.7 \times 10^{-1}$	$3.6 \pm 2.6 \times 10^{-1}$	$4.9 \pm 2.1 \times 10^{-1}$
	P	$4.3 \pm 2.0 \times 10^{0}$	6.6×10^{0}	$2.3 \pm 1.3 \times 10^{0}$	$1.1 \pm 0.0 \times 10^{0}$	$5.4 \pm 3.4 \times 10^{0}$	$1.0 \pm 0.9 \times 10^{1}$	$1.6 \pm 1.1 \times 10^{1}$	$4.0 \pm 1.7 \times 10^{0}$	$2.8 \pm 2.3 \times 10^{0}$
PHE	G	$5.4 \pm 2.1 \times 10^{-1}$	1.7×10^{0}	$7.4 \pm 1.8 \times 10^{-1}$	$3.3 \pm 2.1 \times 10^{-1}$	$5.4 \pm 2.1 \times 10^{-1}$	$1.3 \pm 0.9 \times 10^{0}$	$1.0 \pm 0.6 \times 10^{0}$	$6.1 \pm 4.8 \times 10^{-1}$	$1.2 \pm 0.7 \times 10^{0}$
	P	$7.2 \pm 3.2 \times 10^{0}$	5.1×10^{0}	$3.9 \pm 1.7 \times 10^{0}$	$2.4 \pm 0.1 \times 10^{0}$	$7.2 \pm 3.2 \times 10^{0}$	$1.9 \pm 1.3 \times 10^{1}$	$2.0 \pm 8.5 \times 10^{1}$	$8.3 \pm 3.0 \times 10^{0}$	$5.4 \pm 3.8 \times 10^{0}$
ANT	G	$7.3 \pm 3.0 \times 10^{-2}$	2.5×10^{-1}	$1.0 \pm 0.3 \times 10^{-1}$	$5.5 \pm 3.0 \times 10^{-2}$	$8.7 \pm 3.4 \times 10^{-1}$	$1.9 \pm 1.3 \times 10^{-1}$	$1.5 \pm 0.9 \times 10^{-1}$	$8.4 \pm 5.8 \times 10^{-2}$	$2.4 \pm 1.7 \times 10^{-1}$
	P	$8.9 \pm 3.8 \times 10^{-1}$	6.3×10^{-1}	$5.0 \pm 2.8 \times 10^{-1}$	$3.2 \pm 0.0 \times 10^{-1}$	$9.0 \pm 0.5 \times 10^{0}$	$2.9 \pm 2.2 \times 10^{0}$	$2.8 \pm 0.9 \times 10^{0}$	$9.8 \pm 4.1 \times 10^{-1}$	$8.0 \pm 6.9 \times 10^{-1}$
FLA	G	$7.4 \pm 1.2 \times 10^{-2}$	1.1×10^{-1}	$7.6 \pm 3.7 \times 10^{-2}$	$2.2 \pm 0.8 \times 10^{-2}$	$1.6 \pm 0.7 \times 10^{-1}$	$2.0 \pm 2.5 \times 10^{-1}$	$1.3 \pm 0.5 \times 10^{-1}$	$7.8 \pm 7.0 \times 10^{-2}$	$3.4 \pm 2.4 \times 10^{-2}$
	P	$3.1 \pm 1.7 \times 10^{0}$	3.1×10^{0}	$2.7 \pm 0.7 \times 10^{0}$	$2.8 \pm 0.0 \times 10^{0}$	$1.5 \pm 0.5 \times 10^{0}$	$8.3 \pm 2.8 \times 10^{0}$	$7.8 \pm 0.5 \times 10^{0}$	$2.8 \pm 0.8 \times 10^{0}$	$2.6 \pm 0.7 \times 10^{0}$
PYR	G	$5.4 \pm 1.1 \times 10^{-2}$	1.4×10^{-1}	$5.5 \pm 2.5 \times 10^{-2}$	$3.1 \pm 1.4 \times 10^{-2}$	$1.4 \pm 0.6 \times 10^{-1}$	$2.0 \pm 3.0 \times 10^{-1}$	$1.5 \pm 1.1 \times 10^{-1}$	$9.8 \pm 9.7 \times 10^{-2}$	$3.4 \pm 2.5 \times 10^{-2}$
	P	$2.5 \pm 1.3 \times 10^{0}$	2.6×10^{0}	$2.6 \pm 0.9 \times 10^{0}$	$2.7 \pm 0.0 \times 10^{0}$	$7.2 \pm 4.3 \times 10^{0}$	$7.0 \pm 2.5 \times 10^{0}$	$6.6 \pm 0.5 \times 10^{0}$	$1.9 \pm 0.4 \times 10^{0}$	$2.2 \pm 0.5 \times 10^{0}$
BaA	G	$8.5 \pm 4.0 \times 10^{-3}$	1.5×10^{-2}	$5.9 \pm 2.4 \times 10^{-3}$	$7.1 \pm 3.8 \times 10^{-3}$	$2.0 \pm 1.0 \times 10^{-1}$	$3.3 \pm 3.5 \times 10^{-2}$	$2.9 \pm 0.7 \times 10^{-2}$	$1.4 \pm 1.3 \times 10^{-2}$	$4.5 \pm 1.9 \times 10^{-3}$
	P	$6.0 \pm 4.3 \times 10^{-1}$	4.9×10^{-1}	$4.1 \pm 0.8 \times 10^{-1}$	$4.7 \pm 0.3 \times 10^{-1}$	$6.8 \pm 4.8 \times 10^{0}$	$1.5 \pm 0.6 \times 10^{0}$	$1.4 \pm 0.4 \times 10^{0}$	$4.3 \pm 1.2 \times 10^{-1}$	$4.7 \pm 1.0 \times 10^{-1}$
CHR	G	$1.6 \pm 0.3 \times 10^{-2}$	3.2×10^{-2}	$1.2 \pm 0.5 \times 10^{-2}$	$1.0 \pm 0.5 \times 10^{-2}$	$3.6 \pm 0.5 \times 10^{-2}$	$4.0 \pm 4.0 \times 10^{-2}$	$3.9 \pm 2.2 \times 10^{-2}$	$2.9 \pm 2.9 \times 10^{-2}$	$8.2 \pm 4.7 \times 10^{-3}$
	P	$5.9 \pm 4.0 \times 10^{-1}$	5.6×10^{-1}	$4.1 \pm 0.5 \times 10^{-1}$	$5.5 \pm 0.7 \times 10^{-1}$	$1.2 \pm 0.7 \times 10^{0}$	$1.6 \pm 0.5 \times 10^{0}$	$1.5 \pm 0.3 \times 10^{0}$	$4.9 \pm 2.3 \times 10^{-1}$	$5.2 \pm 1.4 \times 10^{-1}$
BbF	G	$7.5 \pm 6.7 \times 10^{-3}$	1.7×10^{-2}	$3.1 \pm 4.0 \times 10^{-3}$	$2.2 \pm 3.3 \times 10^{-3}$	$5.6 \pm 0.8 \times 10^{-2}$	$1.1 \pm 0.6 \times 10^{-2}$	$8.6 \pm 8.8 \times 10^{-3}$	$1.1 \pm 1.0 \times 10^{-2}$	$4.9 \pm 3.1 \times 10^{-3}$
	P	$6.2 \pm 4.4 \times 10^{-1}$	4.3×10^{-1}	$4.0 \pm 0.6 \times 10^{-1}$	$5.4 \pm 0.8 \times 10^{-1}$	$1.1 \pm 0.7 \times 10^{0}$	$1.4 \pm 0.6 \times 10^{0}$	$1.5 \pm 0.5 \times 10^{0}$	$4.7 \pm 1.3 \times 10^{-1}$	$4.5 \pm 0.0 \times 10^{-1}$
BkF	G	$2.7 \pm 2.4 \times 10^{-3}$	6.0×10^{-3}	$7.2 \pm 1.0 \times 10^{-3}$	$1.5 \pm 1.8 \times 10^{-3}$	$1.3 \pm 1.1 \times 10^{-2}$	$1.6 \pm 1.4 \times 10^{-2}$	$4.1 \pm 3.6 \times 10^{-3}$	$1.6 \pm 2.6 \times 10^{-2}$	$2.5 \pm 1.1 \times 10^{-3}$
	P	$4.3 \pm 3.5 \times 10^{-1}$	4.4×10^{-1}	$3.1 \pm 0.3 \times 10^{-1}$	$4.7 \pm 0.5 \times 10^{-1}$	$1.0 \pm 0.4 \times 10^{0}$	$1.1 \pm 0.3 \times 10^{0}$	$1.0 \pm 0.2 \times 10^{0}$	$3.0 \pm 0.4 \times 10^{-1}$	$3.6 \pm 0.6 \times 10^{-1}$
BaP	G	$4.5 \pm 3.9 \times 10^{-3}$	7.7×10^{-3}	$7.7 \pm 0.0 \times 10^{-3}$	$5.0 \pm 3.7 \times 10^{-3}$	$1.2 \pm 1.1 \times 10^{-2}$	$8.2 \pm 1.2 \times 10^{-3}$	$6.7 \pm 5.8 \times 10^{-3}$	$1.3 \pm 1.1 \times 10^{-2}$	$2.5 \pm 1.7 \times 10^{-3}$
	P	$4.0 \pm 2.9 \times 10^{-1}$	5.1×10^{-1}	$4.1 \pm 1.1 \times 10^{-1}$	$6.3 \pm 0.8 \times 10^{-1}$	$9.5 \pm 6.3 \times 10^{-1}$	$1.4 \pm 0.4 \times 10^{0}$	$1.2 \pm 0.3 \times 10^{0}$	$2.8 \pm 0.4 \times 10^{-1}$	$3.9 \pm 0.0 \times 10^{-1}$
IcdP	G	$2.5 \pm 2.6 \times 10^{-3}$	4.5×10^{-3}	$6.6 \pm 4.6 \times 10^{-3}$	$1.9 \pm 2.2 \times 10^{-3}$	$2.0 \pm 0.6 \times 10^{-2}$	$1.1 \pm 0.9 \times 10^{-2}$	$1.3 \pm 0.6 \times 10^{-2}$	$4.1 \pm 4.8 \times 10^{-3}$	$2.3 \pm 2.0 \times 10^{-3}$
	P	$4.1 \pm 3.4 \times 10^{-1}$	3.6×10^{-1}	$3.4 \pm 0.2 \times 10^{-1}$	$6.0 \pm 1.0 \times 10^{-1}$	$9.0 \pm 5.8 \times 10^{-1}$	$1.0 \pm 0.3 \times 10^{0}$	$8.2 \pm 0.3 \times 10^{-2}$	$2.5 \pm 0.5 \times 10^{-1}$	$2.9 \pm 0.9 \times 10^{-1}$

(continued)

Table 5.3 (continued)

PAH		Horsebean	Peanut	Soybean	Cotton	Rice	Wheat	Rape	Sesame	Corn
DahA	G	$4.4 \pm 5.0 \times 10^{-3}$	2.3×10^{-3}	$1.5 \pm 1.4 \times 10^{-3}$	$4.2 \pm 8.4 \times 10^{-4}$	$4.8 \pm 6.2 \times 10^{-3}$	$1.2 \pm 1.8 \times 10^{-3}$	$3.2 \pm 2.8 \times 10^{-3}$	$3.8 \pm 6.0 \times 10^{-3}$	$1.9 \pm 3.2 \times 10^{-4}$
	P	$1.4 \pm 1.0 \times 10^{-1}$	8.3×10^{-2}	$7.3 \pm 0.9 \times 10^{-1}$	$6.5 \pm 0.5 \times 10^{-2}$	$1.5 \pm 0.4 \times 10^{-1}$	$2.5 \pm 0.9 \times 10^{-1}$	$3.6 \pm 2.5 \times 10^{-1}$	$1.3 \pm 0.2 \times 10^{-1}$	$8.4 \pm 1.1 \times 10^{-2}$
BghiP	G	$3.5 \pm 3.2 \times 10^{-3}$	5.9×10^{-3}	$4.4 \pm 2.6 \times 10^{-3}$	$2.3 \pm 2.1 \times 10^{-4}$	$1.7 \pm 0.4 \times 10^{-2}$	$1.4 \pm 1.5 \times 10^{-2}$	$1.5 \pm 0.9 \times 10^{-2}$	$4.6 \pm 6.0 \times 10^{-3}$	$1.4 \pm 1.5 \times 10^{-3}$
	P	$3.6 \pm 2.8 \times 10^{-1}$	2.9×10^{-1}	$2.9 \pm 0.2 \times 10^{-1}$	$4.9 \pm 0.7 \times 10^{-1}$	$6.7 \pm 4.3 \times 10^{-1}$	$8.6 \pm 2.6 \times 10^{-1}$	$8.5 \pm 1.1 \times 10^{-1}$	$2.3 \pm 0.3 \times 10^{-1}$	$2.6 \pm 0.1 \times 10^{-1}$
Total	G	$\mathbf{1.8 \pm 0.2 \times 10^1}$	$\mathbf{2.8 \times 10^1}$	$\mathbf{2.5 \pm 0.7 \times 10^1}$	$\mathbf{1.6 \pm 0.6 \times 10^1}$	$\mathbf{4.4 \pm 1.2 \times 10^1}$	$\mathbf{4.6 \pm 0.5 \times 10^1}$	$\mathbf{3.6 \pm 1.6 \times 10^1}$	$\mathbf{8.1 \pm 3.6 \times 10^0}$	$\mathbf{1.8 \pm 0.5 \times 10^1}$
	P	$\mathbf{2.6 \pm 1.4 \times 10^1}$	$\mathbf{2.6 \times 10^1}$	$\mathbf{1.7 \pm 0.7 \times 10^1}$	$\mathbf{1.4 \pm 0.0 \times 10^1}$	$\mathbf{4.3 \pm 0.8 \times 10^1}$	$\mathbf{6.6 \pm 4.2 \times 10^1}$	$\mathbf{7.8 \pm 3.1 \times 10^1}$	$\mathbf{2.4 \pm 0.9 \times 10^1}$	$\mathbf{1.9 \pm 1.1 \times 10^1}$

Data listed are results from each burning cycle in mg/kg. Modified from Shen et al. (2011) with permission of American Chemical Society

Table 5.4 EFs of PAHs in gaseous (G) and particulate-bound (P) phases for crop residues in the flaming phase

PAH		Horsebean	Peanut	Soybean	Cotton	Rice	Wheat	Rape	Sesame	Corn
NAP	G	$2.9 \pm 2.0 \times 10^{1}$	2.2×10^{1}	$1.4 \pm 1.2 \times 10^{1}$	$2.8 \pm 2.1 \times 10^{1}$	$4.9 \pm 2.1 \times 10^{1}$	$4.8 \pm 1.7 \times 10^{1}$	$5.4 \pm 0.9 \times 10^{1}$	$1.2 \pm 0.1 \times 10^{1}$	$3.3 \pm 1.0 \times 10^{1}$
	P	$5.0 \pm 2.3 \times 10^{-1}$	4.5×10^{-1}	$2.3 \pm 0.5 \times 10^{-1}$	$1.3 \pm 0.8 \times 10^{-1}$	$6.7 \pm 2.3 \times 10^{-1}$	$5.9 \pm 4.9 \times 10^{-1}$	$5.7 \pm 6.9 \times 10^{0}$	$1.8 \pm 2.5 \times 10^{0}$	$3.1 \pm 1.8 \times 10^{-1}$
ACY	G	$3.2 \pm 1.0 \times 10^{0}$	3.6×10^{0}	$3.9 \pm 2.6 \times 10^{0}$	$6.5 \pm 0.1 \times 10^{0}$	$7.6 \pm 4.1 \times 10^{0}$	$7.2 \pm 2.9 \times 10^{0}$	$1.1 \pm 1.2 \times 10^{1}$	$1.4 \pm 1.2 \times 10^{0}$	$6.0 \pm 1.7 \times 10^{0}$
	P	$4.6 \pm 4.5 \times 10^{-1}$	2.7×10^{-1}	$1.6 \pm 0.7 \times 10^{-1}$	$1.4 \pm 0.3 \times 10^{-1}$	$4.6 \pm 0.7 \times 10^{-1}$	$8.7 \pm 0.9 \times 10^{-1}$	$1.2 \pm 0.2 \times 10^{0}$	$3.0 \pm 1.8 \times 10^{-1}$	$1.6 \pm 0.1 \times 10^{-1}$
ACE	G	$1.9 \pm 0.9 \times 10^{-1}$	2.5×10^{-1}	$1.8 \pm 0.9 \times 10^{-1}$	$1.7 \pm 0.1 \times 10^{-1}$	$5.5 \pm 3.1 \times 10^{-1}$	$4.0 \pm 2.1 \times 10^{-1}$	$5.7 \pm 1.4 \times 10^{-1}$	$2.5 \pm 2.7 \times 10^{-1}$	$3.3 \pm 0.4 \times 10^{-1}$
	P	$7.3 \pm 6.0 \times 10^{0}$	1.1×10^{1}	$4.3 \pm 2.2 \times 10^{0}$	$2.2 \pm 1.1 \times 10^{0}$	$8.8 \pm 6.0 \times 10^{0}$	$1.2 \pm 1.3 \times 10^{1}$	$2.9 \pm 7.6 \times 10^{1}$	$3.6 \pm 2.3 \times 10^{0}$	$2.5 \pm 1.8 \times 10^{0}$
FLO	G	$4.7 \pm 2.8 \times 10^{-1}$	5.2×10^{-1}	$6.2 \pm 3.1 \times 10^{-1}$	$6.0 \pm 2.0 \times 10^{-1}$	$9.8 \pm 6.5 \times 10^{-1}$	$1.0 \pm 0.5 \times 10^{0}$	$1.3 \pm 0.1 \times 10^{0}$	$5.5 \pm 4.5 \times 10^{-1}$	$9.9 \pm 1.9 \times 10^{-1}$
	P	$9.3 \pm 8.5 \times 10^{0}$	1.5×10^{1}	$3.8 \pm 1.3 \times 10^{0}$	$3.1 \pm 0.7 \times 10^{0}$	$1.0 \pm 0.8 \times 10^{1}$	$1.6 \pm 1.5 \times 10^{1}$	$3.8 \pm 2.2 \times 10^{1}$	$5.3 \pm 3.9 \times 10^{0}$	$2.9 \pm 1.2 \times 10^{0}$
PHE	G	$5.2 \pm 3.1 \times 10^{-1}$	6.7×10^{-1}	$7.9 \pm 3.5 \times 10^{-1}$	$5.9 \pm 3.8 \times 10^{-1}$	$1.4 \pm 1.4 \times 10^{0}$	$1.1 \pm 1.0 \times 10^{0}$	$1.7 \pm 0.4 \times 10^{0}$	$1.1 \pm 0.9 \times 10^{0}$	$1.7 \pm 0.5 \times 10^{0}$
	P	$1.6 \pm 1.5 \times 10^{0}$	7.7×10^{0}	$5.6 \pm 1.1 \times 10^{0}$	$7.1 \pm 0.3 \times 10^{0}$	$1.3 \pm 0.4 \times 10^{1}$	$2.5 \pm 0.2 \times 10^{1}$	$4.0 \pm 1.4 \times 10^{1}$	$1.0 \pm 0.7 \times 10^{1}$	$5.9 \pm 4.0 \times 10^{0}$
ANT	G	$8.0 \pm 5.4 \times 10^{-2}$	8.1×10^{-2}	$1.1 \pm 0.5 \times 10^{-1}$	$1.1 \pm 0.7 \times 10^{-1}$	$1.8 \pm 1.6 \times 10^{-1}$	$1.5 \pm 0.9 \times 10^{-1}$	$2.4 \pm 0.6 \times 10^{-1}$	$1.6 \pm 1.3 \times 10^{-1}$	$2.7 \pm 0.9 \times 10^{-1}$
	P	$2.1 \pm 2.0 \times 10^{0}$	8.4×10^{-1}	$7.6 \pm 2.6 \times 10^{-1}$	$9.9 \pm 1.3 \times 10^{-1}$	$2.1 \pm 0.7 \times 10^{0}$	$3.6 \pm 2.9 \times 10^{0}$	$4.6 \pm 0.2 \times 10^{0}$	$1.1 \pm 0.9 \times 10^{0}$	$8.3 \pm 2.5 \times 10^{-1}$
FLA	G	$9.7 \pm 0.6 \times 10^{-2}$	1.0×10^{-1}	$1.0 \pm 0.6 \times 10^{-1}$	$5.5 \pm 0.5 \times 10^{-2}$	$2.3 \pm 1.7 \times 10^{-1}$	$1.5 \pm 1.6 \times 10^{-1}$	$3.7 \pm 1.5 \times 10^{-1}$	$1.7 \pm 1.3 \times 10^{-1}$	$8.4 \pm 0.9 \times 10^{-2}$
	P	$6.9 \pm 7.4 \times 10^{0}$	3.1×10^{0}	$3.5 \pm 0.9 \times 10^{0}$	$5.4 \pm 0.4 \times 10^{0}$	$9.1 \pm 5.7 \times 10^{0}$	$1.0 \pm 0.4 \times 10^{0}$	$1.4 \pm 0.5 \times 10^{0}$	$3.2 \pm 1.8 \times 10^{0}$	$3.6 \pm 0.4 \times 10^{0}$
PYR	G	$7.1 \pm 3.6 \times 10^{-2}$	9.2×10^{-2}	$7.6 \pm 4.0 \times 10^{-2}$	$7.1 \pm 0.8 \times 10^{-2}$	$3.6 \pm 3.3 \times 10^{-1}$	$1.5 \pm 2.2 \times 10^{-1}$	$5.0 \pm 2.8 \times 10^{-1}$	$2.6 \pm 2.8 \times 10^{-1}$	$1.1 \pm 0.3 \times 10^{-1}$
	P	$5.1 \pm 5.2 \times 10^{0}$	2.4×10^{0}	$3.3 \pm 1.1 \times 10^{0}$	$4.8 \pm 0.0 \times 10^{0}$	$8.6 \pm 0.1 \times 10^{0}$	$8.5 \pm 0.1 \times 10^{0}$	$1.1 \pm 0.3 \times 10^{1}$	$2.2 \pm 0.1 \times 10^{0}$	$3.0 \pm 0.3 \times 10^{0}$
BaA	G	$8.1 \pm 2.3 \times 10^{-3}$	1.1×10^{-2}	$7.1 \pm 1.2 \times 10^{-3}$	$1.6 \pm 1.4 \times 10^{-2}$	$4.5 \pm 4.8 \times 10^{-2}$	$4.8 \pm 6.0 \times 10^{-2}$	$1.1 \pm 0.4 \times 10^{-1}$	$8.1 \pm 3.6 \times 10^{-2}$	$3.1 \pm 0.5 \times 10^{-2}$
	P	$1.5 \pm 1.6 \times 10^{0}$	4.4×10^{-1}	$5.7 \pm 1.1 \times 10^{-1}$	$6.9 \pm 0.9 \times 10^{-1}$	$1.6 \pm 0.9 \times 10^{0}$	$1.9 \pm 0.6 \times 10^{0}$	$2.3 \pm 0.2 \times 10^{0}$	$4.8 \pm 2.2 \times 10^{-1}$	$6.1 \pm 1.2 \times 10^{-1}$
CHR	G	$2.1 \pm 1.0 \times 10^{-2}$	1.6×10^{-2}	$1.7 \pm 0.7 \times 10^{-2}$	$2.0 \pm 1.7 \times 10^{-2}$	$6.9 \pm 6.2 \times 10^{-2}$	$5.4 \pm 7.4 \times 10^{-2}$	$1.1 \pm 3.1 \times 10^{-2}$	$8.1 \pm 9.0 \times 10^{-2}$	$3.1 \pm 2.4 \times 10^{-2}$
	P	$1.6 \pm 1.8 \times 10^{0}$	4.7×10^{-1}	$5.8 \pm 0.6 \times 10^{-1}$	$7.9 \pm 1.9 \times 10^{-1}$	$1.3 \pm 0.9 \times 10^{0}$	$2.0 \pm 1.1 \times 10^{0}$	$2.6 \pm 0.3 \times 10^{0}$	$5.9 \pm 0.4 \times 10^{-1}$	$6.5 \pm 0.5 \times 10^{-1}$
BbF	G	$8.1 \pm 0.2 \times 10^{-3}$	1.5×10^{-2}	$3.2 \pm 6.4 \times 10^{-3}$	$3.0 \pm 5.2 \times 10^{-3}$	$1.2 \pm 1.1 \times 10^{-2}$	$1.5 \pm 2.1 \times 10^{-2}$	$1.7 \pm 3.0 \times 10^{-2}$	$3.0 \pm 1.7 \times 10^{-2}$	$1.2 \pm 2.3 \times 10^{-2}$
	P	$1.4 \pm 1.4 \times 10^{0}$	4.2×10^{-1}	$5.1 \pm 1.5 \times 10^{-1}$	$7.6 \pm 4.3 \times 10^{-1}$	$1.3 \pm 0.6 \times 10^{0}$	$1.7 \pm 0.7 \times 10^{0}$	$3.0 \pm 0.6 \times 10^{0}$	$6.7 \pm 3.6 \times 10^{-1}$	$6.3 \pm 2.1 \times 10^{-1}$
BkF	G	$8.1 \pm 1.4 \times 10^{-3}$	8.6×10^{-3}	$1.4 \pm 2.8 \times 10^{-2}$	$4.2 \pm 7.2 \times 10^{-3}$	$1.4 \pm 1.3 \times 10^{-2}$	$1.3 \pm 1.7 \times 10^{-2}$	$6.9 \pm 0.1 \times 10^{-4}$	$1.2 \pm 1.0 \times 10^{-1}$	$6.8 \pm 2.8 \times 10^{-3}$
	P	$9.9 \pm 1.2 \times 10^{-1}$	4.1×10^{-1}	$3.9 \pm 1.1 \times 10^{-1}$	$6.5 \pm 0.9 \times 10^{-1}$	$1.2 \pm 0.6 \times 10^{0}$	$1.5 \pm 0.6 \times 10^{0}$	$1.7 \pm 0.5 \times 10^{0}$	$3.6 \pm 1.1 \times 10^{-1}$	$4.8 \pm 1.0 \times 10^{-1}$
BaP	G	$1.3 \pm 1.5 \times 10^{-2}$	1.1×10^{-4}	$1.8 \pm 0.3 \times 10^{-2}$	$9.5 \pm 1.7 \times 10^{-3}$	$3.5 \pm 1.9 \times 10^{-2}$	$2.5 \pm 1.1 \times 10^{-2}$	$9.0 \pm 0.2 \times 10^{-3}$	$3.4 \pm 1.6 \times 10^{-2}$	$3.5 \pm 2.6 \times 10^{-3}$
	P	$8.2 \pm 8.4 \times 10^{-1}$	4.9×10^{-1}	$5.4 \pm 3.5 \times 10^{-1}$	$9.0 \pm 0.3 \times 10^{-1}$	$1.9 \pm 1.4 \times 10^{0}$	$1.6 \pm 0.6 \times 10^{0}$	$2.9 \pm 0.3 \times 10^{0}$	$4.3 \pm 1.6 \times 10^{-1}$	$5.0 \pm 0.2 \times 10^{-1}$
IcdP	G	$2.9 \pm 5.0 \times 10^{-3}$	1.0×10^{-4}	$5.4 \pm 3.2 \times 10^{-3}$	$9.8 \pm 1.7 \times 10^{-4}$	$1.6 \pm 1.9 \times 10^{-2}$	$6.4 \pm 1.1 \times 10^{-3}$	$2.7 \pm 1.8 \times 10^{-2}$	$1.9 \pm 1.6 \times 10^{-2}$	$4.6 \pm 2.6 \times 10^{-3}$
	P	$8.8 \pm 0.1 \times 10^{-1}$	3.9×10^{-1}	$4.1 \pm 1.4 \times 10^{-1}$	$7.2 \pm 0.6 \times 10^{-1}$	$1.1 \pm 0.7 \times 10^{0}$	$1.1 \pm 0.2 \times 10^{0}$	$1.6 \pm 0.3 \times 10^{0}$	$3.7 \pm 1.8 \times 10^{-1}$	$4.3 \pm 2.2 \times 10^{-1}$

(continued)

Table 5.4 (continued)

PAH		Horsebean	Peanut	Soybean	Cotton	Rice	Wheat	Rape	Sesame	Corn
DahA	G	$3.7 \pm 5.4 \times 10^{-3}$	1.0×10^{-4}	$2.1 \pm 3.2 \times 10^{-3}$	$1.0 \pm 0.0 \times 10^{-4}$	$1.2 \pm 2.1 \times 10^{-2}$	$1.6 \pm 3.2 \times 10^{-3}$	$1.6 \pm 2.5 \times 10^{-2}$	$2.0 \pm 1.7 \times 10^{-2}$	$4.9 \pm 0.7 \times 10^{-3}$
	P	$3.5 \pm 3.4 \times 10^{-1}$	6.0×10^{-2}	$8.9 \pm 3.9 \times 10^{-2}$	$8.1 \pm 3.0 \times 10^{-2}$	$2.1 \pm 1.2 \times 10^{-1}$	$3.1 \pm 1.7 \times 10^{-1}$	$7.3 \pm 3.9 \times 10^{-1}$	$1.9 \pm 0.2 \times 10^{-1}$	$1.1 \pm 0.4 \times 10^{-1}$
BghiP	G	$3.5 \pm 3.1 \times 10^{-3}$	1.1×10^{-2}	$4.3 \pm 2.5 \times 10^{-3}$	$3.4 \pm 0.6 \times 10^{-3}$	$2.1 \pm 1.8 \times 10^{-2}$	$1.1 \pm 2.1 \times 10^{-2}$	$2.4 \pm 2.1 \times 10^{-2}$	$1.8 \pm 1.6 \times 10^{-2}$	$3.1 \pm 4.4 \times 10^{-3}$
	P	$7.8 \pm 8.5 \times 10^{-1}$	3.3×10^{-1}	$3.8 \pm 0.9 \times 10^{-1}$	$6.1 \pm 3.5 \times 10^{-1}$	$8.3 \pm 3.7 \times 10^{-1}$	$1.1 \pm 0.4 \times 10^{0}$	$1.5 \pm 0.5 \times 10^{0}$	$3.5 \pm 1.6 \times 10^{-1}$	$3.5 \pm 1.3 \times 10^{-1}$
Total	G	$\mathbf{3.3 \pm 0.2 \times 10^{1}}$	$\mathbf{2.8 \times 10^{1}}$	$\mathbf{2.0 \pm 1.5 \times 10^{1}}$	$\mathbf{3.6 \pm 2.1 \times 10^{1}}$	$\mathbf{6.1 \pm 2.7 \times 10^{1}}$	$\mathbf{5.9 \pm 2.0 \times 10^{1}}$	$\mathbf{7.0 \pm 1.0 \times 10^{1}}$	$\mathbf{1.6 \pm 0.8 \times 10^{1}}$	$\mathbf{4.3 \pm 1.1 \times 10^{1}}$
	P	$\mathbf{5.6 \pm 5.4 \times 10^{1}}$	$\mathbf{4.3 \times 10^{1}}$	$\mathbf{2.5 \pm 0.8 \times 10^{1}}$	$\mathbf{2.9 \pm 0.1 \times 10^{1}}$	$\mathbf{6.1 \pm 0.5 \times 10^{1}}$	$\mathbf{8.7 \pm 6.1 \times 10^{1}}$	$\mathbf{1.6 \pm 0.4 \times 10^{2}}$	$\mathbf{3.0 \pm 1.9 \times 10^{1}}$	$\mathbf{2.3 \pm 2.1 \times 10^{1}}$

Data listed are results from each burning cycle in mg/kg. Modified from Shen et al. (2011) with permission of American Chemical Society

Table 5.5 EFs of PAHs in gaseous (G) and particulate-bound (P) phases for crop residues in the smoldering phase

PAH		Horsebean	Peanut	Soybean	Cotton	Rice	Wheat	Rape	Sesame	Corn
NAP	G	$1.0 \pm 0.5 \times 10^{1}$	6.6×10^{1}	$2.5 \pm 0.6 \times 10^{1}$	$9.4 \pm 6.9 \times 10^{0}$	$3.2 \pm 1.5 \times 10^{1}$	$3.1 \pm 1.5 \times 10^{1}$	$2.3 \pm 1.6 \times 10^{1}$	$5.3 \pm 1.6 \times 10^{0}$	$1.1 \pm 0.3 \times 10^{1}$
	P	$2.3 \pm 1.0 \times 10^{-1}$	3.2×10^{-2}	$6.3 \pm 0.3 \times 10^{-2}$	$3.5 \pm 0.1 \times 10^{-2}$	$2.9 \pm 2.1 \times 10^{-1}$	$3.8 \pm 4.7 \times 10^{-1}$	$4.9 \pm 4.9 \times 10^{-1}$	$2.2 \pm 1.7 \times 10^{-1}$	$1.9 \pm 2.0 \times 10^{-1}$
ACY	G	$2.2 \pm 1.1 \times 10^{0}$	7.6×10^{0}	$5.0 \pm 0.8 \times 10^{0}$	$2.2 \pm 1.9 \times 10^{0}$	$3.5 \pm 2.1 \times 10^{0}$	$6.7 \pm 1.5 \times 10^{0}$	$4.3 \pm 2.9 \times 10^{0}$	$8.0 \pm 2.4 \times 10^{-1}$	$1.5 \pm 0.4 \times 10^{0}$
	P	$1.6 \pm 0.4 \times 10^{-1}$	9.0×10^{-2}	$7.3 \pm 5.5 \times 10^{-2}$	$2.6 \pm 0.2 \times 10^{-2}$	$1.4 \pm 0.3 \times 10^{-1}$	$3.3 \pm 2.6 \times 10^{-1}$	$4.8 \pm 0.5 \times 10^{-1}$	$1.4 \pm 0.4 \times 10^{-1}$	$1.3 \pm 1.3 \times 10^{-1}$
ACE	G	$1.2 \pm 0.5 \times 10^{-1}$	3.7×10^{-1}	$2.3 \pm 0.5 \times 10^{-1}$	$8.0 \pm 7.0 \times 10^{-2}$	$2.8 \pm 0.8 \times 10^{-1}$	$3.3 \pm 1.0 \times 10^{-1}$	$2.1 \pm 1.1 \times 10^{-1}$	$1.4 \pm 0.9 \times 10^{-1}$	$1.1 \pm 0.4 \times 10^{-1}$
	P	$2.9 \pm 1.6 \times 10^{0}$	2.8×10^{0}	$1.6 \pm 1.3 \times 10^{0}$	$3.7 \pm 1.3 \times 10^{-1}$	$2.9 \pm 2.6 \times 10^{0}$	$5.8 \pm 6.0 \times 10^{0}$	$1.0 \pm 1.1 \times 10^{1}$	$1.4 \pm 0.8 \times 10^{0}$	$2.5 \pm 3.3 \times 10^{0}$
FLO	G	$3.7 \pm 2.1 \times 10^{-1}$	1.0×10^{0}	$6.7 \pm 1.2 \times 10^{-1}$	$2.1 \pm 1.9 \times 10^{-1}$	$6.5 \pm 2.1 \times 10^{-1}$	$1.1 \pm 0.5 \times 10^{0}$	$6.0 \pm 4.4 \times 10^{-1}$	$3.6 \pm 2.8 \times 10^{-1}$	$2.5 \pm 0.6 \times 10^{-1}$
	P	$3.3 \pm 0.9 \times 10^{0}$	4.0×10^{0}	$1.7 \pm 1.6 \times 10^{0}$	$4.6 \pm 0.8 \times 10^{-1}$	$3.0 \pm 1.7 \times 10^{0}$	$7.6 \pm 5.9 \times 10^{0}$	$1.2 \pm 1.1 \times 10^{1}$	$3.6 \pm 0.8 \times 10^{0}$	$3.0 \pm 3.3 \times 10^{0}$
PHE	G	$5.0 \pm 2.8 \times 10^{-1}$	2.2×10^{0}	$8.1 \pm 2.4 \times 10^{-1}$	$3.0 \pm 2.7 \times 10^{-1}$	$6.4 \pm 2.0 \times 10^{-1}$	$1.5 \pm 1.1 \times 10^{0}$	$8.9 \pm 8.4 \times 10^{-1}$	$5.5 \pm 3.9 \times 10^{-1}$	$4.1 \pm 1.9 \times 10^{0}$
	P	$5.4 \pm 1.0 \times 10^{0}$	4.5×10^{0}	$3.3 \pm 2.5 \times 10^{0}$	$8.7 \pm 0.2 \times 10^{-1}$	$6.8 \pm 0.3 \times 10^{0}$	$1.8 \pm 1.2 \times 10^{1}$	$1.6 \pm 1.1 \times 10^{1}$	$8.0 \pm 1.0 \times 10^{0}$	$5.8 \pm 0.7 \times 10^{0}$
ANT	G	$6.4 \pm 3.6 \times 10^{-2}$	3.2×10^{-1}	$1.1 \pm 0.3 \times 10^{-1}$	$4.8 \pm 4.1 \times 10^{-2}$	$1.6 \pm 3.8 \times 10^{-1}$	$2.3 \pm 1.9 \times 10^{-1}$	$1.3 \pm 1.3 \times 10^{-1}$	$7.1 \pm 3.7 \times 10^{-2}$	$6.3 \pm 0.4 \times 10^{-1}$
	P	$6.6 \pm 0.7 \times 10^{-1}$	5.8×10^{-1}	$4.2 \pm 3.5 \times 10^{-1}$	$9.9 \pm 0.5 \times 10^{-2}$	$1.2 \pm 0.3 \times 10^{0}$	$2.9 \pm 2.4 \times 10^{0}$	$2.4 \pm 2.1 \times 10^{0}$	$9.7 \pm 3.5 \times 10^{-1}$	$9.0 \pm 0.4 \times 10^{-1}$
FLA	G	$5.3 \pm 1.1 \times 10^{-2}$	1.2×10^{-1}	$6.9 \pm 3.2 \times 10^{-2}$	$1.6 \pm 0.9 \times 10^{-2}$	$1.0 \pm 0.3 \times 10^{-1}$	$2.5 \pm 3.3 \times 10^{-1}$	$7.0 \pm 0.5 \times 10^{-2}$	$6.0 \pm 5.8 \times 10^{-2}$	$9.7 \pm 7.6 \times 10^{-2}$
	P	$2.4 \pm 0.5 \times 10^{0}$	3.3×10^{0}	$2.5 \pm 0.8 \times 10^{0}$	$2.0 \pm 0.2 \times 10^{0}$	$6.0 \pm 2.8 \times 10^{0}$	$7.9 \pm 2.4 \times 10^{0}$	$6.4 \pm 2.1 \times 10^{0}$	$2.7 \pm 0.4 \times 10^{0}$	$2.2 \pm 1.4 \times 10^{0}$
PYR	G	$3.8 \pm 1.5 \times 10^{-2}$	1.7×10^{-1}	$4.9 \pm 2.1 \times 10^{-2}$	$2.3 \pm 2.0 \times 10^{-2}$	$1.3 \pm 0.7 \times 10^{-1}$	$2.6 \pm 3.8 \times 10^{-1}$	$5.5 \pm 0.4 \times 10^{-1}$	$5.8 \pm 4.9 \times 10^{-2}$	$9.2 \pm 6.1 \times 10^{-2}$
	P	$2.0 \pm 0.5 \times 10^{0}$	2.8×10^{0}	$2.5 \pm 0.9 \times 10^{0}$	$2.0 \pm 0.2 \times 10^{0}$	$5.7 \pm 3.1 \times 10^{0}$	$6.8 \pm 2.3 \times 10^{0}$	$5.4 \pm 1.0 \times 10^{0}$	$1.9 \pm 0.2 \times 10^{0}$	$1.8 \pm 0.9 \times 10^{0}$
BaA	G	$6.8 \pm 2.6 \times 10^{-3}$	1.7×10^{-2}	$6.0 \pm 3.5 \times 10^{-3}$	$5.4 \pm 3.4 \times 10^{-3}$	$3.4 \pm 1.8 \times 10^{-2}$	$2.7 \pm 2.5 \times 10^{-2}$	$1.1 \pm 0.5 \times 10^{-2}$	$6.0 \pm 0.5 \times 10^{-3}$	$2.6 \pm 2.3 \times 10^{-2}$
	P	$4.3 \pm 2.0 \times 10^{-1}$	5.2×10^{-1}	$3.8 \pm 0.9 \times 10^{-1}$	$4.1 \pm 0.9 \times 10^{-1}$	$1.1 \pm 0.5 \times 10^{0}$	$1.4 \pm 0.4 \times 10^{0}$	$1.2 \pm 0.8 \times 10^{0}$	$4.3 \pm 0.8 \times 10^{-1}$	$4.2 \pm 2.5 \times 10^{-1}$
CHR	G	$1.1 \pm 1.7 \times 10^{-2}$	3.9×10^{-2}	$9.9 \pm 5.1 \times 10^{-3}$	$8.3 \pm 8.1 \times 10^{-3}$	$5.3 \pm 3.2 \times 10^{-2}$	$3.4 \pm 2.6 \times 10^{-2}$	$1.3 \pm 0.4 \times 10^{-2}$	$1.5 \pm 1.0 \times 10^{-2}$	$2.6 \pm 2.1 \times 10^{-2}$
	P	$4.1 \pm 1.3 \times 10^{-1}$	6.2×10^{-1}	$3.6 \pm 0.6 \times 10^{-1}$	$4.8 \pm 1.8 \times 10^{-1}$	$1.0 \pm 0.6 \times 10^{0}$	$1.5 \pm 0.3 \times 10^{0}$	$1.2 \pm 0.6 \times 10^{0}$	$4.7 \pm 1.9 \times 10^{-1}$	$4.8 \pm 2.7 \times 10^{-1}$
BbF	G	$6.3 \pm 5.8 \times 10^{-3}$	1.8×10^{-2}	$3.5 \pm 4.0 \times 10^{-3}$	$2.3 \pm 4.6 \times 10^{-3}$	$1.4 \pm 1.2 \times 10^{-2}$	$8.5 \pm 7.9 \times 10^{-3}$	$5.8 \pm 5.6 \times 10^{-3}$	$5.9 \pm 6.8 \times 10^{-3}$	$2.3 \pm 1.6 \times 10^{-2}$
	P	$4.6 \pm 2.7 \times 10^{-1}$	4.6×10^{-1}	$4.0 \pm 0.3 \times 10^{-1}$	$4.7 \pm 1.5 \times 10^{-1}$	$8.8 \pm 2.5 \times 10^{-1}$	$1.4 \pm 0.6 \times 10^{0}$	$1.2 \pm 0.7 \times 10^{0}$	$4.0 \pm 0.4 \times 10^{-1}$	$3.7 \pm 1.3 \times 10^{-1}$
BkF	G	$3.8 \pm 3.5 \times 10^{-3}$	5.3×10^{-3}	$4.9 \pm 6.2 \times 10^{-3}$	$1.1 \pm 2.2 \times 10^{-3}$	$1.2 \pm 1.0 \times 10^{-2}$	$1.7 \pm 1.8 \times 10^{-2}$	$5.4 \pm 4.7 \times 10^{-3}$	$2.8 \pm 3.3 \times 10^{-3}$	$1.7 \pm 1.5 \times 10^{-2}$
	P	$3.3 \pm 1.8 \times 10^{-1}$	4.6×10^{-1}	$3.2 \pm 0.0 \times 10^{-1}$	$4.2 \pm 1.1 \times 10^{-1}$	$8.3 \pm 6.0 \times 10^{-1}$	$1.0 \pm 0.3 \times 10^{0}$	$8.9 \pm 4.6 \times 10^{-1}$	$3.0 \pm 0.0 \times 10^{-1}$	$3.1 \pm 0.7 \times 10^{-1}$
BaP	G	$5.4 \pm 7.6 \times 10^{-3}$	1.1×10^{-2}	$8.4 \pm 6.6 \times 10^{-3}$	$4.4 \pm 6.5 \times 10^{-3}$	$2.3 \pm 0.4 \times 10^{-2}$	$8.6 \pm 1.0 \times 10^{-3}$	$1.4 \pm 2.0 \times 10^{-3}$	$7.5 \pm 7.9 \times 10^{-3}$	$2.4 \pm 2.2 \times 10^{-2}$
	P	$3.2 \pm 1.8 \times 10^{-1}$	5.3×10^{-1}	$4.0 \pm 0.0 \times 10^{-1}$	$5.5 \pm 1.4 \times 10^{-1}$	$1.2 \pm 0.9 \times 10^{0}$	$1.4 \pm 0.4 \times 10^{0}$	$8.1 \pm 0.0 \times 10^{-1}$	$2.2 \pm 0.0 \times 10^{-1}$	$3.3 \pm 0.0 \times 10^{-1}$
IcdP	G	$2.4 \pm 2.1 \times 10^{-3}$	6.2×10^{-3}	$8.2 \pm 6.7 \times 10^{-3}$	$2.4 \pm 3.2 \times 10^{-3}$	$1.0 \pm 1.4 \times 10^{-2}$	$1.3 \pm 1.1 \times 10^{-2}$	$4.4 \pm 3.9 \times 10^{-3}$	$1.0 \pm 0.0 \times 10^{-4}$	$1.5 \pm 0.1 \times 10^{-2}$
	P	$3.3 \pm 2.1 \times 10^{-1}$	3.6×10^{-1}	$3.5 \pm 0.4 \times 10^{-1}$	$5.7 \pm 1.7 \times 10^{-1}$	$7.7 \pm 4.6 \times 10^{-1}$	$1.0 \pm 0.4 \times 10^{0}$	$6.4 \pm 0.2 \times 10^{-2}$	$2.1 \pm 0.0 \times 10^{-1}$	$2.2 \pm 0.4 \times 10^{-1}$

(continued)

Table 5.5 (continued)

PAH		Horsebean	Peanut	Soybean	Cotton	Rice	Wheat	Rape	Sesame	Corn
DahA	G	$4.8 \pm 8.4 \times 10^{-3}$	3.2×10^{-3}	$1.4 \pm 2.2 \times 10^{-3}$	$5.9 \pm 1.2 \times 10^{-4}$	$1.3 \pm 1.1 \times 10^{-3}$	$1.1 \pm 1.3 \times 10^{-3}$	$1.2 \pm 2.0 \times 10^{-3}$	$1.0 \pm 0.0 \times 10^{-4}$	$1.3 \pm 2.2 \times 10^{-3}$
	P	$9.0 \pm 6.4 \times 10^{-2}$	9.4×10^{-2}	$7.4 \pm 0.5 \times 10^{-2}$	$6.1 \pm 0.8 \times 10^{-2}$	$1.1 \pm 0.5 \times 10^{-1}$	$2.4 \pm 0.6 \times 10^{-1}$	$2.9 \pm 2.8 \times 10^{-1}$	$1.1 \pm 0.6 \times 10^{-1}$	$7.3 \pm 4.2 \times 10^{-2}$
BghiP	G	$3.6 \pm 3.6 \times 10^{-3}$	4.5×10^{-3}	$5.0 \pm 2.9 \times 10^{-3}$	$2.3 \pm 3.2 \times 10^{-3}$	$1.7 \pm 0.7 \times 10^{-2}$	$1.5 \pm 1.6 \times 10^{-2}$	$3.8 \pm 3.3 \times 10^{-3}$	$1.2 \pm 2.4 \times 10^{-3}$	$1.1 \pm 0.9 \times 10^{-2}$
	P	$2.8 \pm 1.7 \times 10^{-1}$	3.0×10^{-1}	$2.8 \pm 1.2 \times 10^{-1}$	$4.6 \pm 1.4 \times 10^{-1}$	$5.9 \pm 4.4 \times 10^{-1}$	$8.1 \pm 2.1 \times 10^{-1}$	$6.9 \pm 3.7 \times 10^{-1}$	$1.9 \pm 0.3 \times 10^{-1}$	$2.2 \pm 0.5 \times 10^{-1}$
Total	G	$\mathbf{1.4 \pm 0.6 \times 10^{1}}$	$\mathbf{7.8 \times 10^{1}}$	$\mathbf{3.1 \pm 0.7 \times 10^{1}}$	$\mathbf{1.2 \pm 0.9 \times 10^{1}}$	$\mathbf{3.7 \pm 1.7 \times 10^{1}}$	$\mathbf{4.2 \pm 1.3 \times 10^{1}}$	$\mathbf{2.9 \pm 2.0 \times 10^{1}}$	$\mathbf{7.3 \pm 2.4 \times 10^{0}}$	$\mathbf{1.3 \pm 0.4 \times 10^{1}}$
	P	$\mathbf{1.9 \pm 0.6 \times 10^{1}}$	$\mathbf{2.1 \times 10^{1}}$	$\mathbf{1.5 \pm 0.8 \times 10^{1}}$	$\mathbf{9.3 \pm 1.2 \times 10^{0}}$	$\mathbf{3.2 \pm 0.6 \times 10^{1}}$	$\mathbf{5.8 \pm 3.3 \times 10^{1}}$	$\mathbf{6.0 \pm 4.1 \times 10^{1}}$	$\mathbf{2.1 \pm 0.4 \times 10^{1}}$	$\mathbf{1.9 \pm 1.7 \times 10^{1}}$

Data listed are results from each burning cycle in mg/kg. Modified from Shen et al. (2011) with permission of American Chemical Society

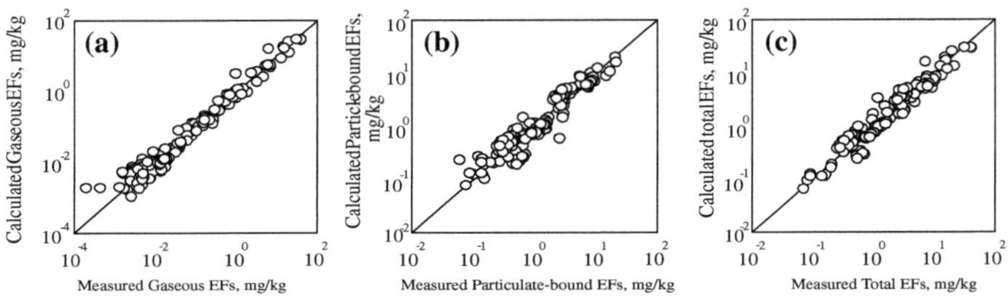

Fig. 5.9 Comparison between the measured and calculated EFs of gaseous (**a**), particulate-bound (**b**), and total (**c**) PAHs for crop residue burning. The calculation was based on a regression model with moisture and MCE as independent variables. The results are presented in log-scale. Adapted from Shen et al. (2011) with permission of American Chemical Society

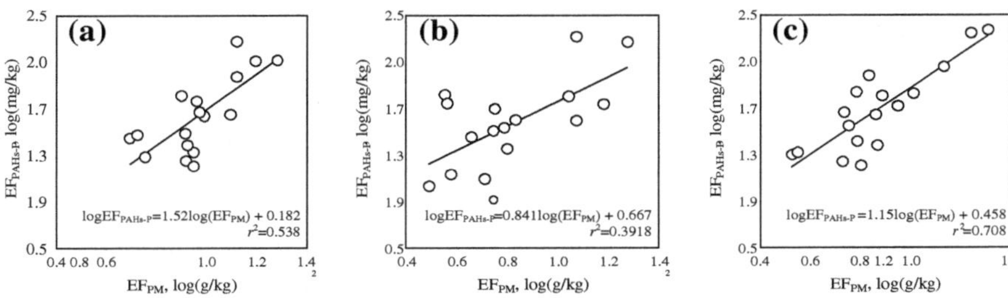

Fig. 5.10 Relationship between PAHs and PM from crop residue burning in the whole burning cycle (**a**), flaming (**b**) and smoldering phases (**c**). Adapted from Shen et al. (2011) with permission of American Chemical Society

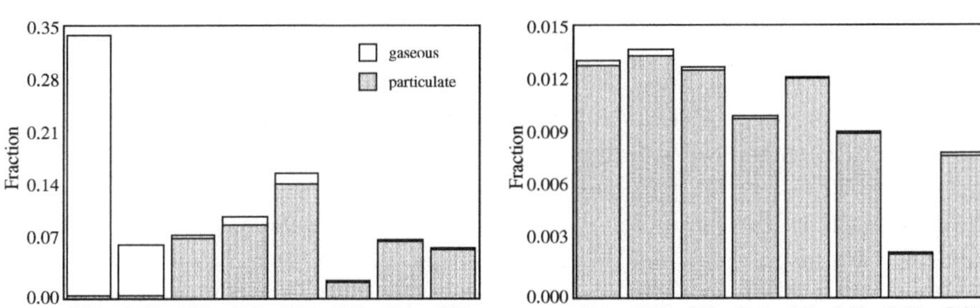

Fig. 5.11 Composition profile of PAH emission from crop residues burning in the cooking stove during a whole burning cycle. Adapted from Shen et al. (2011) with permission of American Chemical Society

(76 ± 5 %), followed by ACY (15 ± 4 %). In particulate phase, PHE (20 ± 12 %), FLA (11 ± 7 %), FLO (11 ± 7 %) and PYR (10 ± 7 %) dominated. There was no significant difference in the PAH profile between flaming and smoldering phases ($p = 0.121$), and low molecular weight PAHs with 2–3 rings made up 80 ± 4 and 78 ± 7 % of the total in flaming and smoldering phases, respectively.

Table 5.6 Comparison of parent PAH isomer ratios from this study and the literature reported values for crop residue

	This study	Stove burning	Chamber study	Open-fire
ANT/(ANT + PHE)	0.12 ± 0.01	0.2	0.18–0.25	0.17–0.25
FLA/(FLA + PYR)	0.53 ± 0.03	0.51–0.80	0.50–0.53	0.34–0.53
BaA/(BaA + CHR)	0.48 ± 0.02	0.46	0.46–0.53	0.39–0.50
IcdP/(IcdP + BghiP)	0.54 ± 0.02	0.31–0.50	0.46–0.52	0.39–0.94
BbF/(BbF + BkF)	0.55 ± 0.03	0.50–0.65	0.28–0.41	0.35–0.80
BaP/(BaP + BghiP)	0.60 ± 0.05	0.23–0.67	0.56–0.83	0.43–0.98

Modified from Shen et al. (2011) with permission of American Chemical Society

Several commonly used isomer ratios including ANT/(ANT + PHE), FLA/(FLA + PYR), BaA/(BaA + CHR), IcdP/(IcdP + BghiP), BbF/(BbF + BkF), and BaP/(BaP + BghiP) for crop residue derived in our present study and those reported in the literature from stove burning (Kim Oanh et al. 2005; Sheesley et al. 2003), chamber study (Lu et al. 2009; Zhang et al. 2008; Dhammapala et al. 2007a, b; Keshtkar and Ashbaugh 2007) and simulated open fire burning (Hays et al. 2005; Jenkins et al. 1996a, b) are provided in Table 5.6. These ratios varied largely between and within different burning situations. Our measured ratios are generally similar to those of crop residue burning in stoves or chambers, with a few exceptions, but are significantly different from those from open-field crop residue combustions.

5.2.3 *Size Distribution of Particulate Phase PAHs*

The size distributions of particulate phase PAHs between different crop straws were similar in general (Fig. 5.12). On average, over 80 % could be found in fine $PM_{2.5}$. Most of PAHs (54 ± 11 %) were present in PM with diameter between 0.7–2.1 μm. Distribution of individual PAH compounds between fine (<2.1 μm) and coarse (2.1–10 μm) particles are also illustrated in Fig. 5.12. It is clearly shown that the higher the molecular weight, the stronger the association with finer particles, similar to that found in coal emission. About 60 % of particulate phase NAP was associated with particles larger than 2.1 μm, while 81–90 % of 4- to 6-ring PAHs from PYR to BghiP bound to particles with diameter less than 2.1 μm. Since high molecular weight PAHs tended to be present in fine PM fraction (Allen et al. 1996; Hays et al. 2003; Venkataraman et al. 1999), difference in normalized composition profiles of particle-bound PAHs in different size fractions could be expected. As shown in Fig. 5.13, the contribution of high molecular weight PAHs increased in fine PM fractions.

In both flaming and smoldering phases, emitted particle-bound PAHs were also mainly present in fine particles (Fig. 5.14). The mass fractions of total PAHs in fine particles in the flaming phase was 74 ± 9 %, significantly lower than that in the smoldering phase (82 ± 7 %) ($p = 0.019$). It has been mentioned above that,

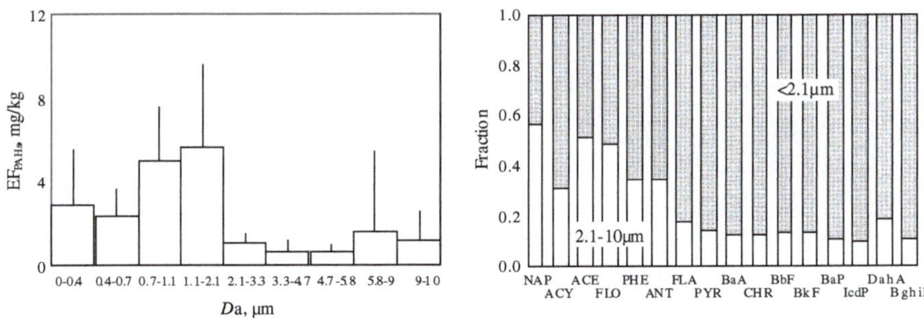

Fig. 5.12 Size distribution of particulate phase PAHs emitted from crop residue burning (*left panel*) and relative distribution of 16 individual PAH compounds between fine (<2.1 μm) and coarse (2.1–10 μm) particles (*right panel*). The means and standard derivations of EFs of PAHs associated with PM with different sizes from 17 burning experiments are shown. Adapted from Shen et al. (2011) with permission of American Chemical Society

Fig. 5.13 Normalized composition profile of particulate phase PAHs in each size fraction

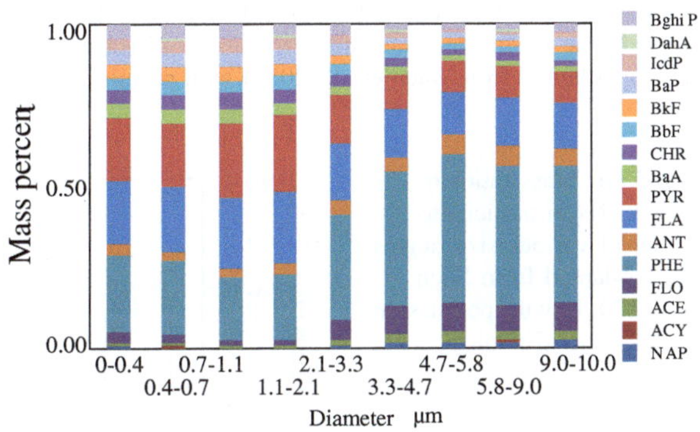

Fig. 5.14 Size distributions of particulate phase PAHs from flaming (*blank column*) and smoldering (*filled column*) phases of crop residues burning. Adapted from Shen et al. (2011) with permission of American Chemical Society

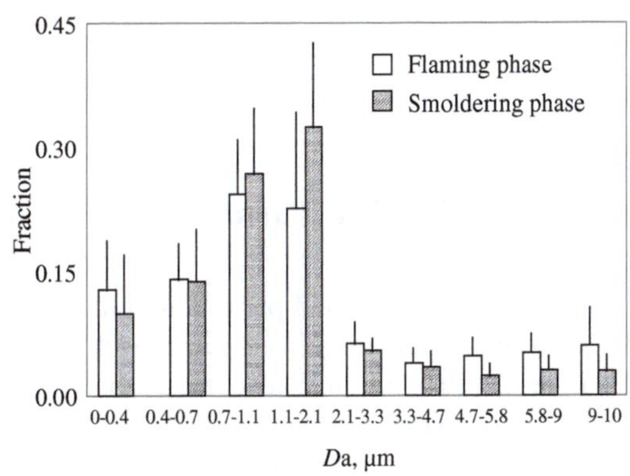

particles emitted during the smoldering phase were smaller than those during the flaming phase and this is the main reason causing the difference in size distribution of particle-bound PAHs.

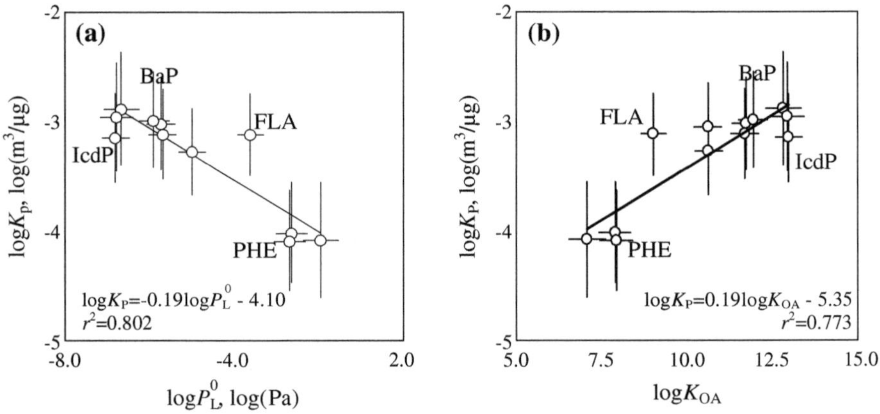

Fig. 5.15 Dependence of log(K_p) on log(P_L^0) (**a**) and log(K_{OA}) (**b**). P_L^0 and K_{OA} were calculated based on the measured temperatures and equations established by Odabasi et al. (2006). PAH compounds of concern due to high abundance and/or toxicity, like PHE, FLA, BaP, and IcdP, are labeled. The means and standard derivations of measured *Kp* from 17 burning experiments are shown. Adapted from Shen et al. (2011) with permission of American Chemical Society

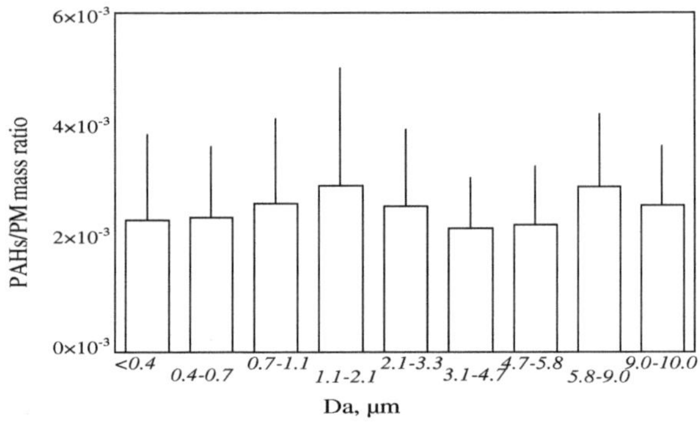

Fig. 5.16 Mass ratios of total PAHs in particulate matter of various size ranges (D_a). Adapted from Shen et al. (2011) with permission of American Chemical Society

5.2.4 Gas-Particle Partitioning

Total EFs of gaseous and particulate phase PAHs were 27 ± 13 and 35 ± 23 mg/kg, respectively. Similar to the approach in analysis of gas-particle partitioning of PAHs from residential coal combustion, the partitioning coefficients were calculated and probable mechanism(s) controlling the partitioning was investigated. As shown in Fig. 5.15, the dependence of K_P on P_L^0 and K_{OA} was well demonstrated. An extremely shallow slope of −0.19 (−0.23 to −0.15, 95 % confidence interval) was derived from the relationship between K_P and P_L^0. K_P positively correlated with K_{OA} significantly. The mass ratios of particle-bound PAHs did not show any dependence on particle size (Fig. 5.16). Moveover, the correlation between PAHs and OC ($r = 0.563$, $p = 0.0093$) was stronger than that between PAHs and EC ($r = 0.376$, $p = 0.0685$). All of these indicated that the partitioning was mainly controlled by absorption.

5.3 Residential Wood Combustion

5.3.1 EFs and Influencing Factor

The measured results are provided in Tables 5.7 and 5.8. EF_{PAHs} for fuel wood log ranged from 3.2 to 32.7 mg/kg, with a mean of 12.7 ± 7.0 mg/kg. For the brushwood, the EF_{PAHs} ranged from 27.1 to 160 mg/kg, with a mean and standard deviation of 86.7 ± 67.6 mg/kg. The later was significantly higher than the former, which coincided with that in PM emission. Positive correlations between EF_{PAHs} and EF_{PM} and between EF_{PAHs} and EF_{CO} could be found (Fig. 5.17).

Total EFs of the 16 priority PAHs for fuel wood, brushwood, and bamboo were 12.1 ± 6.7, 79.7 ± 60.4, and 31.4 ± 16.6 mg/kg, respectively, accounting for approximately 95 ± 3 % of the EF_{PAHs}. Although the 12 non-U.S. EPA priority PAHs contributed only a small fraction to the EF_{PAH28}, it is worth noting that most of these PAHs are carcinogenic, and hence should be given more attention. As most added non-priority PAHs are high molecular weight ones which are preferable to be in particulate phase, it is expected that the increase in EFs of total 28 PAHs would be much obvious in particulate phase (Fig. 5.18). For example, the EFs of total gaseous and particulate phase 16 priority PAHs for the brushwood were 53.4 ± 22.7 and 25.1 ± 13.4 mg/kg, respectively, while the sum of 28 PAHs had EFs of 54.5 ± 23.1 and 30.8 ± 17.1 mg/kg for gaseous and particulate phases, respectively.

The EFs were negatively correlated with MCE, and significantly positively correlated with fuel moisture (Fig. 5.19). EFs of wood fuel with moisture larger than 35 % were between 15.3–32.7 mg/kg while for fuels with moisture lower than 10 %, the measured EFs were between 3.2 and 10.7 mg/kg. This might be due to an appreciable amount of energy is needed to vaporize the water when burning high moisture fuels, leading to reduced combustion efficiencies and enhanced pollutant emissions The relationship between EF_{PAHs} and other quantified fuel properties was not significant ($p > 0.05$).

5.3.2 Composition Profile and Isomer Ratio

The composition profile of parent PAHs from residential wood combustion was dominated by 2- and 3-ring PAHs (from NAP to ANT), accounting for 73.6 % of the total. NAP (41.0 ± 12.6 %) and PHE (16.9 ± 5.6 %), followed by FLA

Table 5.7 Average emission factors (mg/kg) of parent PAHs for residential wood combustions of different fuel types

	White poplar	Elm	Locust	Maple	Fir	Larch	Water Chinese fir
NAP	5.5×10^{0}	4.9×10^{0}	2.4×10^{0}	3.3×10^{0}	4.2×10^{0}	2.8×10^{0}	3.2×10^{0}
ACY	1.5×10^{0}	1.7×10^{0}	4.9×10^{-1}	8.4×10^{-1}	1.1×10^{0}	1.0×10^{0}	8.5×10^{-1}
ACE	8.8×10^{-2}	1.1×10^{-1}	5.7×10^{-2}	8.1×10^{-2}	5.0×10^{-2}	5.2×10^{-2}	4.4×10^{-2}
FLO	3.6×10^{-1}	5.1×10^{-1}	1.6×10^{-1}	2.9×10^{-1}	2.5×10^{-1}	2.8×10^{-1}	2.5×10^{-1}
PHE	1.7×10^{0}	2.7×10^{0}	9.2×10^{-1}	1.6×10^{0}	1.9×10^{0}	1.4×10^{0}	1.8×10^{0}
ANT	2.3×10^{-1}	3.9×10^{-1}	1.3×10^{-1}	2.0×10^{-1}	2.7×10^{-1}	1.7×10^{-1}	2.1×10^{-1}
FLU	1.0×10^{0}	1.1×10^{0}	4.3×10^{-1}	8.4×10^{-1}	1.2×10^{0}	4.6×10^{-1}	9.2×10^{-1}
PYR	9.2×10^{-1}	9.8×10^{-1}	3.7×10^{-1}	6.5×10^{-1}	9.2×10^{-1}	4.0×10^{-1}	7.8×10^{-1}
RET	6.7×10^{-2}	4.4×10^{-2}	7.5×10^{-2}	6.2×10^{-2}	7.4×10^{-2}	2.1×10^{-1}	1.3×10^{-1}
BcP	7.4×10^{-2}	5.4×10^{-2}	2.1×10^{-2}	5.1×10^{-2}	4.1×10^{-2}	2.1×10^{-2}	4.3×10^{-2}
CcdP	2.9×10^{-1}	1.2×10^{-1}	6.2×10^{-2}	9.4×10^{-2}	5.0×10^{-2}	3.8×10^{-2}	2.3×10^{-2}
BaA	2.7×10^{-1}	2.1×10^{-1}	8.0×10^{-2}	2.5×10^{-1}	1.3×10^{-1}	7.0×10^{-2}	1.3×10^{-1}
CHR	2.5×10^{-1}	2.0×10^{-1}	8.6×10^{-2}	2.3×10^{-1}	1.3×10^{-1}	8.4×10^{-2}	1.5×10^{-1}
BbF	1.3×10^{-1}	1.1×10^{-1}	3.2×10^{-2}	1.5×10^{-1}	6.6×10^{-2}	3.9×10^{-2}	8.0×10^{-2}
BkF	1.3×10^{-1}	9.8×10^{-2}	3.0×10^{-2}	1.3×10^{-1}	6.9×10^{-2}	3.9×10^{-2}	7.5×10^{-2}
BeP	7.8×10^{-2}	6.5×10^{-2}	2.0×10^{-2}	8.7×10^{-2}	4.1×10^{-2}	2.6×10^{-2}	4.8×10^{-2}
BaP	1.3×10^{-1}	8.4×10^{-2}	2.7×10^{-2}	1.2×10^{-1}	4.8×10^{-2}	2.6×10^{-2}	5.0×10^{-2}
PER	2.5×10^{-2}	1.6×10^{-2}	6.3×10^{-3}	2.7×10^{-2}	1.4×10^{-2}	8.5×10^{-3}	8.5×10^{-3}
IcdP	8.6×10^{-2}	6.0×10^{-2}	2.0×10^{-2}	8.3×10^{-2}	3.4×10^{-2}	2.2×10^{-2}	3.4×10^{-2}
DahA	1.3×10^{-2}	8.0×10^{-3}	3.5×10^{-3}	1.3×10^{-2}	4.8×10^{-3}	1.8×10^{-3}	4.3×10^{-3}
BghiP	6.6×10^{-2}	4.9×10^{-2}	1.5×10^{-2}	6.9×10^{-2}	3.1×10^{-2}	2.1×10^{-2}	3.1×10^{-2}
DacP	2.2×10^{-2}	1.1×10^{-2}	5.0×10^{-3}	1.5×10^{-2}	5.9×10^{-3}	5.2×10^{-3}	3.2×10^{-3}
DalP	1.5×10^{-2}	9.1×10^{-3}	3.4×10^{-3}	1.3×10^{-2}	5.1×10^{-3}	2.4×10^{-3}	3.5×10^{-3}
DaeF	9.6×10^{-3}	6.2×10^{-3}	2.8×10^{-3}	1.2×10^{-2}	3.8×10^{-3}	1.7×10^{-3}	2.8×10^{-3}

(continued)

Table 5.7 (continued)

	White poplar	Elm	Locust	Maple	Fir	Larch	Water Chinese fir
Coron	3.6×10^{-2}	2.3×10^{-2}	8.2×10^{-3}	2.3×10^{-2}	1.0×10^{-2}	1.0×10^{-2}	8.8×10^{-3}
DaeP	9.4×10^{-3}	8.3×10^{-3}	7.5×10^{-3}	8.9×10^{-3}	1.6×10^{-3}	4.4×10^{-3}	2.9×10^{-3}
DaiP	9.7×10^{-3}	4.7×10^{-3}	1.1×10^{-3}	1.4×10^{-2}	7.9×10^{-4}	1.9×10^{-3}	2.3×10^{-3}
DahP	2.1×10^{-3}	7.1×10^{-4}	2.8×10^{-4}	2.1×10^{-3}	7.4×10^{-5}	3.5×10^{-4}	2.2×10^{-4}

	Cypress	Oak	Chinese pine	Willow	Paulownia tomentosa	Toon	White birch
NAP	3.4×10^{0}	3.1×10^{0}	2.1×10^{0}	1.1×10^{0}	3.2×10^{0}	2.7×10^{0}	6.5×10^{0}
ACY	6.1×10^{-1}	5.9×10^{-1}	5.5×10^{-1}	1.9×10^{-1}	5.6×10^{-1}	4.1×10^{-1}	1.6×10^{0}
ACE	5.4×10^{-2}	5.4×10^{-2}	4.5×10^{-2}	1.7×10^{-2}	4.7×10^{-2}	3.2×10^{-2}	1.2×10^{-1}
FLO	2.4×10^{-1}	2.2×10^{-1}	1.8×10^{-1}	7.6×10^{-2}	1.3×10^{-1}	1.1×10^{-1}	4.0×10^{-1}
PHE	1.3×10^{0}	1.3×10^{0}	1.2×10^{0}	7.2×10^{-1}	1.0×10^{0}	8.2×10^{-1}	2.0×10^{0}
ANT	1.6×10^{-1}	1.6×10^{-1}	1.4×10^{-1}	8.1×10^{-2}	1.1×10^{-1}	8.4×10^{-2}	2.7×10^{-1}
FLU	6.2×10^{-1}	6.4×10^{-1}	6.6×10^{-1}	3.7×10^{-1}	4.9×10^{-1}	3.5×10^{-1}	1.1×10^{0}
PYR	5.0×10^{-1}	5.3×10^{-1}	5.5×10^{-1}	2.9×10^{-1}	4.1×10^{-1}	2.9×10^{-1}	1.0×10^{0}
RET	1.4×10^{-1}	1.3×10^{-1}	3.4×10^{-1}	9.3×10^{-2}	1.6×10^{-1}	4.8×10^{-2}	8.1×10^{-2}
BcP	2.8×10^{-2}	3.6×10^{-2}	3.9×10^{-2}	1.6×10^{-2}	4.3×10^{-2}	2.0×10^{-2}	7.5×10^{-2}
CcdP	6.0×10^{-2}	1.3×10^{-1}	2.1×10^{-2}	2.6×10^{-2}	1.3×10^{-1}	2.7×10^{-2}	3.5×10^{-1}
BaA	1.1×10^{-1}	1.5×10^{-1}	1.1×10^{-1}	4.6×10^{-2}	1.5×10^{-1}	7.0×10^{-2}	3.0×10^{-1}
CHR	1.2×10^{-1}	1.7×10^{-1}	1.4×10^{-1}	6.3×10^{-2}	1.8×10^{-1}	8.0×10^{-2}	3.0×10^{-1}
BbF	6.1×10^{-2}	7.0×10^{-2}	6.9×10^{-2}	2.3×10^{-2}	1.0×10^{-1}	4.1×10^{-2}	1.8×10^{-1}
BkF	5.3×10^{-2}	6.1×10^{-2}	7.1×10^{-2}	2.6×10^{-2}	8.6×10^{-2}	4.2×10^{-2}	1.9×10^{-1}
BeP	3.6×10^{-2}	4.3×10^{-2}	3.8×10^{-2}	1.8×10^{-2}	6.1×10^{-2}	2.9×10^{-2}	1.0×10^{-1}
BaP	4.7×10^{-2}	6.6×10^{-2}	3.7×10^{-2}	1.9×10^{-2}	7.7×10^{-2}	3.1×10^{-2}	1.9×10^{-1}
PER	1.0×10^{-2}	1.3×10^{-2}	7.2×10^{-3}	6.0×10^{-3}	1.3×10^{-2}	7.8×10^{-3}	3.2×10^{-2}
IcdP	3.6×10^{-2}	4.0×10^{-2}	2.3×10^{-2}	1.1×10^{-2}	5.8×10^{-2}	2.2×10^{-2}	1.4×10^{-1}

(continued)

Table 5.7 (continued)

	Cypress	Oak	Chinese pine	Willow	Paulownia tomentosa	Toon	White birch
DahA	5.0×10^{-3}	6.9×10^{-3}	3.1×10^{-3}	2.8×10^{-3}	9.4×10^{-3}	3.3×10^{-3}	1.3×10^{-2}
BghiP	2.8×10^{-2}	3.3×10^{-2}	2.0×10^{-2}	1.1×10^{-2}	4.7×10^{-2}	2.1×10^{-2}	1.1×10^{-1}
DacP	5.8×10^{-3}	8.2×10^{-3}	1.8×10^{-3}	1.4×10^{-3}	9.7×10^{-3}	3.4×10^{-3}	3.3×10^{-2}
DalP	6.2×10^{-3}	6.4×10^{-3}	2.0×10^{-3}	1.7×10^{-3}	8.1×10^{-3}	2.5×10^{-3}	2.2×10^{-2}
DaeF	5.0×10^{-3}	5.1×10^{-3}	1.5×10^{-3}	2.0×10^{-3}	8.1×10^{-3}	2.2×10^{-3}	1.5×10^{-2}
Coron	1.3×10^{-2}	1.6×10^{-2}	3.9×10^{-3}	3.9×10^{-3}	2.5×10^{-2}	7.4×10^{-3}	6.8×10^{-2}
DaeP	5.3×10^{-3}	7.6×10^{-3}	1.5×10^{-3}	8.8×10^{-4}	9.3×10^{-3}	2.9×10^{-3}	1.7×10^{-2}
DaiP	4.9×10^{-3}	5.6×10^{-3}	3.9×10^{-4}	2.4×10^{-4}	6.7×10^{-3}	1.1×10^{-3}	2.1×10^{-2}
DahP	1.1×10^{-3}	6.7×10^{-4}	4.3×10^{-4}	0.0×10^{0}	1.3×10^{-3}	6.0×10^{-4}	2.6×10^{-3}

	Ribbed birch	Paulownia elongata	Black poplar	China aspen	Chinaberry	Jujube tree	Persimmon tree
NAP	7.0×10^{0}	1.4×10^{1}	4.5×10^{0}	6.0×10^{0}	1.3×10^{1}	6.4×10^{0}	6.6×10^{0}
ACY	1.5×10^{0}	3.3×10^{0}	1.2×10^{0}	1.2×10^{0}	2.4×10^{0}	2.3×10^{0}	2.2×10^{0}
ACE	8.9×10^{-2}	3.8×10^{-1}	1.1×10^{-1}	8.7×10^{-2}	2.6×10^{-1}	1.3×10^{-1}	1.4×10^{-1}
FLO	3.6×10^{-1}	8.8×10^{-1}	5.0×10^{-1}	3.4×10^{-1}	6.2×10^{-1}	5.8×10^{-1}	5.5×10^{-1}
PHE	1.8×10^{0}	5.6×10^{0}	5.1×10^{0}	2.4×10^{0}	3.1×10^{0}	2.7×10^{0}	2.4×10^{0}
ANT	2.2×10^{-1}	1.0×10^{0}	8.8×10^{-1}	3.4×10^{-1}	4.9×10^{-1}	3.9×10^{-1}	3.4×10^{-1}
FLU	1.0×10^{0}	2.2×10^{0}	2.3×10^{0}	1.4×10^{0}	1.3×10^{0}	1.6×10^{0}	1.1×10^{0}
PYR	9.3×10^{-1}	1.9×10^{0}	1.9×10^{0}	1.2×10^{0}	1.1×10^{0}	1.5×10^{0}	9.7×10^{-1}
RET	6.3×10^{-2}	5.4×10^{-2}	1.5×10^{-1}	7.7×10^{-2}	1.5×10^{-1}	3.9×10^{-2}	7.7×10^{-2}
BcP	6.8×10^{-2}	1.5×10^{-1}	1.6×10^{-1}	7.9×10^{-2}	8.1×10^{-2}	1.4×10^{-1}	7.5×10^{-2}
CcdP	1.4×10^{-1}	7.0×10^{-1}	6.8×10^{-1}	3.2×10^{-1}	3.7×10^{-1}	5.2×10^{-1}	2.9×10^{-1}
BaA	2.3×10^{-1}	6.0×10^{-1}	7.5×10^{-1}	3.7×10^{-1}	3.6×10^{-1}	4.7×10^{-1}	2.9×10^{-1}
CHR	2.1×10^{-1}	5.5×10^{-1}	7.3×10^{-1}	3.1×10^{-1}	4.3×10^{-1}	4.2×10^{-1}	2.6×10^{-1}
BbF	1.4×10^{-1}	2.5×10^{-1}	3.3×10^{-1}	2.3×10^{-1}	1.5×10^{-1}	3.2×10^{-1}	1.6×10^{-1}

(continued)

Table 5.7 (continued)

	Ribbed birch	Paulownia elongata	Black poplar	China aspen	Chinaberry	Jujube tree	Persimmon tree
BkF	1.3×10^{-1}	2.1×10^{-1}	2.4×10^{-1}	2.0×10^{-1}	1.3×10^{-1}	3.8×10^{-1}	1.7×10^{-1}
BeP	7.9×10^{-2}	1.4×10^{-1}	1.8×10^{-1}	1.2×10^{-1}	8.3×10^{-2}	1.9×10^{-1}	1.0×10^{-1}
BaP	1.3×10^{-1}	2.7×10^{-1}	3.5×10^{-1}	2.0×10^{-1}	1.5×10^{-1}	3.5×10^{-1}	1.9×10^{-1}
PER	2.1×10^{-2}	4.2×10^{-2}	5.0×10^{-2}	3.3×10^{-2}	2.4×10^{-2}	5.8×10^{-2}	3.2×10^{-2}
IcdP	1.0×10^{-1}	1.8×10^{-1}	2.6×10^{-1}	1.5×10^{-1}	9.3×10^{-2}	2.7×10^{-1}	1.2×10^{-1}
DahA	9.0×10^{-3}	2.4×10^{-2}	4.4×10^{-2}	2.0×10^{-2}	1.5×10^{-2}	2.8×10^{-2}	1.3×10^{-2}
BghiP	8.3×10^{-2}	1.1×10^{-1}	1.6×10^{-1}	1.1×10^{-1}	6.5×10^{-2}	2.1×10^{-1}	9.7×10^{-2}
DacP	1.7×10^{-2}	6.7×10^{-2}	6.1×10^{-2}	3.6×10^{-2}	2.8×10^{-2}	6.3×10^{-2}	2.8×10^{-2}
DalP	1.3×10^{-2}	3.4×10^{-2}	4.8×10^{-2}	2.6×10^{-2}	1.6×10^{-2}	4.6×10^{-2}	2.4×10^{-2}
DaeF	1.1×10^{-2}	2.7×10^{-2}	3.4×10^{-2}	2.0×10^{-2}	1.3×10^{-2}	3.3×10^{-2}	1.5×10^{-2}
Coron	4.5×10^{-2}	7.6×10^{-2}	8.7×10^{-2}	5.1×10^{-2}	3.4×10^{-2}	1.4×10^{-1}	6.1×10^{-2}
DaeP	1.0×10^{-2}	2.9×10^{-2}	3.8×10^{-2}	2.6×10^{-2}	1.3×10^{-2}	3.3×10^{-2}	1.5×10^{-2}
DaiP	1.3×10^{-2}	2.6×10^{-2}	3.0×10^{-2}	1.9×10^{-2}	1.2×10^{-2}	3.5×10^{-2}	1.3×10^{-2}
DahP	2.0×10^{-3}	5.9×10^{-3}	1.1×10^{-2}	3.1×10^{-3}	1.4×10^{-3}	3.7×10^{-3}	2.6×10^{-3}

	Mulberry tree	Peach tree	Lespedeza	Buxus sinica	Holly	Bamboo
NAP	6.6×10^{0}	6.2×10^{0}	5.9×10^{1}	2.5×10^{1}	1.2×10^{1}	1.4×10^{1}
ACY	1.7×10^{0}	1.6×10^{0}	1.3×10^{1}	6.2×10^{0}	2.4×10^{0}	3.4×10^{0}
ACE	1.3×10^{-1}	1.2×10^{-1}	3.4×10^{-1}	1.8×10^{-1}	7.1×10^{-2}	9.4×10^{-2}
FLO	5.2×10^{-1}	4.6×10^{-1}	2.6×10^{0}	1.3×10^{0}	5.4×10^{-1}	6.7×10^{-1}
PHE	2.3×10^{0}	2.1×10^{0}	1.2×10^{1}	7.7×10^{0}	3.7×10^{0}	3.5×10^{0}
ANT	3.4×10^{-1}	2.6×10^{-1}	2.3×10^{0}	1.2×10^{0}	4.9×10^{-1}	5.3×10^{-1}
FLU	8.9×10^{-1}	8.4×10^{-1}	1.5×10^{1}	7.2×10^{0}	2.3×10^{0}	2.8×10^{0}
PYR	7.4×10^{-1}	7.5×10^{-1}	1.4×10^{1}	7.0×10^{0}	1.9×10^{0}	2.6×10^{0}
RET	5.7×10^{-2}	5.7×10^{-2}	3.9×10^{-2}	9.5×10^{-2}	4.2×10^{-2}	8.1×10^{-2}

(continued)

Table 5.7 (continued)

	Mulberry tree	Peach tree	Lespedeza	Buxus sinica	Holly	Bamboo
BcP	4.5×10^{-2}	5.1×10^{-2}	1.1×10^{0}	5.7×10^{-1}	1.4×10^{-1}	1.9×10^{-1}
CcdP	1.6×10^{-1}	2.4×10^{-1}	6.5×10^{0}	1.3×10^{0}	1.7×10^{-1}	6.7×10^{-1}
BaA	1.8×10^{-1}	2.3×10^{-1}	4.9×10^{0}	2.2×10^{0}	5.2×10^{-1}	6.6×10^{-1}
CHR	1.7×10^{-1}	2.0×10^{-1}	3.5×10^{0}	1.6×10^{0}	5.0×10^{-1}	5.7×10^{-1}
BbF	9.2×10^{-2}	1.2×10^{-1}	3.6×10^{0}	1.6×10^{0}	3.9×10^{-1}	4.6×10^{-1}
BkF	8.0×10^{-2}	1.2×10^{-1}	3.4×10^{0}	1.6×10^{0}	4.0×10^{-1}	5.9×10^{-1}
BeP	5.3×10^{-2}	7.3×10^{-2}	1.9×10^{0}	8.7×10^{-1}	2.4×10^{-1}	3.1×10^{-1}
BaP	7.6×10^{-2}	1.3×10^{-1}	4.4×10^{0}	1.7×10^{0}	3.5×10^{-1}	6.2×10^{-1}
PER	1.3×10^{-2}	2.6×10^{-2}	6.7×10^{-1}	2.7×10^{-1}	6.3×10^{-2}	1.0×10^{-1}
IcdP	4.9×10^{-2}	9.7×10^{-2}	4.0×10^{0}	1.6×10^{0}	3.1×10^{-1}	5.2×10^{-1}
DahA	7.5×10^{-3}	9.0×10^{-3}	3.2×10^{-1}	1.3×10^{-1}	2.6×10^{-2}	3.9×10^{-2}
BghiP	4.0×10^{-2}	7.7×10^{-2}	3.2×10^{0}	1.4×10^{0}	2.8×10^{-1}	4.7×10^{-1}
DacP	1.1×10^{-2}	2.2×10^{-2}	9.5×10^{-1}	2.2×10^{-1}	4.4×10^{-2}	1.1×10^{-1}
DalP	7.8×10^{-3}	1.3×10^{-2}	4.2×10^{-1}	1.9×10^{-1}	3.9×10^{-2}	5.6×10^{-2}
DaeF	7.6×10^{-3}	8.9×10^{-3}	3.6×10^{-1}	1.6×10^{-1}	2.8×10^{-2}	5.4×10^{-2}
Coron	1.9×10^{-2}	4.3×10^{-2}	2.2×10^{0}	8.3×10^{-1}	1.5×10^{-1}	3.6×10^{-1}
DaeP	9.3×10^{-3}	9.8×10^{-3}	3.8×10^{-1}	1.9×10^{-1}	3.8×10^{-2}	1.1×10^{-1}
DaiP	6.0×10^{-3}	1.0×10^{-2}	3.7×10^{-1}	1.2×10^{-1}	2.9×10^{-2}	7.3×10^{-2}
DahP	8.2×10^{-4}	2.0×10^{-3}	4.9×10^{-2}	1.5×10^{-2}	4.1×10^{-3}	8.8×10^{-3}

For each fuel, combustion experiments were done in triplicate. Adapted from Shen et al. (2012a) with permission of American Chemical Society

Table 5.8 Standard derivations in triplicate measurements on emission factors (mg/kg) of parent PAHs for residential wood combustions of different fuel types

	White poplar	Elm	Locust	Maple	Fir	Larch	Water Chinese fir
NAP	2.8×10^{0}	2.2×10^{0}	1.3×10^{0}	1.0×10^{0}	3.4×10^{0}	1.4×10^{0}	7.6×10^{-1}
ACY	7.2×10^{-1}	7.7×10^{-1}	2.4×10^{-1}	2.8×10^{-1}	8.4×10^{-1}	3.8×10^{-1}	2.7×10^{-1}
ACE	3.6×10^{-2}	4.5×10^{-2}	3.2×10^{-2}	4.2×10^{-2}	3.4×10^{-2}	1.9×10^{-2}	1.4×10^{-2}
FLO	1.6×10^{-1}	2.2×10^{-1}	4.7×10^{-2}	1.3×10^{-1}	1.6×10^{-1}	1.1×10^{-1}	8.8×10^{-2}
PHE	5.3×10^{-1}	1.1×10^{0}	1.3×10^{-1}	6.6×10^{-1}	5.6×10^{-1}	3.4×10^{-1}	5.9×10^{-1}
ANT	9.5×10^{-2}	1.6×10^{-1}	2.8×10^{-2}	7.3×10^{-2}	1.1×10^{-1}	4.2×10^{-2}	7.9×10^{-2}
FLU	4.4×10^{-1}	5.2×10^{-1}	1.9×10^{-2}	2.8×10^{-1}	3.6×10^{-1}	7.1×10^{-2}	1.9×10^{-1}
PYR	3.6×10^{-1}	3.7×10^{-1}	3.2×10^{-2}	2.3×10^{-1}	3.3×10^{-1}	5.8×10^{-2}	2.0×10^{-1}
RET	1.4×10^{-2}	1.3×10^{-2}	7.1×10^{-2}	4.3×10^{-2}	2.6×10^{-2}	2.1×10^{-1}	8.8×10^{-2}
BcP	4.1×10^{-2}	9.9×10^{-3}	3.2×10^{-3}	1.8×10^{-2}	1.8×10^{-3}	1.7×10^{-3}	7.4×10^{-3}
CcdP	2.8×10^{-1}	8.0×10^{-3}	3.2×10^{-2}	5.3×10^{-2}	7.3×10^{-3}	4.9×10^{-3}	3.2×10^{-3}
BaA	1.7×10^{-1}	4.5×10^{-2}	2.1×10^{-2}	1.1×10^{-1}	1.6×10^{-2}	1.7×10^{-3}	2.9×10^{-2}
CHR	1.3×10^{-1}	4.3×10^{-2}	2.2×10^{-2}	9.1×10^{-2}	5.6×10^{-3}	3.4×10^{-3}	2.7×10^{-2}
BbF	6.1×10^{-2}	1.3×10^{-2}	1.2×10^{-2}	6.8×10^{-2}	6.8×10^{-3}	4.5×10^{-3}	2.3×10^{-2}
BkF	7.5×10^{-2}	1.6×10^{-2}	7.8×10^{-3}	6.8×10^{-2}	2.5×10^{-3}	8.7×10^{-3}	1.5×10^{-2}
BeP	3.7×10^{-2}	1.0×10^{-2}	6.4×10^{-3}	4.2×10^{-2}	1.2×10^{-3}	2.3×10^{-3}	1.5×10^{-2}
BaP	9.6×10^{-2}	8.3×10^{-3}	1.1×10^{-2}	6.6×10^{-2}	5.2×10^{-3}	3.7×10^{-3}	2.6×10^{-2}
PER	1.7×10^{-2}	2.5×10^{-3}	1.7×10^{-3}	1.7×10^{-2}	1.8×10^{-3}	2.1×10^{-3}	3.0×10^{-3}
IcdP	6.4×10^{-2}	2.8×10^{-3}	8.6×10^{-3}	4.4×10^{-2}	4.3×10^{-3}	5.4×10^{-3}	1.7×10^{-2}
DahA	1.0×10^{-2}	1.5×10^{-3}	1.7×10^{-3}	6.3×10^{-3}	1.3×10^{-3}	6.8×10^{-4}	1.5×10^{-3}
BghiP	4.0×10^{-2}	5.3×10^{-3}	6.5×10^{-3}	3.6×10^{-2}	1.4×10^{-3}	5.0×10^{-3}	1.4×10^{-2}
DacP	2.3×10^{-2}	1.8×10^{-3}	2.5×10^{-3}	8.6×10^{-3}	3.4×10^{-3}	3.7×10^{-3}	3.3×10^{-3}
DalP	1.1×10^{-2}	7.8×10^{-4}	2.2×10^{-3}	8.2×10^{-3}	4.9×10^{-3}	9.6×10^{-4}	3.7×10^{-3}
DaeF	8.1×10^{-3}	1.1×10^{-3}	1.6×10^{-3}	7.1×10^{-3}	2.8×10^{-3}	4.5×10^{-4}	2.0×10^{-3}
Coron	3.0×10^{-2}	4.3×10^{-3}	3.9×10^{-3}	1.2×10^{-2}	6.3×10^{-3}	3.5×10^{-3}	8.2×10^{-3}

(continued)

Table 5.8 (continued)

	White poplar	Elm	Locust	Maple	Fir	Larch	Water Chinese fir
DaeP	7.5×10^{-3}	5.6×10^{-4}	9.1×10^{-3}	4.7×10^{-3}	1.0×10^{-3}	2.2×10^{-3}	1.8×10^{-3}
DaiP	1.0×10^{-2}	3.2×10^{-4}	1.8×10^{-3}	1.1×10^{-2}	6.9×10^{-4}	2.0×10^{-3}	2.7×10^{-3}
DahP	2.8×10^{-3}	1.7×10^{-4}	4.8×10^{-4}	1.9×10^{-3}	1.3×10^{-4}	2.1×10^{-4}	1.4×10^{-4}
	Cypress	**Oak**	**Chinese pine**	**Willow**	**Paulownia tomentosa**	**Toon**	**White birch**
NAP	8.6×10^{-1}	1.3×10^{0}	1.6×10^{0}	1.3×10^{0}	2.0×10^{0}	1.9×10^{0}	5.3×10^{0}
ACY	1.1×10^{-1}	1.6×10^{-1}	3.0×10^{-1}	2.4×10^{-1}	1.7×10^{-1}	2.7×10^{-1}	1.1×10^{0}
ACE	1.1×10^{-2}	2.3×10^{-2}	2.1×10^{-2}	2.0×10^{-2}	2.2×10^{-2}	2.0×10^{-2}	8.4×10^{-2}
FLO	2.1×10^{-2}	7.9×10^{-2}	8.9×10^{-2}	8.5×10^{-2}	4.1×10^{-2}	5.9×10^{-2}	2.6×10^{-1}
PHE	1.1×10^{-2}	5.4×10^{-1}	4.1×10^{-1}	4.8×10^{-1}	2.0×10^{-1}	1.8×10^{-1}	7.9×10^{-1}
ANT	1.1×10^{-2}	7.8×10^{-2}	4.6×10^{-2}	4.3×10^{-2}	2.4×10^{-2}	2.1×10^{-2}	9.3×10^{-2}
FLU	5.6×10^{-2}	2.1×10^{-1}	2.0×10^{-1}	1.9×10^{-1}	6.1×10^{-2}	4.5×10^{-2}	6.7×10^{-2}
PYR	9.7×10^{-2}	1.6×10^{-1}	1.1×10^{-1}	1.4×10^{-1}	7.3×10^{-2}	8.3×10^{-2}	8.4×10^{-2}
RET	4.9×10^{-2}	1.0×10^{-1}	8.1×10^{-2}	1.7×10^{-2}	2.1×10^{-2}	2.2×10^{-2}	2.3×10^{-2}
BcP	6.0×10^{-3}	9.1×10^{-3}	1.5×10^{-2}	1.2×10^{-2}	1.3×10^{-2}	3.7×10^{-3}	3.2×10^{-3}
CcdP	3.0×10^{-2}	6.1×10^{-3}	3.5×10^{-3}	8.7×10^{-3}	6.6×10^{-2}	6.6×10^{-3}	1.1×10^{-1}
BaA	2.7×10^{-2}	3.0×10^{-2}	6.2×10^{-2}	2.9×10^{-2}	5.8×10^{-2}	8.1×10^{-3}	1.8×10^{-2}
CHR	2.4×10^{-2}	2.6×10^{-2}	7.5×10^{-2}	4.1×10^{-2}	5.0×10^{-2}	6.0×10^{-3}	9.5×10^{-3}
BbF	1.1×10^{-2}	1.4×10^{-2}	5.3×10^{-2}	1.0×10^{-2}	2.2×10^{-2}	2.5×10^{-3}	1.6×10^{-2}
BkF	8.2×10^{-3}	1.3×10^{-2}	5.3×10^{-2}	1.3×10^{-2}	2.1×10^{-2}	5.7×10^{-3}	2.8×10^{-2}
BeP	6.4×10^{-3}	7.8×10^{-3}	2.6×10^{-2}	9.6×10^{-3}	1.3×10^{-2}	4.0×10^{-3}	9.7×10^{-3}
BaP	1.0×10^{-2}	1.6×10^{-2}	2.6×10^{-2}	1.1×10^{-2}	3.0×10^{-2}	4.0×10^{-3}	1.6×10^{-2}
PER	1.9×10^{-3}	1.2×10^{-4}	3.6×10^{-3}	2.7×10^{-3}	4.4×10^{-3}	7.7×10^{-4}	1.7×10^{-3}
IcdP	6.8×10^{-3}	7.7×10^{-3}	1.6×10^{-2}	5.4×10^{-3}	2.3×10^{-2}	2.8×10^{-3}	5.3×10^{-3}
DahA	1.7×10^{-3}	1.4×10^{-3}	1.9×10^{-3}	2.1×10^{-3}	5.1×10^{-3}	1.3×10^{-3}	2.7×10^{-3}
BghiP	5.3×10^{-3}	4.9×10^{-3}	1.5×10^{-2}	5.3×10^{-3}	1.7×10^{-2}	2.6×10^{-3}	1.5×10^{-2}
DacP	1.2×10^{-3}	3.1×10^{-3}	6.3×10^{-4}	4.4×10^{-4}	5.3×10^{-3}	6.4×10^{-4}	4.1×10^{-3}

(continued)

Table 5.8 (continued)

	Cypress	Oak	Chinese pine	Willow	Paulownia tomentosa	Toon	White birch
DalP	1.2×10^{-3}	2.7×10^{-3}	1.8×10^{-3}	3.9×10^{-4}	6.1×10^{-4}	5.5×10^{-4}	3.9×10^{-3}
DaeF	1.1×10^{-3}	1.1×10^{-3}	2.2×10^{-3}	3.4×10^{-4}	2.2×10^{-3}	8.6×10^{-4}	3.2×10^{-3}
Coron	3.4×10^{-3}	2.0×10^{-3}	2.9×10^{-3}	2.5×10^{-3}	8.2×10^{-3}	2.8×10^{-3}	5.6×10^{-3}
DaeP	2.2×10^{-3}	5.9×10^{-4}	2.3×10^{-3}	7.5×10^{-4}	3.5×10^{-3}	1.0×10^{-3}	1.2×10^{-3}
DaiP	1.5×10^{-3}	1.6×10^{-3}	3.4×10^{-4}	3.5×10^{-4}	4.3×10^{-3}	5.4×10^{-4}	4.4×10^{-3}
DahP	7.1×10^{-4}	7.9×10^{-4}	4.2×10^{-4}	0.0×10^{0}	5.6×10^{-4}	5.2×10^{-4}	2.2×10^{-3}

	Ribbed birch	Paulownia elongata	Black poplar	China aspen	Chinaberry	Jujube tree	Persimmon tree
NAP	9.5×10^{-1}	1.5×10^{0}	5.4×10^{0}	1.6×10^{0}	4.5×10^{0}	4.3×10^{0}	1.9×10^{0}
ACY	1.9×10^{-1}	1.8×10^{0}	4.5×10^{-1}	4.2×10^{-1}	8.6×10^{-1}	1.2×10^{0}	8.2×10^{-1}
ACE	1.3×10^{-2}	1.8×10^{-1}	2.9×10^{-2}	2.2×10^{-2}	6.6×10^{-2}	4.9×10^{-2}	5.1×10^{-2}
FLO	5.7×10^{-2}	7.4×10^{-1}	1.5×10^{-1}	1.0×10^{-1}	1.7×10^{-1}	2.6×10^{-1}	2.0×10^{-1}
PHE	3.2×10^{-1}	6.4×10^{0}	3.5×10^{0}	7.5×10^{-1}	8.7×10^{-1}	1.0×10^{0}	8.6×10^{-1}
ANT	4.7×10^{-2}	1.2×10^{0}	7.1×10^{-1}	1.1×10^{-1}	2.3×10^{-1}	1.8×10^{-1}	1.4×10^{-1}
FLU	3.1×10^{-1}	2.3×10^{0}	9.4×10^{-1}	5.2×10^{-1}	4.3×10^{-1}	6.3×10^{-1}	3.9×10^{-1}
PYR	2.1×10^{-1}	2.0×10^{0}	8.2×10^{-1}	4.4×10^{-1}	4.1×10^{-1}	6.0×10^{-1}	3.7×10^{-1}
RET	6.9×10^{-2}	1.9×10^{-2}	3.3×10^{-2}	1.0×10^{-1}	8.9×10^{-2}	9.7×10^{-3}	4.0×10^{-2}
BcP	1.6×10^{-2}	1.7×10^{-1}	9.6×10^{-2}	3.9×10^{-2}	3.8×10^{-2}	5.3×10^{-2}	2.6×10^{-2}
CcdP	4.8×10^{-2}	8.2×10^{-1}	2.8×10^{-1}	1.7×10^{-1}	3.1×10^{-1}	3.7×10^{-1}	1.4×10^{-1}
BaA	9.3×10^{-2}	6.7×10^{-1}	4.4×10^{-1}	2.2×10^{-1}	1.6×10^{-1}	2.0×10^{-1}	1.0×10^{-1}
CHR	7.9×10^{-2}	5.8×10^{-1}	4.8×10^{-1}	1.8×10^{-1}	1.6×10^{-1}	1.3×10^{-1}	7.9×10^{-2}
BbF	5.7×10^{-2}	2.5×10^{-1}	7.6×10^{-2}	1.7×10^{-1}	6.0×10^{-2}	1.4×10^{-1}	4.0×10^{-2}
BkF	4.6×10^{-2}	1.9×10^{-1}	4.4×10^{-2}	1.3×10^{-1}	6.4×10^{-2}	1.5×10^{-1}	5.2×10^{-2}
BeP	3.6×10^{-2}	1.3×10^{-1}	6.6×10^{-2}	8.5×10^{-2}	3.5×10^{-2}	6.1×10^{-2}	3.0×10^{-2}
BaP	5.3×10^{-2}	2.9×10^{-1}	1.4×10^{-1}	1.4×10^{-1}	8.6×10^{-2}	1.6×10^{-1}	6.5×10^{-2}
PER	1.0×10^{-2}	3.7×10^{-2}	1.5×10^{-2}	2.3×10^{-2}	1.2×10^{-2}	2.7×10^{-2}	1.1×10^{-2}
IcdP	4.0×10^{-2}	1.9×10^{-1}	9.1×10^{-2}	1.0×10^{-1}	5.4×10^{-2}	1.1×10^{-1}	4.0×10^{-2}

(continued)

Table 5.8 (continued)

	Ribbed birch	Paulownia elongata	Black poplar	China aspen	Chinaberry	Jujube tree	Persimmon tree
DahA	4.4×10^{-3}	2.8×10^{-2}	3.6×10^{-2}	1.6×10^{-2}	2.3×10^{-3}	1.8×10^{-2}	4.9×10^{-3}
BghiP	3.3×10^{-2}	1.0×10^{-1}	3.6×10^{-2}	7.2×10^{-2}	2.9×10^{-2}	7.1×10^{-2}	2.7×10^{-2}
DacP	5.0×10^{-3}	7.5×10^{-2}	2.7×10^{-2}	2.3×10^{-2}	2.1×10^{-2}	4.9×10^{-2}	1.2×10^{-2}
DalP	5.1×10^{-3}	3.4×10^{-2}	2.8×10^{-2}	2.0×10^{-2}	1.0×10^{-2}	2.3×10^{-2}	1.2×10^{-2}
DaeF	4.4×10^{-3}	2.6×10^{-2}	2.1×10^{-2}	1.4×10^{-2}	7.1×10^{-3}	1.1×10^{-2}	1.8×10^{-3}
Coron	1.5×10^{-2}	8.2×10^{-2}	1.2×10^{-2}	3.0×10^{-2}	2.8×10^{-2}	6.5×10^{-2}	2.3×10^{-2}
DaeP	2.1×10^{-3}	3.1×10^{-2}	2.2×10^{-2}	2.5×10^{-2}	2.1×10^{-3}	1.2×10^{-2}	5.4×10^{-3}
DaiP	4.2×10^{-3}	1.9×10^{-2}	1.3×10^{-2}	1.7×10^{-2}	3.0×10^{-3}	1.9×10^{-2}	4.7×10^{-3}
DahP	1.2×10^{-3}	1.5×10^{-3}	1.3×10^{-2}	3.6×10^{-3}	2.2×10^{-4}	1.7×10^{-3}	1.8×10^{-3}

	Mulberry tree	Peach tree	Lespedeza	Buxus sinica	Holly	Bamboo
NAP	2.0×10^{0}	3.8×10^{0}	4.4×10^{1}	2.2×10^{1}	4.4×10^{0}	8.1×10^{0}
ACY	8.8×10^{-1}	8.9×10^{-1}	9.0×10^{0}	3.8×10^{0}	4.8×10^{-1}	5.0×10^{-1}
ACE	3.4×10^{-2}	5.9×10^{-2}	2.2×10^{-1}	1.0×10^{-1}	6.9×10^{-3}	2.4×10^{-2}
FLO	1.8×10^{-1}	2.3×10^{-1}	1.8×10^{0}	7.4×10^{-1}	9.3×10^{-2}	6.5×10^{-2}
PHE	5.4×10^{-1}	8.4×10^{-1}	7.8×10^{0}	2.6×10^{0}	7.2×10^{-1}	3.3×10^{-1}
ANT	1.1×10^{-1}	1.1×10^{-1}	1.7×10^{0}	4.9×10^{-1}	8.8×10^{-2}	1.1×10^{-1}
FLU	5.6×10^{-2}	1.6×10^{-1}	1.3×10^{1}	3.6×10^{0}	4.0×10^{-1}	1.5×10^{0}
PYR	3.0×10^{-2}	1.8×10^{-1}	1.1×10^{0}	3.7×10^{0}	2.3×10^{-1}	1.9×10^{0}
RET	1.9×10^{-2}	4.1×10^{-2}	1.2×10^{-3}	7.3×10^{-2}	1.6×10^{-2}	8.1×10^{-2}
BcP	2.8×10^{-3}	1.2×10^{-2}	8.3×10^{-1}	3.4×10^{-1}	2.1×10^{-2}	1.9×10^{-1}
CcdP	1.4×10^{-2}	5.5×10^{-2}	6.9×10^{0}	8.8×10^{-1}	1.0×10^{-1}	1.0×10^{0}
BaA	2.3×10^{-2}	5.7×10^{-2}	4.1×10^{0}	1.3×10^{0}	9.3×10^{-2}	7.1×10^{-1}
CHR	2.3×10^{-2}	4.4×10^{-2}	2.7×10^{0}	8.5×10^{-1}	8.5×10^{-2}	5.6×10^{-1}
BbF	1.5×10^{-2}	3.0×10^{-2}	2.6×10^{0}	1.1×10^{0}	6.5×10^{-2}	5.7×10^{-1}
BkF	9.9×10^{-3}	3.4×10^{-2}	2.7×10^{0}	1.0×10^{0}	6.8×10^{-3}	7.6×10^{-1}
BeP	5.2×10^{-3}	1.9×10^{-2}	1.4×10^{0}	5.0×10^{-1}	2.8×10^{-2}	3.9×10^{-1}1

(continued)

Table 5.8 (continued)

	Mulberry tree	Peach tree	Lespedeza	Buxus sinica	Holly	Bamboo
BaP	1.1×10^{-2}	4.7×10^{-2}	3.6×10^{0}	1.1×10^{0}	3.7×10^{-2}	8.3×10^{-1}
PER	3.2×10^{-3}	8.7×10^{-3}	5.9×10^{-1}	1.5×10^{-1}	8.6×10^{-3}	1.4×10^{-1}
IcdP	6.1×10^{-3}	3.1×10^{-2}	3.3×10^{0}	1.1×10^{0}	4.1×10^{-2}	7.6×10^{-1}
DahA	1.2×10^{-3}	4.4×10^{-3}	2.7×10^{-1}	6.0×10^{-2}	5.1×10^{-3}	5.7×10^{-2}
BghiP	2.1×10^{-3}	2.7×10^{-2}	2.6×10^{0}	8.8×10^{-1}	4.3×10^{-2}	6.7×10^{-1}
DacP	9.8×10^{-4}	4.2×10^{-3}	9.2×10^{-1}	1.4×10^{-1}	1.6×10^{-2}	1.8×10^{-1}
DalP	2.0×10^{-3}	3.8×10^{-3}	3.1×10^{-1}	9.5×10^{-2}	1.1×10^{-2}	8.7×10^{-2}
DaeF	3.6×10^{-4}	9.2×10^{-4}	2.8×10^{-1}	1.1×10^{-1}	4.6×10^{-3}	8.5×10^{-2}
Coron	1.8×10^{-3}	1.4×10^{-2}	1.7×10^{0}	5.6×10^{-1}	6.3×10^{-2}	5.9×10^{-1}
DaeP	1.7×10^{-3}	1.4×10^{-3}	4.1×10^{-1}	1.3×10^{-1}	4.6×10^{-3}	1.8×10^{-1}
DaiP	2.4×10^{-3}	2.7×10^{-3}	2.5×10^{-1}	6.9×10^{-2}	4.5×10^{-4}	1.2×10^{-1}
DahP	5.7×10^{-4}	1.4×10^{-3}	1.5×10^{-2}	7.7×10^{-3}	1.1×10^{-3}	1.3×10^{-2}

Adapted from Shen et al. (2012a) with permission of American Chemical Society

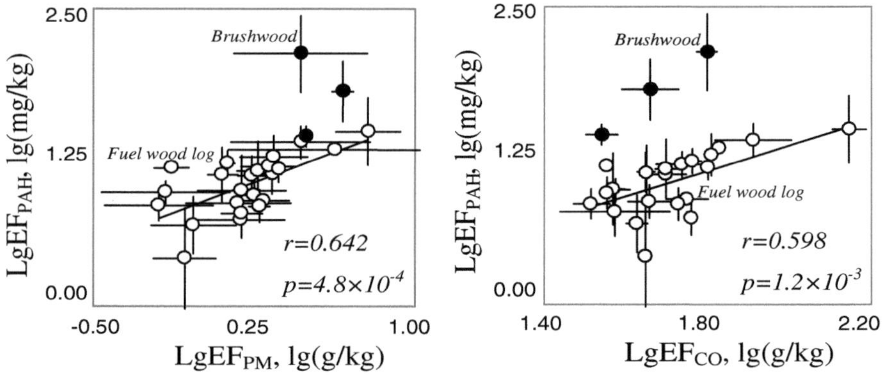

Fig. 5.17 Relationship between EF_{PAHs} and EF_{PM} (*left*) and between EF_{PAHs} and EF_{CO} (*right*) for wood burned in residential stove in the present study. Adapted from Shen et al. (2012a) with permission of American Chemical Society

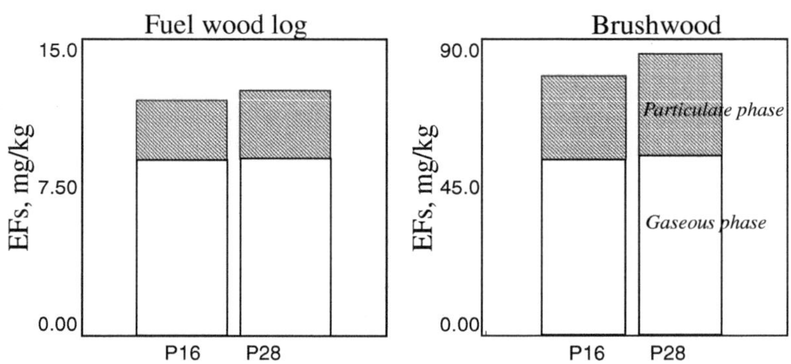

Fig. 5.18 Comparisons of gaseous and particulate phase total EFs between EF_{PAH16} and EF_{PAH28}

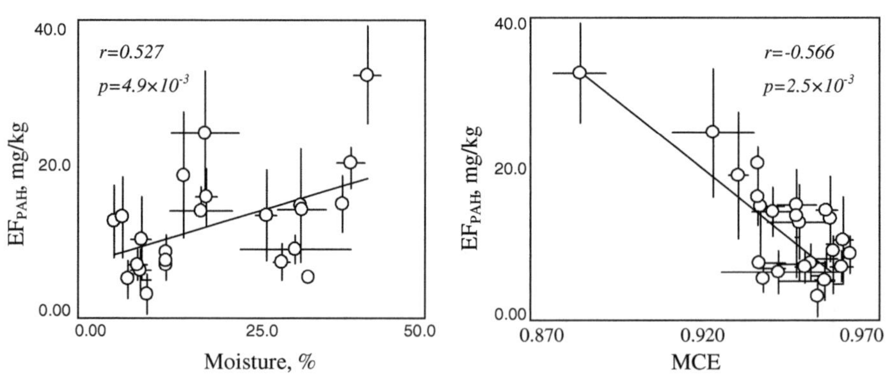

Fig. 5.19 Dependence of EF_{PAHs} on fuel moisture (*left*) and MCE (*right*) for wood. Modified from Shen et al. (2012a) with permission of American Chemical Society

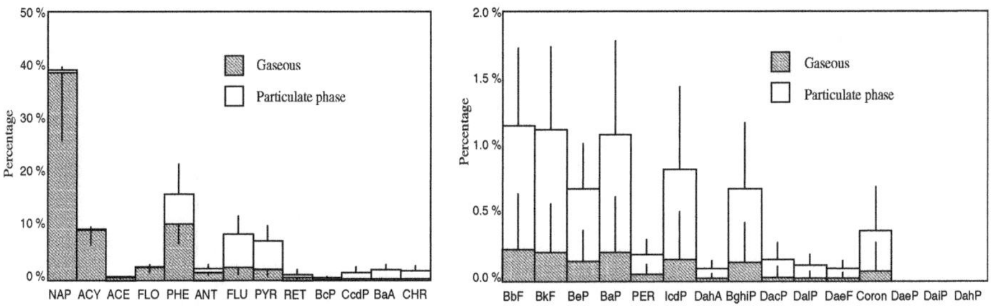

Fig. 5.20 Composition profile of PAH emitted from residential wood combustion. Modified from Shen et al. (2012a) with permission of American Chemical Society

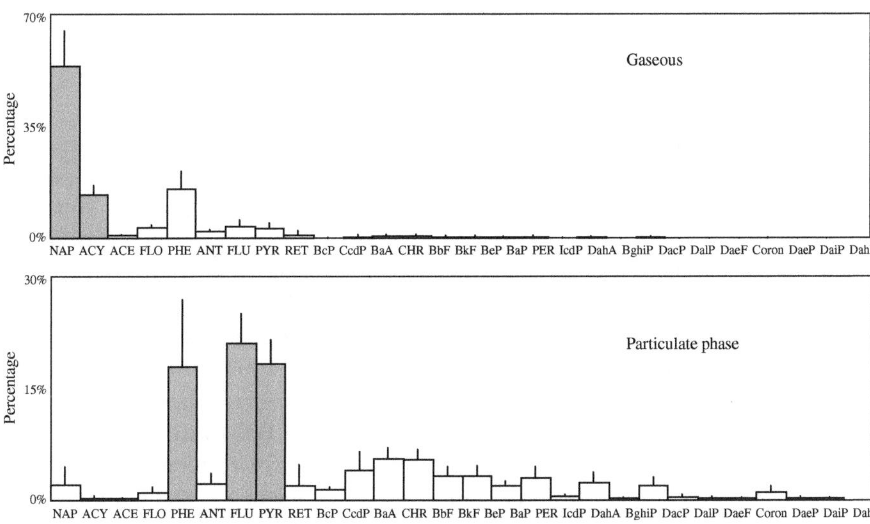

Fig. 5.21 Normalized mass percents of PAH individuals in gaseous and particulate phases. Modified from Shen et al. (2012a) with permission of American Chemical Society

(9.1 ± 3.4 %) and PYR (7.8 ± 2.9 %) were the dominated species (Fig. 5.20). In gaseous phase, NAP contributed about 54 % and ACY and PHE comprised to 14 and 15 % of total gaseous PAHs, respectively. In particulate phase, the mass percents of PHE, FLA and PYR were 18, 21 and 18 %, respectively (Fig. 5.21). No significant difference was found in the profile between fuel wood and brushwood (Fig. 5.22).

Calculated isomer ratios for ANT/(ANT + PHE), FLA/(FLA + PYR), BaA/(BaA + CHR), IcdP/(IcdP + BghiP), BbF/(BbF + BkF), and BaP/(BaP + BghiP) were 0.12 ± 0.02, 0.54 ± 0.02, 0.50 ± 0.04, 0.55 ± 0.03, 0.51 ± 0.03, and 0.62 ± 0.04, respectively. They were within the previously reported ranges of 0.10–0.30, 0.43–0.74, 0.39–0.56, 0.16–0.69, 0.35–0.51, and 0.38–0.78 for these six pairs in general (Kim Oanh et al. 1999, 2005; Hedberg et al. 2002).

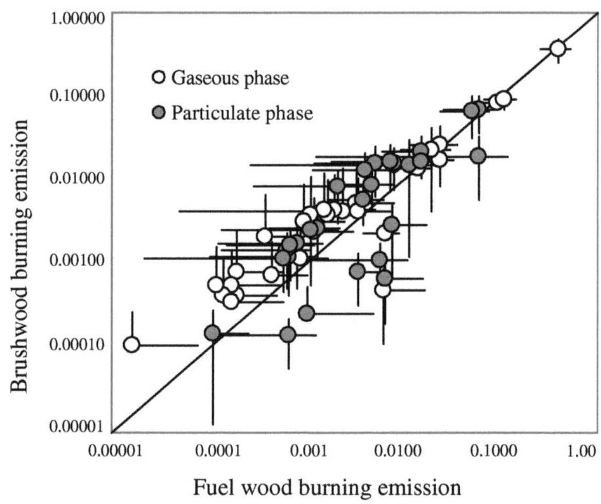

Fig. 5.22 Comparison of mass percent of PAH individual to the total in emissions between brushwood and fuel wood combustion

In addition to the 6 commonly used isomer ratios, BeP/(BeP + BaP) was also calculated in this study. Although BeP is not among the 16 U.S. EPA priority PAHs, it is sometimes reported in field measurements together with the 16 PAHs. The BeP/(BaP + BeP) ration has been used as an indicator of the degree of photo-degradation in ambient air because BaP degrades faster than BeP in the atmosphere (Wang et al. 2007; Luo et al. 2006). For example, in ambient air during the summer and winter periods, the ratio of BeP/(BaP + BeP) was reported to be at 0.42–0.75 and 0.28–0.44, respectively (Wang et al. 2007; Ding et al. 2011). Enhanced degradation under high temperature in summer resulted in more faster degradation of BaP and subsequent high BeP/(BaP + BeP) ratio in comparison with that in cold winter. Measured BeP/(BaP + BeP) for residential wood combustion in our study varied from 0.29 to 0.54, with a mean and standard deviation of 0.41 ± 0.06. The result was similar to that of 0.51 ± 0.12, ranging from 0.33 to 0.71 reported in the literature (Jenkins et al. 1996a, b; Kim Oanh et al. 2005; Rogge et al. 1998; Schauer et al. 2001).

5.3.3 Size Distribution of Particulate Phase PAHs

Figure 5.23 shows the normalized mass percent of particulate phase PAHs in PM fractions with diameter of 2.1–10.0, 1.1–2.1, 0.4–1.1, and <0.4 μm. Because of much high laboratory analysis work, we combined the nine fractions into these four fractions. It appears that the distributions were similar in emissions between fuel wood and bamboo, which were different from that of brushwood. PAHs associated with $PM_{0.4-1.1}$ contributed 39.4 ± 15.4 and 39.4 ± 4.7 % of the total for fuel wood and bamboo, respectively. The second largest size fractions were those associated with $PM_{0.4}$ accounting for 26.7 ± 7.7 and 21.8 ± 21.4 %,

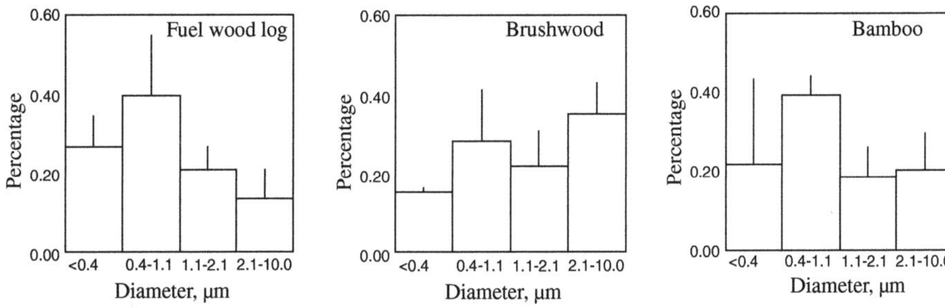

Fig. 5.23 Normalized mass percents of PAHs in different PM size fractions to the total mass of particulate phase PAHs in emissions from the fuel wood, brushwood and bamboo burnings. Reprinted from Biomass and Bioenergy 55, Shen et al., Emission and size distribution of particle-bound polycyclic aromatic hydrocarbons from residential wood combustion in rural China, 141–147, with permission from Elsevier

respectively. For brushwood, however, PAHs associated with $PM_{0.4-1.1}$ was only 28.0 ± 13.0 %, much less than those for fuel wood and bamboo. Meantime, percent of PAHs associated with course $PM_{2.1-10}$ was 35.0 ± 7.8 % for brushwood, significantly higher than that from fuel wood (13.3 ± 7.5 %) and bamboo (20.3 ± 9.2 %). Overall, there were over 86.7, 65.0 and 79.9 % of total particulate phase PAHs bound in fine $PM_{2.5}$. Such a difference in PAH size fraction was consistent with that in the size distribution of PM, which also showed that PM emitted from fuel wood combustion was finer than those from burning of brushwood. Calculated MMAD of particle-bound PAHs for fuel wood, brushwood, and bamboo were 0.75 (0.38–1.6 as range), 1.4 (0.65–1.9) and 0.92 (0.49–1.4) µm, respectively.

Differences in fuel properties and combustion conditions can result in variation in size distributions of PM, consequently the distribution of particle-bound PAHs (Hays et al. 2003; Purvis et al. 2000; Venkataraman et al. 2002). It is believed that fine particle-bound PAHs could be influenced by the combustion temperature, as high temperature often results in strong air convection and sufficient oxygen for combustion. In this study, there was no significant difference in the recorded temperature among the combustions of various fuels. Hence, the possible impact of combustion temperature was not further discussed.

As shown in Fig. 5.24, the mass percents of PAHs associated with $PM_{2.1}$ were positively correlated with fuel density and moisture, and negatively correlated with MCE ($p < 0.05$). The impacts of other factors tested in this study were not insignificant ($p > 0.05$). This can be partly explained by the change in the size distributions of PM, to which the PAHs associated through absorption and/or adsorption. It has been previously found that the combustions of wood fuels with relative higher moisture under lower MCE produced higher mass percents of finer particles. As a result, more PAHs in fine PM could be expected under high moisture and low MCE.

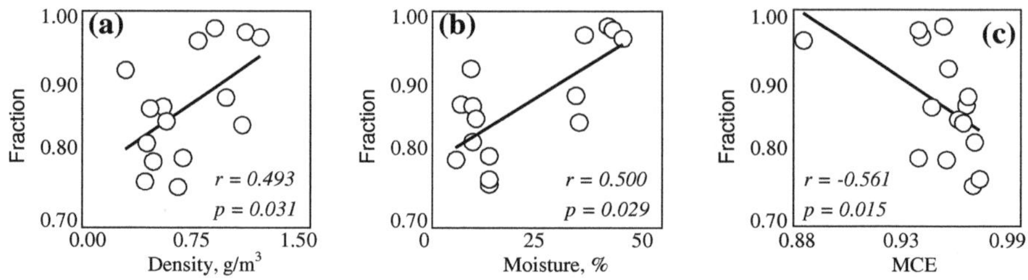

Fig. 5.24 Relationship between mass fractions of $PM_{2.1}$-bound PAHs and fuel density (**a**), moisture (**b**), and combustion efficiency (**c**) from residential wood combustion. Data shown are results from fuel wood combustion. Reprinted from Biomass and Bioenergy 55, Shen et al., Emission and size distribution of particle-bound polycyclic aromatic hydrocarbons from residential wood combustion in rural China, 141–147, with permission from Elsevier

Size distributions of individual PAHs usually follow a trend that higher molecular weight (MW) compounds tend to be present in finer particles, while lower MW ones prefer to partition to coarser PM. The same tendency was also revealed in this study. For example, for fuel wood combustion, there were 24.2 ± 8.9 and 26.7 ± 0.8 % of total particle-bound NAP and PHE in $PM_{0.4}$, while $PM_{0.4}$-bound BaP and IcdP made up 32.2 ± 2.8 and 33.0 ± 2.9 % of the total. MMAD values for individual PAHs varied from 0.47 to 1.0 μm, decreasing generally with the increase of PAH MW. As such, although the composition profiles of PAHs in PM size fractions were similar in general (Fig. 5.25), dominated by PHE, FLA and PYR, slight difference could be expected. Higher MW PAH preferentially segregated to finer PM, leading to relatively higher normalized mass percents of these HMW PAHs in small particle. The difference could be more clearly shown in the Fig. 5.26 using log-transformed scale. PAHs with MW > 228 made up to 18, 25, 33, and 32 % of the total PAHs in $PM_{2.1-10}$, $PM_{1.1-2.1}$, $PM_{0.4-1.1}$, and $PM_{0.4}$, respectively.

Since high MW PAHs tended to occur in fine PM in comparison with low MW ones, and high MW PAHs often had relatively high TEF values, it is expected that the size distribution of BaP*eq* should be different from that of the total PAHs. Because the TEF values for the other 12 PAHs identified in this study were not available, the comparison includes only P16. Figure 5.27 compares the normalized size distributions of P16 and calculated BaP*eq* from fuel wood combustion as an example. The toxicity equivalency factors used were a group of data suggested by Nisbet and LaGoy (1992) which was believed to be a better set (Petry et al. 1996). For P16, total mass concentrations in $PM_{2.1-10}$, $PM_{1.1-2.1}$, $PM_{0.4-1.1}$, and $PM_{0.4}$ made up 13.6 ± 8.1, 21.0 ± 7.8, 39.1 ± 17.8, and 26.4 ± 9.9 % of the total particulate phase PAHs, respectively, while the mass percentages of BaP*eq* in these fractions were 8.4 ± 5.5, 17.7 ± 6.6, 42.9 ± 17.8, and 31.0 ± 9.9 %, respectively. The differences between mass percentages of P16 and those of BaP*eq* in the same size fraction were statistically significant ($p < 0.05$).

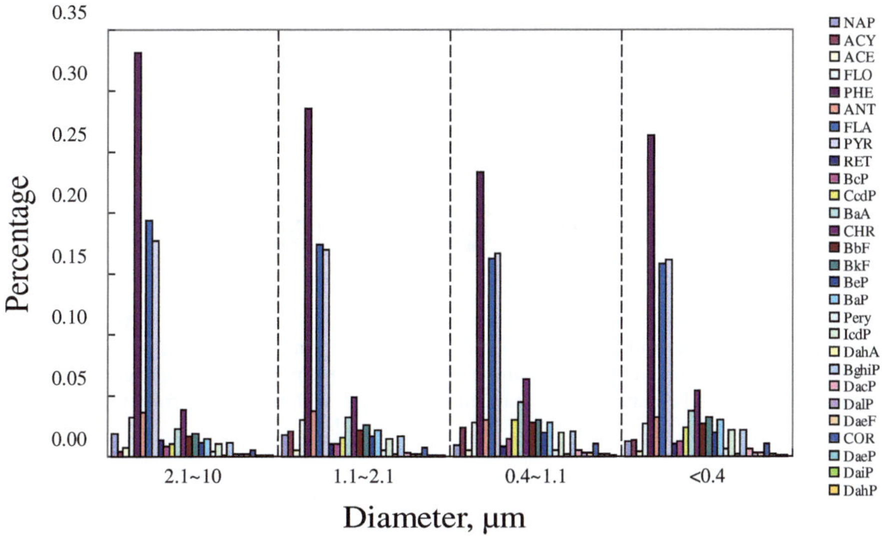

Fig. 5.25 Composition profiles of PAHs in different PM size fractions in emission from residential wood combustion

Fig. 5.26 Comparison of mass percent of PAH individual between fine PM and coarse PM in emissions from residential wood combustion

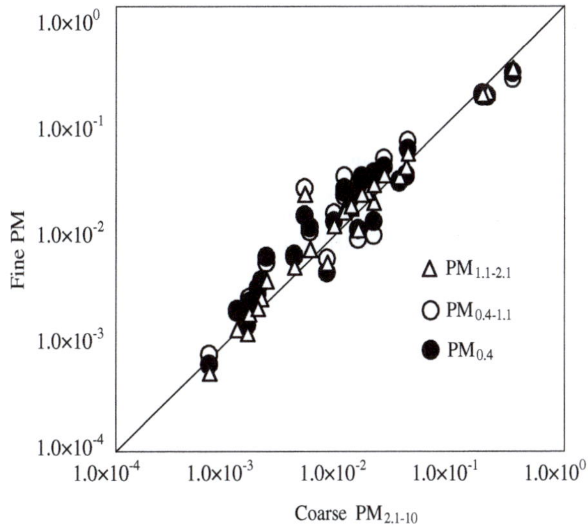

5.3.4 Gas-Particle Partitioning

The calculated partitioning coefficient was plotted against P_L^0 and K_{OA} in Fig. 5.28. Significantly positive correlation between *Kp* and K_{OA} ($r = 0.919$) and a shallow slope of -0.39 ± 0.06 indicated that in emissions from residential wood

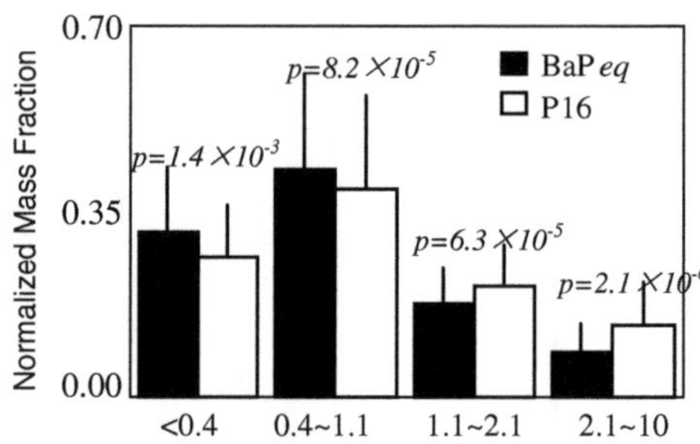

Fig. 5.27 Size distributions of total concentration of 16 U.S. EPA priority PAHs (P16) and calculated BaP*eq* from residential fuel wood combustion. Reprinted from Biomass and Bioenergy 55, Shen et al., Emission and size distribution of particle-bound polycyclic aromatic hydrocarbons from residential wood combustion in rural China, 141–147, with permission from Elsevier

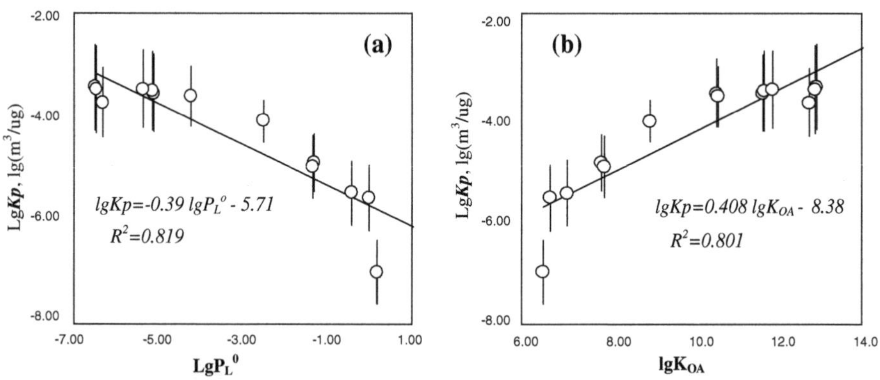

Fig. 5.28 Dependence of K_p on $\log P_L^0$ (**a**) and K_{OA} (**b**) for PAHs in emissions from residential wood combustion. Modified from Shen et al. (2012a, b) with permission of American Chemical Society

combustion, absorption rather than adsorption dominated the gas-particle partitioning of freshly emitted PAHs, which was the same to that in emissions from the coal and crop residue burning.

5.4 Fuel Comparison

5.4.1 Emission Factor

PAH emissions from coal combustion highly depend on the coal property and form. Generally, bituminous produced higher emissions than anthracite, and raw chunk had much higher emissions than briquette coals. Average EF_{PAHs} for crop

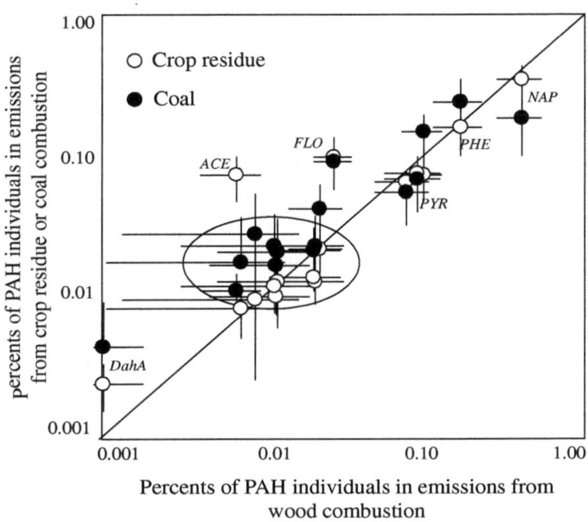

Fig. 5.29 Comparison of normalized mass percents of PAH individuals in emissions from wood, crop residue and coal combustions

residue was 62.7 mg/kg, comparable to 79.7 mg/kg, both of which were significantly higher than that of 12.1 mg/kg for fuel wood log. Particulate phase PAHs contributed only 0.41, 0.60 and 0.16 % of total PAHs in emissions from crop residue, brushwood and fuel wood combustions.

5.4.2 Composition Profile

The composition profiles for different solid fuel types were similar in general with the domination of NAP, PHE, FLA and PYR. Slight difference could be identified, as shown in Fig. 5.29. The profiles for crop residue and wood were very similar but crop residue burning had relatively higher emissions of ACE and FLO. In coal combustion, the mass percents of high molecular weight PAHs were higher than those in emissions from biomass burning. The total mass percents of PAHs with MW larger than 228 (from BaA to BghiP) were 15.6, 9.5 and 8.2 % on average for coal, crop residue and wood, respectively. Since most high molecular weight PAHs had more toxic, the burning of coal would cause more adverse health outcome.

The isomer ratios for three fuels are compared in Table 5.9. The means were not significant different among these three fuel types. In fact, these three fuels are often classified into the same group in PAH source apportionment. But, it should be noted that relatively large variations were found within fuels, particularly for coal. Thus, the use of a specific one value in source apportionment may cause considerable bias.

Table 5.9 Comparison of isomer ratios between crop residue, wood and coal

	ANT/(ANT + PHE)	FLA/(FLA + PYR)	BaA/(BaA + CHR)	IcdP/(IcdP + BghiP)	BbF/(BbF + BkF)	BaP/(BaP + BghiP)
Wood	0.12 ± 0.02	0.54 ± 0.02	0.50 ± 0.04	0.55 ± 0.03	0.51 ± 0.03	0.62 ± 0.04
Crop residue	0.12 ± 0.01	0.53 ± 0.03	0.48 ± 0.02	0.54 ± 0.02	0.55 ± 0.03	0.60 ± 0.05
Coal	0.17 ± 0.07	0.56 ± 0.02	0.48 ± 0.13	0.58 ± 0.16	0.54 ± 0.06	0.58 ± 0.16

Data shown are means and standard deviations from all combustion tests

5.4.3 Evaluation of Retene as a Marker

In source apportionment of PAHs, some compounds are often used as a maker for a specific source by assuming that the compound is only or mainly emitted from the given source while the contributions of other sources are minor. For example, Retene (1-methyl-7-isopropylphenanthrene) has been proposed as a marker for softwood (McDonald et al. 2000; Ramdahl 1983; Schauer et al. 2001; Simoneit 2002). Its use can be frequently found in many studies (Li et al. 2009; McDonald et al. 2000; Schauer et al. 1996, 2001). Perylene is another compounds sometimes used as a maker since it can not only form from the incomplete combustion burning but also through biosynthesis process, and thus by comparing the difference in environmental behaviors like concentration, vertical distribution and temporal- spatial distributions, between perylene and other PAHs it is able to analyze the potential source of biosynthesis (Grice et al. 2009; Suzuki et al. 2010; Silliman et al. 1998; Wilcke et al. 2002). The use of perylene can be often found in studies on PAHs in sediment (Bachtel et al. 2007; Bouloubassi et al. 2001; Silliman et al. 1998; Venkatesan, 1988). Another PAH of interest is Coronone. Coronene was first found in vehicle emission, and had high abundance in the area where vehicle emission was the dominated PAH source, like in a tunnel. It was considered as a potential marker for vehicle emission (Gordon and Bryan 1973; Gordon 1976; Greenberg et al. 1981; Harkov et al. 1984; Sawicki et al. 1962). However, with more and more report of coronene on non-vehicle sources, and variable levels in the vehicle source (Freeman and Cattell 1990; Hasegawa and Hibino 2011; Mastral et al. 2003), it seems that the use of coronene as a marker has been less mentioned nowadays.

In our present study, we analyzed emission factor of retene (EF_{RET}) in the source samples from residential combustions of wood, crop residue and coal, and thus it provided an opportunity to compare the RET emission among there sources so as to evaluate the use of it as a marker for softwood. The results are provided in Table 5.9, together with collected data from the literature.

For the crop residues, EF_{RET} ranged from 0.083 ± 0.007 (*pea straw*) to 0.37 ± 0.14 (*rape straw*) mg/kg, with a mean and standard derivation of 0.14 ± 0.10 mg/kg. For wood, EF_{RET} varied dramatically among different tree species, ranging from 0.016 ± 0.006 (*China Aspen*) to 0.34 ± 0.08 mg/kg

(*Chinese Pine*). Of the wood types burned, five were softwood. EF_{RET} for softwood (0.20 ± 0.11 mg/kg) were significantly higher than for hardwood (0.075 ± 0.043 mg/kg) ($p < 0.05$), and among the five softwood species, the highest EF_{RET}s were measured for pine (0.34 ± 0.08 mg/kg) and larch (0.29 ± 0.22 mg/kg), followed by cypress (0.14 ± 0.05 mg/kg) and water Chinese fir species (0.14 ± 0.09 mg/kg). EF_{RET} for the different coals measured varied much more dramatically, from 2.2 ± 1.5 (anthracite briquette, Beijing) to 187 ± 113 (bituminous chunk, Yulin) mg/kg since coal properties and form varied obviously among five coals. It was previously reported in the literature that EF_{RET} could be as high as 101 ± 2 and 282 ± 45 mg/kg, respectively, for lignite and sub-bituminous coals (Oros and Simoneit, 2000) (Table 5.10).

In addition to significant differences among fuel types (crop residues, wood, or coal), EF_{RET} varied significantly within fuel types, which can be attributed to the differences in fuel properties and combustion conditions. For crop residues, significantly negative correlations between EF_{RET} and fuel moisture ($r = -0{,}596$, $p = 0.002$) and between EF_{RET} and MCE ($r = -0{,}415$, $p = 0.027$) were found. For coal, EF_{RET} was found to be positively correlated with VM content ($r = 0.900$, $p = 0.019$). But for wood materials, when softwood and hardwood were burned, there was no significant correlation between EF_{RET} and moisture, VM content, and MCE ($p > 0.05$). Although EF_{RET} were significantly different among the hardwood and softwood tested, there was no significant difference between hardwood and softwood in fuel properties, including density, moisture, elemental and proximate analysis, and combustion conditions.

Phenanthrene was selected as a representative compound to retene since their molecular structures are similar, and it is one most abundant and widely reported PAH. Figure 5.30 illustrated the relationship between EF_{RET} and EF_{PHE} for different fuel types. Significantly positively correlations were found in emissions from the burning of crop residues and coal ($p < 0.05$), but there were no significant correlations for residential hardwood or softwood ($p > 0.05$). Although EF_{RET} for both crop residues and coals were correlated with EF_{PHE} and EF_{PM}, the difference between crop residues and coal was clear. Although EF_{RET} for both crop residues and coals were correlated with EF_{PHE}, the difference between crop residues and coal was clear. At the same levels of EF_{RET}, EF_{PHE} for crop residue was much higher than those for coal. The ratios of EF_{RET}/EF_{PHE} for wood were comparable to that for crop residue, and they were lower than that for coal.

Retene was considered as a biomarker for the combustion of conifer fuel containing abundant diterpenoid resin acids, the thermal degradation of which usually leads to the formation of retene. In the present study, we compared the retene emissions among different solid fuels. Although relatively high EF_{RET} values were found for the softwood fuels of *Chinese pine* (0.29 ± 0.22 mg/kg) and *larch* (0.34 ± 0.08 mg/kg), EF_{RET} for the other three softwood fuels (0.13 ± 0.09, 0.14 ± 0.05, and 0.07 ± 0.03 mg/kg for *redwood*, *cypress*, and *fir*, respectively) were not significantly higher than those of many hardwood and crop residues

Table 5.10 Emission Factors of RET and ratios of RET/PHE for various solid fuels measured in this study and reported in the literature

Fuel type	Combustion facility	EF, mg/kg	RET/PHE	References
Softwood				
Pine wood	FP	0.70		Simoneit et al. (1993)
Pine wood	FP	9.90	0.60	Schauer et al. (2001)
Pine wood	FP	0.68	1.45	Rogge et al. (1998)
Pine	FP	0.49		Gullett et al. (2003)
Pine (Ponderosa pine, pinion pine)	FP	1.79	0.17	McDonald et al. (2000)
Penderosa pine	FP	18.9	66.04	Fine et al. (2004a, b)
Pinyon pine	FP	43.9	40.00	Fine et al. (2004a, b)
Esatern white pine	FP	45.9	34.94	Fine et al. (2001)
Loblolly pine	FP	16.8	11.33	Fine et al. (2002)
Slash pine	FP	9.01	48.27	Fine et al. (2002)
Eastern hemlock	FP	6.31	22.82	Fine et al. (2001)
White spruce	FP	6.52	22.81	Fine et al. (2004a, b)
Balsam fir	FP	3.79	10.16	Fine et al. (2001)
Douglas fir	FP	+		Fine et al. (2004a, b)
Apache pine		25.00	12.06	Oros et al. (2001a, b)
Eastern white pine		25.339	6.04	Oros et al. (2001a, b)
Lodgepole pine		4.165	2.82	Oros et al. (2001a, b)
Montezuma pine		25.043	3.26	Oros et al. (2001a, b)
Pacific pine		7.797	10.49	Oros et al. (2001a, b)
Western white pine		34.717	17.35	Oros et al. (2001a, b)
California redwood		0.274	0.43	Oros et al. (2001a, b)
Douglas fir		6.51	4.40	Oros et al. (2001a, b)
Moutain hemlock		0.573	0.07	Oros et al. (2001a, b)
Noble fir		0.963	2.43	Oros et al. (2001a, b)
Pacific silver fir		3.842	0.51	Oros et al. (2001a, b)
Sitka spruce		2.634	3.68	Oros et al. (2001a, b)
Pinus pinaster	WS	4.18	8.17	Goncalves et al. (2010)
Loblolly pine	WS	3.39	6.64	Fine et al. (2004a, b)
Douglas fir	WS	1.70	2.06	Fine et al. (2004a, b)
Pine	WS	0.84[a]	6.46	Bari et al. (2009)
Spruce	WS	1113[a]	1.64	Ramdahl (1983)
Spruce	WS	16[a]	0.19	Ramdahl (1983)
Spruce	boiler	1420[a]	0.29	Ramdahl (1983)
Spruce	boiler	360[a]	0.11	Ramdahl (1983)
Pine wood	furance	82[a]	1.46	Kozinski and Saade (1998)
Pine	WS	15.4[b]	148.08	Goncalves et al. (2011)
Pine	WS	4.12[b]	20.70	Goncalves et al. (2011)
Pine	FP	1.54[b]	39.59	Goncalves et al. (2011)
Pine	FP	6.43[b]	17.33	Goncalves et al. (2011)

(continued)

Table 5.10 (continued)

Fuel type	Combustion facility	EF, mg/kg	RET/PHE	References
Douglas fir	WS	3.52	4.34	Fine et al. (2004a, b)
Fir		0.07	0.04	This study
Larch		0.29	0.27	This study
Water Chinese fir		0.13	0.07	This study
Cypress		0.14	0.10	This study
Chinese pine		0.34	0.28	This study
Hardwood				
Oak	FP	0.10		Simoneit et al. (1993)
Oak	FP	2.33	0.25	Schauer et al. (2001)
Eucalyptus	FP	0.15	0.02	Schauer et al. (2001)
Oak	FP	0.11	0.37	Rogge et al. (1998)
Mixed oak birch aspen	FP	0.50	0.03	McDonald et al. (2000)
Oak	FP	ND		Gullett et al. (2003)
White oak_hw	FP	+		Fine et al. (2004a, b)
Sugar maple	FP	+		Fine et al. (2004a, b)
Black oak	FP	+		Fine et al. (2004a, b)
American beech	FP	+		Fine et al. (2004a, b)
Black cherry	FP	+		Fine et al. (2004a, b)
Quaking aspen	FP	+		Fine et al. (2004a, b)
Red maple	FP	+		Fine et al. (2001)
Nirthern red oak	FP	+		Fine et al. (2001)
Paper birch	FP	+		Fine et al. (2001)
Yellow polar	FP	+		Fine et al. (2002)
White ash	FP	+		Fine et al. (2002)
Sweetgum	FP	+		Fine et al. (2002)
Mockernut hickory	FP	+		Fine et al. (2002)
Oak	WS	ND		Gullett et al. (2003)
Eucalyptus	WS	0.09	3.84	Goncalves et al. (2010)
Quercus suber_cork oak	WS	0.17	1.19	Goncalves et al. (2010)
Acacia lognifolia_Acacia	WS	0.04	22.37	Goncalves et al. (2010)
Birch	WS	<0.1		Hedberg et al. (2002)
Mixed oak birch aspen	WS	0.02	0.00	McDonald et al. (2000)
Red maple	WS	0.00	0.01	Fine et al. (2004a, b)
White oak	WS	0.03	0.09	Fine et al. (2004a, b)
White oak	WS	0.01	0.04	Fine et al. (2004a, b)
Sugar maple	WS	0.01	0.04	Fine et al. (2004a, b)
Beech	WS	0.17[a]	4.25	Bari et al. (2009)
Golden wattle	WS	0.16[b]	0.95	Goncalves et al. (2011)
Golden wattle	WS	0.44[b]	1.92	Goncalves et al. (2011)
Golden wattle	FP	0.28[b]	4.07	Goncalves et al. (2011)
Golden wattle	FP	0.049[b]	0.28	Goncalves et al. (2011)
Oak	WS	0.15[b]	1.31	Goncalves et al. (2011)

(continued)

Table 5.10 (continued)

Fuel type	Combustion facility	EF, mg/kg	RET/PHE	References
Oak	WS	0.36[b]	0.90	Goncalves et al. (2011)
Oak	FP	0.023[b]	0.04	Goncalves et al. (2011)
Oak	FP	0.035[b]	0.31	Goncalves et al. (2011)
Eucalypt	WS	0.10[b]	1.45	Goncalves et al. (2011)
Eucalypt	WS	0.12[b]	0.41	Goncalves et al. (2011)
Eucalypt	FP	0.084[b]	0.95	Goncalves et al. (2011)
Chinese white poplar		0.07	0.04	This study
Elm		0.04	0.02	This study
Yellow locust		0.07	0.09	This study
Maple		0.06	0.04	This study
Bamboo		0.08	0.02	This study
OAK		0.13	0.12	This study
Willow		0.09	0.16	This study
Paulownia tomentosa		0.16	0.17	This study
Toon		0.05	0.06	This study
White birch		0.08	0.05	This study
Lespedeza		0.04	0.00	This study
Holly		0.09	0.01	This study
Buxus sinica		0.04	0.01	This study
Ribbed birch		0.02	0.01	This study
Paulownia elongata		0.05	0.02	This study
Black poplar		0.15	0.04	This study
China aspen		0.02	0.01	This study
Chinaberry		0.15	0.06	This study
Jujube tree		0.04	0.02	This study
Persimmon tree		0.08	0.04	This study
Mulberry tree		0.06	0.02	This study
Peach tree		0.06	0.03	This study
Crop residue				
Rice		0.90	3.26	Hays et al. (2005)
Wheat		0.01	0.48	Hays et al. (2005)
Broomcorn		0.10	0.05	This study
Pea straw		0.08	0.02	This study
Horsebean		0.10	0.01	This study
Peanut straw		0.13	0.02	This study
Soybean straw		0.05	0.01	This study
Cotton stalk		0.05	0.02	This study
Rice straw		0.12	0.01	This study
Wheat straw		0.21	0.01	This study
Rape straw		0.37	0.02	This study
Seasame straw		0.17	0.02	This study
Corn straw		0.18	0.03	This study

(continued)

Table 5.10 (continued)

Fuel type	Combustion facility	EF, mg/kg	RET/PHE	References
Coal				
Lignite		101.00	0.78	Oros and Simoneit (2000)
Sub-bituminous		282.00	1.86	Oros and Simoneit (2000)
Briquette-anthricate		2.17	1.76	This study
Briquette-bituminous		3.68	1.29	This study
Chunk-bituminous		30.15	1.45	This study
Chunk-bituminous		186.90	6.10	This study
Chunk-bituminous		30.31	3.07	This study
Synthetic logs				
Synthetic logs (pine mountain)	FP	2.72	5.91	Rogge et al. (1998)
Synthetic log	FP	0.39	0.05	McDonald et al. (2000)
Artificial log	FP	10.96		Gullett et al. (2003)

[a] Unit: ug/m^3; [b] Unit: mg/g OC; *ND* Not detected; +: detected but not quantified
FP Fireplace, *WD* Woodstove
Adapted from Shen et al. (2012b) with permission of American Chemical Society

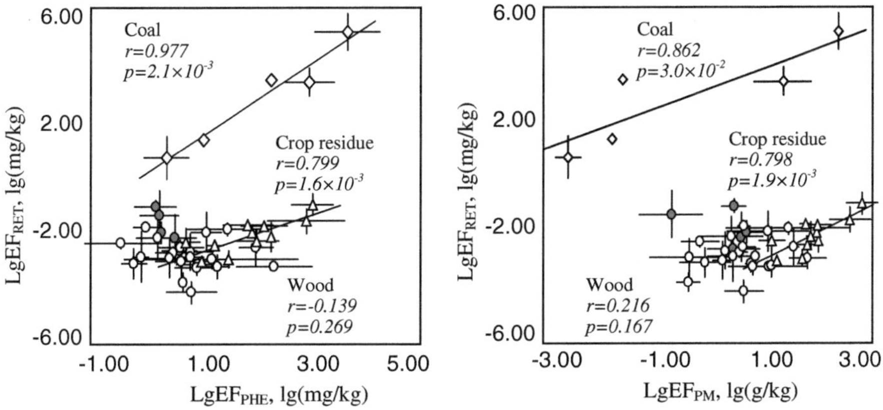

Fig. 5.30 Relationship between log-transformed EF_{RET} and EF_{PHE} (*left panel*) and between log EF_{RET} and log EF_{PM} (*right panel*) for different fuel types. Modified from Shen et al. (2012b) with permission of American Chemical Society

(Fig. 5.31). Much high retene emission factors were found for coal. Fine et al. (2004a, b) reported unquantifiable levels of RET emitted from *Douglas fir* combustion and suggested that the emission of *Douglas fir* may be different from other softwood species. In fact, the occurrence of retene had been reported in the literature for sources other than softwood burning (Hays et al. 2005; Hedberg et al.

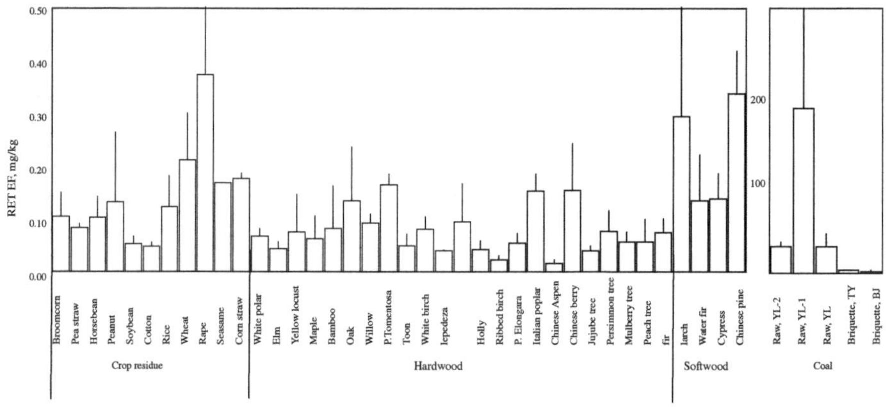

Fig. 5.31 Comparison of EF_{RET} for crop residue, harwood, softwood and coal from residential combustions in our study. Adapted from Shen et al. (2012b) with permission of American Chemical Society

2002; Bari et al. 2009; Oros and Simoneit 2000). In addition to the thermal degradation of diterpenoid resin acids, other formation mechanisms of retene have been reported in the literature. For example, it has been suggested that retene can be formed during the maturation of phyllocladane and/or kaurane-type compounds and abiogenic cyclisation and rearrangement of bicyclic terpenoids (Romero-Sarmiento et al. 2010). All of these suggest that retene is not a unique biomarker for conifer emissions and that the emissions of retene from coal combustion, in particular, should be taken into account.

5.5 Summary

EFs of 16 priority PAHs for the crop residue ranged from 23.6 to 142 mg/kg, with a mean and standard deviation of 62.1 ± 34.6 mg/kg. For the brushwood and fuel wood log, the total EFs of 28 PAHs were 86.7 ± 67.6 (27.1–160) and 12.7 ± 7.0 (3.2–32.7) mg/kg, respectively, of which over 90 % were 16 priority PAHs. For the coal, EF_{PAHs} ranged from 6.25 to 253 mg/kg. In general, PAHs emissions for coals were higher than that for crop residue and wood. The composition profiles were similar among these three fuel types. But slight differences can be identified. In emissions from coal combustion, the mass percents of high molecular weight PAHs to total PAHs appear to be higher than those from crop residue and wood. Since these high molecular weight ones usually have more toxic effects, relatively high emissions of these PAHs from coal combustion suggest that the use of coal in residential cooking would cause much serious detrimental health outcome, thus requiring more strict control strategies.

Low molecular weight PAHs are mostly in gaseous phase while high molecular weight ones prefer to be bound to particle. The mass percentages of fine particle bound PAHs increase with the increase of molecular weight. The gas-particle partitioning of freshly emitted PAHs is mainly controlled by absorption into organics rather than adsorption. The size distribution of particulate phase PAHs is similar to the distribution of co-emitted PM. Most particulate phase PAHs are present in fine particles. The size distributions of particulate phase PAHs for biomass fuels are similar within the fuel type, but different between fuel wood and brushwood and between crop residue and brushwood. For coal, coal property, like the caking property affects the size distribution significantly.

In our present study, by comparing retene emission among different fuel types, the use of retene as a marker for softwood was evaluated. EF_{RET} for crop residue, wood and coal were 0.012–0.45, 0.042–0.47 and 1.1–267 mg/kg, respectively. In emissions from coal and crop residue combustion, retene was positively correlated with other co-emitted PAHs like PHE. But in wood combustion emission, no significant correlation was found between retene and PHE. It was also found that renete was significantly affected by moisture and MCE in crop residue burning, and positively correlated with volatile matter content in coal combustion, but in wood combustion, no any significant correlation was found between retene emission and quantified parameters. As such, it is thought that retene may be not a unique marker for softwood, and other sources, particularly coal combustion, could produce high amounts of retene as well.

References

Allen, J. O., Dookeran, N. M., Smith, K. A., Sarofim, A. F., Taghizadeh, K., & Lafleur, A. (1996). Measurement of polycyclic aromatic hydrocarbons associated with size-segregated atmospheric aerosols in Massachusetts. *Environmental Science and Technology, 30*, 1023–1031.

Bachtel, A., Widera, M., Sachsenhofer, R. F., Gratzer, R., Lucke, A., & Woszczyk, M. (2007). Biomarker and stable carbon isotope systematics of fossil wood from the second Lusatian lignite seem of the Lubstow deposite (Poland). *Organic Geochemistry, 38*, 1850–1864.

Bari, M. A., Baumbach, G., Kuch, B., & Scheffknecht, G. (2009). Wood smoke as a source of particle-phase organic compounds in residential areas. *Atmospheric Environment, 43*, 4722–4732.

Bouloubassi, I., Fillaux, J., & Saliot, A. (2001). Hydrocarbons in surface sediments from the Changjing (Yangtze river) estuary, east China sea. *Marine Pollution Bulletin, 42*, 1335–1346.

Chen, Y. J., Bi, X. H., Mai, B. X., Sheng, G. Y., & Fu, J. M. (2004). Emission characterization of particulate/gaseous phases and size association for polycyclic aromatic hydrocarbons from residential coal combustion. *Fuel, 83*, 781–790.

Chen, Y. J., Sheng, G. Y., Bi, X. H., Feng, Y. L., Mai, B. X., & Fu, J. M. (2005). Emission factors for carbonaceous particles and polycyclic aromatic hydrocarbons from residential coal combustion in China. *Environmental Science and Technology, 39*, 1861–1867.

Dhammapala, R., Claiborn, C., Jimenez, J., Corkill, J., Gullett, B., Simpson, C., et al. (2007a). Emission factors of PAHs, methoxyphenols, levoglucosan, elemental carbon and organic carbon from simulated wheat and Kentucky bluegrass stubble burns. *Atmospheric Environment, 41*, 2660–2669.

Dhammapala, R., Claiborn, C., Simpson, C., & Jimenez, J. (2007b). Emission factors from wheat and Kentucky bluegrass stubble burning: comparison of field and simulated burn experiments. *Atmospheric Environment, 41*, 1512–1520.

Ding, J., Zhong, J., Yang, Y., Li, B., Shen, G., Su, Y., et al. (2011). Occurrence and exposure to polycyclic aromatic hydrocarbons and their derivatives in a rural Chinese home through biomass fuelled cooking. *Environmental Pollution,*. doi:10.1016/j.envpol.2011.10.008

Fine, P. M., Cass, G. R., & Simoneit, B. R. T. (2001). Chemical characterization of fine particle emissions from fireplace combustion of woods grown in the Northeastern United States. *Environmental Science and Technology, 35*, 2665–2675.

Fine, P. M., Cass, G. R., & Simoneit, B. R. T. (2002). Chemical characterization of fine particle emissions from fireplace combustion of woods grown in the Southern United States. *Environmental Science and Technology, 36*, 1442–1451.

Fine, P. M., Cass, G. R., & Simoneit, B. R. T. (2004a). Chemical characterization of fine particle emissions from fireplace combustion of wood stove combustion of prevalent United States tree species. *Environmental Engineering and Science, 21*, 705–721.

Fine, P. M., Cass, G. R., & Simoneit, B. R. T. (2004b). Chemical characterization of fine particle emissions from fireplace combustion of wood types grown in the Midwestern and Western United States. *Environmental Engineering and Science, 21*, 387–409.

Freeman, D. J., & Cattell, F. C. R. (1990). Woodburning as a source of atmospheric polycyclic aromatic hydrocarbons. *Environmental Science and Technology, 24*, 1581–1585.

Goncalves, C., Alves, C., Evtyugina, M., Mirante, F., Pio, C., Caseiro, A., et al. (2010). Characterisation of PM10 emissions from woodstove combustion of common woods grown in Portugal. *Atmospheric Environment, 44*, 4474–4480.

Goncalves, C., Alves, C., Fernandes, A. P., Monteiro, C., Tarelho, L., Evtyugina, M., et al. (2011). Organic compounds in PM2.5 emitted from fireplace and woodstove combustion of typical Portuguese wood species. *Atmospheric Environment, 45*, 4533–4545.

Gordon, R. J. (1976). Distribution of airborne polycyclic aromatic hydrocarbons throuhout Los Angeles. *Environmental Science and Technology, 10*, 370–373.

Gordon, R. J., & Bryan, R. J. (1973). Patterns in airborne polynuclear hydrocarbon concentrations at four Los Angeles sites. *Environmental Science and Technology, 7*, 1050–1053.

Goss, K., & Schwarzenbach, R. P. (1998). Gas/solid and gas/liquid partitioning of organic compounds: Critical evaluation of the interpretation of equilibrium constants. *Environmental Sciecne and Technology, 32*, 2025–2032.

Greenberg, A., Bozzelli, J. W., Cannova, F., Forstner, E., Giorgio, P., Stout, D., et al. (1981). Correlations between leed and coronene concentrations at urban, suburban and industrial sites in New Jersey. *Environmental Science and Technology, 15*, 566–570.

Grice, K., Lu, H., Atahan, P., Asif, M., Hallmann, C., Greenwood, P., et al. (2009). New insights into the origin of perylene in geological samples. *Geochimica et Cosmochimica Acta, 73*, 6531–6543.

Gullett, B. K., Touati, A., & Hays, M. D. (2003). PCDD/F, PCB, HxCBz, PAH, and PM emission factors for fireplaces and woodstove combustion in the San Francisco Bay region. *Environmental Science and Technology, 37*, 1758–1765.

Harkov, R., Greenberg, A., Darack, F., Dalsey, J. M., & Lloy, P. J. (1984). Summertime variations in polycyclic aromatic hydrocarbons at four sites in New Jersey. *Environmental Science and Technology, 18*, 287–291.

Hasegawa, T., & Hibino, T. (2011). Polycyclic aromatic hydrocarbons in the Jurassic-Cretaceous Tetori group, central Japan. *Island Arc, 20*, 23–34.

Hays, M. D., Fine, P. M., Geron, C. D., Kleeman, M. J., & Gullett, B. K. (2005). Open burning of agricultural biomass: physical and chemical properties of particle-phase emissions. *Atmospheric Environment, 39*, 6747–6764.

Hays, M. D., Smith, N. D., Kinsey, J., Dong, Y., & Kariher, P. (2003). Polycyclic aromatic hydrocarbon size distributions in aerosols from appliances of residential wood combustion as determined by direct thermal desorption-GC/MS. *Journal of Aerosol Science, 34*, 1061–1084.

Hedberg, E., Kristensson, A., Ohlsson, M., Johansson, C., Johansson, P., Swietlicki, E., et al. (2002). Chemical and physical characterization of emissions from birch wood combustion in a wood stove. *Atmospheric Environment, 36*, 4823–4837.

Jenkins, B. M., Jones, A. D., Turn, S. Q., & Williams, R. B. (1996a). Particle concentrations, gas-particle partitioning and species intercorrelations for polycyclic aromatic hydrocarbons (PAHs) emitted during biomass burning. *Atmospheric Environment, 30*, 3825–3835.

Jenkins, B. M., Jones, A. D., Turn, S. Q., & Williams, R. B. (1996b). Emission factors for polycyclic aromatic hydrocarbons from biomass burning. *Environmental Science and Technology, 30*, 2462–2469.

Keshtkar, H., & Ashbaugh, L. L. (2007). Size distribution of polycyclic aromatic hydrocarbon particulate emission factors form agricultural burning. *Atmospheric Environment, 41*, 2729–2739.

Kim Oanh, N. T., Albina, D. O., Li, P., & Wang, X. (2005). Emission of particulate matter and polycyclic aromatic hydrocarbons from select cookstove-fuel systems in Asia. *Biomass and Bioenergy, 28*, 579–590.

Kim Oanh, N. T., Reutergardh, L. B., & Dung, N. T. (1999). Emission of polycyclic aromatic hydrocarbons and particulate matter from domestic combustion of selected fuels. *Environmental Science and Technology, 33*, 2703–2709.

Kozinski, J. A., & Saade, R. (1998). Effect of biomass burning on the formation of soot particles and heavy hydrocarbons. An experimental study. *Fuel, 77*, 225–237.

Levendis, Y. A., Atal, A., Carlson, J., Dunayevskiy, Y., & Vouros, P. (1996). Comparative study on the combustion and emissions of waste tire crumb and pulverized coal. *Environmental Science and Technology, 30*, 2742–2754.

Li, Z., Sjodin, A., Porter, E., Patterson, D. G, Jr, Needham, L. L., Lee, S., et al. (2009). Characterization of PM2.5-bound polycyclic aromatic hydrocarbons in Atlantes. *Atmospheric Environment, 43*, 1043–1050.

Liu, W. X., Dou, H., Wei, Z. C., Chang, B., Qiu, W. X., Liu, Y., et al. (2009). Emission characteristics of polycyclic aromatic hydrocarbons from combustion of different residential coals in North China. *Science of the Total Environment, 407*, 1436–1446.

Lohmann, R., & Lammel, G. (2004). Adsorptive and absorptive contributions to the gas-particle partitioning of polycyclic aromatic hydrocarbons: state of knowledge and recommended parametrization for modeling. *Environmental Science and Technology, 38*, 3793–3803.

Lu, H., Zhu, L., & Zhu, N. (2009). Polycyclic aromatic hydrocarbon emissions from straw burning and the influence of combustion parameters. *Atmospheric Environment, 43*, 978–983.

Luo, X., Chen, S., Mai, B., Yang, Q., Sheng, G., & Fu, J. (2006). Polycyclic aromatic hydrocarbons in suspended particulate matter and sediments from the Pearl River Estuary and adjacent coastal areas, China. *Environmental Pollution, 139*, 9–20.

Mastral, A., & Calleän, M. S. (2000). A review on polycyclic aromatic hydrocarbon (PAH) emissions from energy generation. *Environmental Science and Technology, 34*, 3051–3057.

Mastral, A. M., Lopez, J. M., Callen, M. S., Garcia, T., Murillo, R., & Navarro, M. V. (2003). Spatial and temporal PAH concentrations in Zaragoza, Spain. *The Science of the Total Environment, 207*, 111–124.

McDonald, J. D., Zielinska, B., Fugita, E. M., Saebiel, J. C., Chow, J. C., & Watson, J. G. (2000). Fine particle and gaseous emission rates from residential wood combustion. *Environmental Science and Technology, 34*, 2080–2091.

McMeeking, G. R., Kreidenweis, S. M., Baker, S., Carrico, C. M., Chow, J. C., Collett, J. L, Jr, et al. (2009). Emission of trace gases and aerosols during the open combustion of biomass in the laboratory. *Journal of Geophysical Research, 114*, D19210. doi:10.1029/2009JD011836

Mitra, A., Sarofim, A., & Bar-Ziv, E. (1987). The influence of coal type on the evolution of polycyclic aromatic hydrocarbons during coal devolatilization. *Aerosol Science and Technology, 6*, 261–271.

Nisbet, I. C. T., & LaGoy, P. K. (1992). Toxic equivalency factors (TEFs) for polycyclic aromatic hydrocarbons (PAHs). *Regulatory Toxicology and Pharmacology, 16*, 290–300.

Odabasi, M., Cetin, E., & Sofuoglu, A. (2006). Determination of octanol-air partition coefficients and supercooled liquid vapor pressures of PAHs as a function of temperature: Application to gas-particle partitioning in an urban atmosphere. *Atmospheric Environment, 40*, 6615–6625.

Oros, D. R., & Simoneit, B. R. T. (2000). Identification and emission rates of molecular tracers in coal smoke particulate matter. *Fuel, 79*, 515–536.

Oros, D. R., & Simoneit, B. R. T. (2001a). Identification and emission factors of molecular tracers in organic aerosols from biomass burning Part 1. Temperate climate conifers. *Applied Geochemistry, 16*, 1513–1544.

Oros, D. R., & Simoneit, B. R. T. (2001b). Identification and emission factors of molecular tracers in organic aerosols from biomass burning Part 2. Deciduous trees. *Applied Geochemistry, 16*, 1545–1565.

Pankow, J. (1987). Review and comparative analysis of the theories on partitioning between the gas and aerosol particulate phases in the atmosphere. *Atmospheric Environment, 21*, 2275–2283.

Petry, T., Schmid, P., & Schlatter, C. (1996). The use of toxic equivalency factors in assessing occupational and environmental health risk associated with exposure to airborne mixtures of polycyclic aromatic hydrocarbons (PAHs). *Chemosphere, 32*, 639–648.

Purvis, C. R., McCrillis, R. C., & Kariher, P. H. (2000). Fine particulate matter (PM) and organic speciation of fireplace emissions. *Environmental Science and Technology, 34*, 1653–1658.

Ramdahl, T. (1983). Retene—a molecular marker of wood combustion in ambient air. *Nature, 306*, 580–582.

Rogge, W. F., Hildemann, L. M., Mazurek, M. A., Cass, G. R., & Simoneit, B. R. T. (1998). Sources of fine organic aerosol. 9. Pine, oak and synthetic log combustion in residential fireplaces. *Environmental Science and Technology, 32*, 13–22.

Romero-Sarmiento, M.-F., Riboulleau, A., Vecoli, M., & Versteegh, G. J. M. (2010). Occurance of retene in upper Silurian-lower Devonian sediements from North Africa: origin and implications. *Organic Geochemistry, 41*, 302–306.

Sawicki, E., Hauser, T. R., Elbert, W. C., Fox, F. T., & Meeker, J. E. (1962). Polynuclear aromatic hydrocarbon composition of the atmosphere in some large American cities. *American Industrial Hygiene Assocaition Journal, 23*, 137–144.

Schauer, J. J., Rogge, W. F., Hildemann, L. M., Mazurek, M. A., & Cass, G. L. (1996). Source apportionment of airborne particulate matter using organic compounds as tracers. *Atsmospheric Environment, 30*, 3837–3855.

Schauer, J. J., Kleeman, M. J., Cass, G. R., & Simoneit, B. R. T. (2001). Measurement of emissions from air pollution sources 3. C1 through C29 organic compounds from fireplace combustion of wood. *Environmental Science and Technology, 33*, 1716–1728.

Sheesley, R. J., Schauer, J. J., Chowdhury, Z., Cass, G. R., & Simoneit, B. R. T. (2003). Characterization of organic aerosols emitted from the combustion of biomass indigenous to South Asia. *Journal of Geophysical Research, 108*, D94285. doi:10.1029/2002JD002981

Shen, G. F., Wang, W., Yang, Y., Ding, J., Xue, M., Min, Y., Zhu, C., Shen, H., Li, W., Wang, B., Wang, R., Wang, X., Tao, S., Russell, A. (2011) Emissions of PAHs from indoor crop residue burning in a typical rural stove: emission factors, size distributions and gas-particle partitioning. *Environmental Science & Technology, 45*, 1206–1212.

Shen, G. F., Tao, S., Wei, S., Zhang, Y., Wang, R., Wang, B., Li, W., Shen, H., Huang, Y., Chen, Y.C., Chen, H., Yang, Y., Wang, W., Wang, X., Liu, W., Simonich, S. (2012a) Emissions of parent, nitro, and oxygenated polycyclic aromatic hydrocarbons from residential wood combustion in rural China. *Environmental Science & Technology, 46*, 8123–8130.

Shen, G. F., Tao, S., Wei, S., Zhang, Y., Wang, R., Wang, B., Li, W., Shen, H.Z., Huang, Y., Yang, Y., Wang, W., Wang, X., Simonich, S. (2012b) Retene emission from residential solid fuels in China and evaluation of retene as a unique marker for soft wood combustion. *Environmental Science & Technology, 46*, 4666–4672.

Suzuki, N., Yessalina, S., Kikuchi, T. (2010) Probable fungal origin of perylene in late Cretaceous to Paleogene terrestrial sedimentary rocks of northeastern Japan as indicated from stable carbon isotopes. *Organic geochemistry, 41*, 234–241.

Silliman, J. E., Meyers, P. A., & Eadie, B. J. (1998). Perylene: an indicator of alternation processes or precursor materials? *Organical Geochemistry, 29*, 5–7.

Simoneit, B. R. T. (2002). Biomass burning- a review of organic tracers for smoke from incomplete combustion. *Applied Geochemistry, 17*, 129–162.

Simoneit, B. R. T., Rogge, W. F., Mazurek, M. A., Standley, L. J., Hildemann, L. M., & Cass, G. R. (1993). Lignin pyrolysis products, lignans, and resin acids as specific tracers of plant classes in emissions from biomass combustion. *Environmental Science and Technology, 27*, 2533–2541.

Venkataraman, C., Negi, G., Sardar, S. B., & Rastogi, R. (2002). Size distributions of polycyclic aromatic hydrocarbons in aerosol emissions from biofuel combustion. *Journal of Aerosol Science, 33*, 503–518.

Venkataraman, C., Thomas, S., & Kulkarni, P. (1999). Size distributions of polycyclic aromatic hydrocarbons-gas/particle partitioning to urban aerosols. *Journal of Aerosol Science, 30*, 759–770.

Venkatesan, M. L. (1988). Occurrence and possible sources of perylene in marine sediments: a review. *Marine Chemistry, 25*, 1–27.

Wang, G., Kawamura, K., Zhao, X., Li, Q., Niu, H., & Dai, Z. (2007). Idenfication, abundance and seasonal variation of anthropogenic organic aerosols from a mega-city in China. *Atmospheric Environment, 41*, 407–416.

Wang, W., Tao, S., Wang, W., Shen, G., Zhao, J., & Lam, K. (2009). Airborne particulates and polycyclic aromatic hydrocarbons (PAHs) in ambient air in Donghe, Northern China. *Journal of Environmental Science and Health, Part A, 44*, 854–860.

Watson, J. G. (1984). Overview of receptor model principles. *Journal of the Air Pollution Control Association, 34*, 619–623.

Wilcke, W., Krauss, M., & Amelung, W. (2002). Carbon isotope signature of polycyclic aromatic hydrocarbons (PAHs): evidence for different sources in tropical and temperate environments? *Environmental Science and Technology, 36*, 3530–3535.

Wu, S. P., Tao, S., & Liu, W. X. (2006). Particle size distributions of polycyclic aromatic hydrocarbons in rural and urban atmosphere of Tianjin, China. *Chemosphere, 62*, 357–367.

Yunker, M. B., Macdonald, R. W., Vingarzan, R., Mitchell, H. R., Goyette, D., & Sylvestre, S. (2002). PAHs in the Fraser River basin: A critical appraisal PAH ratios as indicators of PAH source and composition. *Organic Geochemistry, 33*, 489–515.

Zhang, Y. X., Schauer, J. J., Zhang, Y., Zeng, L., Wei, Y., Liu, Y., et al. (2008). Characteristics of particulate carbon emissions from real-world Chinese coal combustion. *Environmental Scicence and Technology, 42*, 5068–5073.

Zhang, X. L., Tao, S., Liu, W. X., Yang, Y., Zuo, Q., & Liu, S. Z. (2005). Source diagnostics of polycyclic aromatic hydrocarbons based on species ratios: A multimedia approach. *Environmental Science and Technology, 39*, 9109–9114.

Chapter 6
Nitro- and Oxygenated PAHs

PAH derivatives like nitrated and oxygenated PAHs are of growing concern because these polar derivatives are able to produce more direct adverse health outcome. In the measurement on emissions from crop residue and coal combustions, four oxygenated PAHs (oPAHs) were analyzed, and in the later wood combustion experiment, not only four oPAHs but also 12 nitrated PAHs (nPAHs) were measured. Gas-particle partitioning of the derivatives, size distribution of particle-bound derivatives and the relationship between parent PAHs and the derivatives were discussed.

6.1 Residential Coal Combustion

6.1.1 EFs and Influencing Factor

The total EFs were 0.049 ± 0.009 and 0.29 ± 0.02 mg/kg for the 2 types of honeycomb briquettes and 8.8 ± 4.3, 40 ± 42, and 0.53 ± 0.04 mg/kg for the 3 different raw chunk coals, respectively. 9FO (0.034 ± 0.009 and 0.22 ± 0.01 mg/kg for Beijing and Taiyuan briquettes, respectively) was the dominant specie from briquette combustion, while 9FO (3.3–12 mg/kg) and BZO (4.0–24 mg/kg) were the most abundant compounds for raw chunk coals. EF_{OPAHs} of the tested coals varied 3 orders of magnitude with EF_{OPAHs} of chunk coals higher than those of honeycomb briquettes.

Similar to the approach in the analysis of influencing factors on CPM and PAHs emissions from coal combustion, stepwise regression was undertaken to quantitatively address the influence of fuel properties and combustion conditions on oPAH emissions. The results showed that EF_{OPAH} for coal was significantly affected by heating value, VM, MCE and moisture. F_{OPAH} can be predicted from the regression models with these four factors as independent variables by using the following formula:

$$EF_{OPAH}(\mu g/kg) = a \times M(\%) + b \times VM(\%) + c \times HV(MJ/kg) + d \times MCE + e$$

G. Shen, *Emission Factors of Carbonaceous Particulate Matter and Polycyclic Aromatic Hydrocarbons from Residential Solid Fuel Combustions*, Springer Theses,
DOI: 10.1007/978-3-642-39762-2_6, © Springer-Verlag Berlin Heidelberg 2014

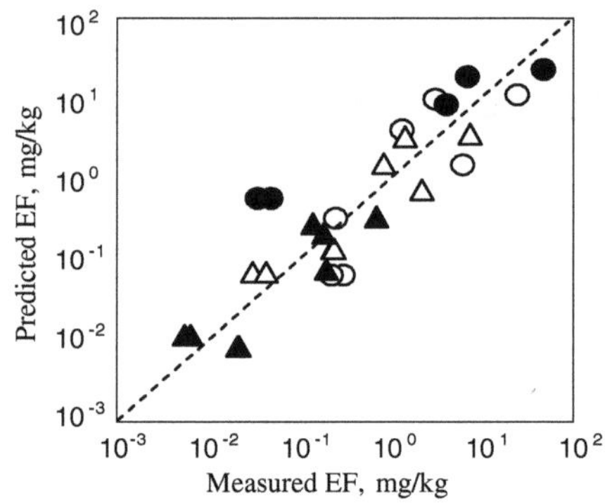

Fig. 6.1 Comparison of measured and predicted emission factors of 9-fluorenone (*open circle*), 9,10-anthraquinone (*open triangle*), benzanthrone (*filled circle*) and Benz[a]anthrane-7,12-dione (*filled triangle*) from residential coal combustions. Reprinted from Shen et al. (2011) with permission of American Chemical Society

Table 6.1 Emission factors (mg/kg) of individual OPAHs from residential coal combustions

Fuel type	9FO	ATQ	BZO	BaAQ
Honeycomb briquette, Beijing	$3.4 \pm 0.9 \times 10^{-2}$	$6.6 \pm 0.6 \times 10^{-3}$	$7.0 \pm 0.1 \times 10^{-3}$	$5.2 \pm 3.6 \times 10^{-4}$
Honeycomb briquette, Taiyuan	$2.2 \pm 0.0 \times 10^{-1}$	$3.2 \pm 0.8 \times 10^{-2}$	$3.6 \pm 0.9 \times 10^{-2}$	$5.2 \pm 0.6 \times 10^{-3}$
Raw chunk, Taiyuan	$3.3 \pm 3.0 \times 10^{0}$	$1.3 \pm 0.8 \times 10^{0}$	$4.0 \pm 0.4 \times 10^{0}$	$1.7 \pm 0.1 \times 10^{-1}$
Raw chunk-A, Yulin	$1.2 \pm 1.3 \times 10^{1}$	$4.0 \pm 3.8 \times 10^{0}$	$2.4 \pm 2.5 \times 10^{1}$	$3.7 \pm 3.5 \times 10^{-1}$
Raw chunk-B, Yulin	$2.3 \pm 0.5 \times 10^{-1}$	$2.1 \pm 0.0 \times 10^{-1}$	$7.2 \pm 0.3 \times 10^{-2}$	$1.9 \pm 0.7 \times 10^{-2}$

Reprinted from Shen et al. (2011) with permission of American Chemical Society

Figure 6.1 compares the estimated EFs with measured ones. They are generally comparable, and about 42–52 % of the total variations in EF_{OPAH} for coals were captured by the regression models Tables 6.1 and 6.2.

6.1.2 Relationship Between oPAHs and Parent PAHs

Parent PAHs generated from solid fuel combustion can be oxygenated to form OPAHs. Since both parent and oxygenated PAHs were affected significantly by the factors like coal VM content and MCE, it is interesting to investigate the relationship between oPAHs and the corresponding parent PAHs. As shown in Fig. 6.2, there were significantly positive correlations ($p < 0.05$) between oPAHs and pPAHs. In fact, oPAHs were also significantly positively correlated with other co-emitted pollutants including PM, OC and EC (Table 6.3).

The linear relationship between log-transformed EFs of oxygenated and parent PAH species may imply the quantitative dependence of the former on the latter.

Table 6.2 Calculated regression coefficients (μg/kg) and R^2 for predicting EF_{OPAH} for residential coal combustion in the improved coal stove

	a (value, *p*)	b (value, *p*)	c (value, *p*)	d (value, *p*)	e (value, *p*)	R^2
9FO	−308, 7.7×10^{-1}	274, 4.7×10^{-1}	37, 9.1×10^{-1}	-6.5×10^4, 2.7×10^{-1}	5.6×10^4, 2.7×10^{-1}	0.444
ATQ	−113, 7.2×10^{-1}	96, 4.0×10^{-1}	21, 8.3×10^{-1}	-2.0×10^4, 2.5×10^{-1}	1.8×10^4, 2.8×10^{-1}	0.508
BZO	−408, 8.5×10^{-1}	498, $5.1. \times 10^{-1}$	37, 9.5×10^{-1}	-1.2×10^5, 2.7×10^{-1}	1.1×10^5, 3.0×10^{-1}	0.415
BaAQ	−18, 5.2×10^{-1}	10, 3.1×10^{-1}	2.9, 7.3×10^{-1}	-1.4×10^3, 3.6×10^{-1}	1.1×10^3, 4.1×10^{-1}	0.522

p values for individual coefficients are also listed. Reprinted from Shen et al. (2011) with permission of American Chemical Society

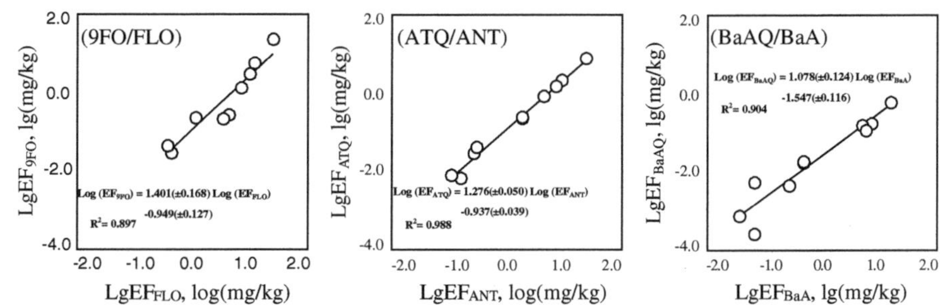

Fig. 6.2 Relationship between the log-transformed EF_{OPAH} and EF_{PAH} from residential coal combustions. The 3 pairs of OPAH/PAH from *left* to *right* are 9FO/FLO, ATQ/ANT, and BaAQ/BaA. Reprinted from Shen et al. (2011) with permission of American Chemical Society

Such a relationship was characterized by defining an oxygenation rate (R_o) in this study as the quantity of a given OPAH emitted per unit quantity of its parent PAH emitted (EF_{OPAH}/EF_{PAH}). For the 3 pairs of OPAH/PAH studied, the calculated R_o were 0.25 ± 0.25, 0.14 ± 0.08, and 0.03 ± 0.02 for 9FO/FLO, ATQ/ANT, and BaAQ/BaA from coal combustion. Moreover, it is interesting to see that the R_o for 9FO/FLO and ATQ/ANT pairs from coal combustion were positively correlated to EF_{FLO} and EF_{ANT}, but for BaAQ, the *Ro* did not show any significant dependence on EF of BaA (Fig. 6.3). Although many factors might influence the formation and emissions of OPAHs and parent PAHs, no significant correlation between *R*o and any coal property or combustion condition was found. The interaction among these factors was likely the reason causing the insignificance.

6.1.3 Gas-Particle Partitioning and Size Distribution

Similar to their parent PAHs, OPAHs are either in gaseous or condensed phases. Generally, those with relatively low molecular weight and high volatility occur dominantly in gaseous phase, while those with relatively high molecular weight

Table 6.3 Correlation coefficients and *p* values between log-transformed EFs of OPAHs and those of CO, OC, PM, and parent PAHs

		FLO	ANT	BaA	CO	OC	BC	PM
9FO	*r*	0.948			0.902	0.912	0.862	0.947
	p	1.5×10^{-5}			1.8×10^{-4}	1.2×10^{-4}	6.7×10^{-4}	1.7×10^{-5}
ATQ	*r*		0.994		0.942	0.969	0.856	0.923
	p		2.8×10^{-9}		2.2×10^{-5}	2.0×10^{-6}	8.0×10^{-4}	6.9×10^{-5}
BZO	*r*				0.903	0.929	0.920	0.981
	p				1.7×10^{-4}	5.1×10^{-5}	8.3×10^{-5}	3.0×10^{-7}
BaAQ	*r*			0.951	0.927	0.951	0.834	0.913
	p			1.2×10^{-5}	5.6×10^{-5}	1.2×10^{-5}	1.4×10^{-3}	1.1×10^{-4}

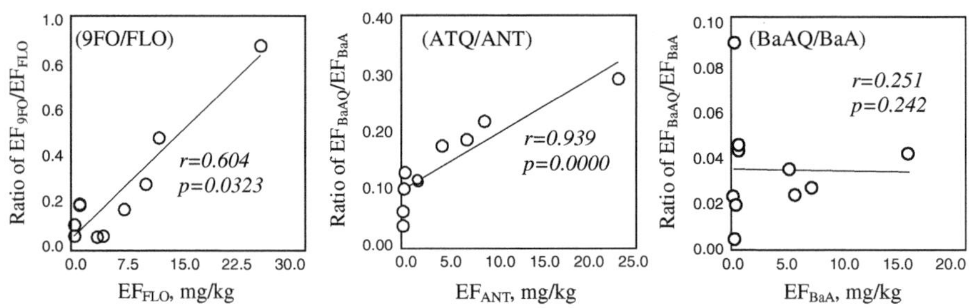

Fig. 6.3 Dependence of oxygenation rates (R_o) on EF_{PAH} for coal. The 3 pairs of OPAH/PAH from *left* to *right* are 9FO/FLO, ATQ/ANT, and BaAQ/BaA. Reprinted from Shen et al. (2011) with permission of American Chemical Society

and low volatility tend to associate with particles. Figure 6.4 compares the calculated partitioning coefficients for oPAHs with those for the corresponding parent ones. The K_P values of 9FO and ATQ were significantly higher than their parent PAHs ($p < 0.05$) and the differences were as high as 0.6–1.3 orders of magnitude, primarily because the vapor pressures of 9FO (7.6×10^{-3} Pa) and ATQ (1.6×10^{-5} Pa) are significantly lower than those of FLO (1.7×10^{-1} Pa) and ANT (4.1×10^{-3} Pa) (Walgraeve et al. 2010; Howard and Meylan 1997; Mackay et al. 1992). For BaA and BaAQ, calculated K_P values of BaAQ (vapor pressure is 4.8×10^{-6} Pa) and those of BaA (vapor pressure is 1.5×10^{-5} Pa) was not significantly different ($p > 0.05$). This might be explained by the fact that high molecular weight PAHs have quite low vapor pressures and are dominantly bound to particles, so differences in K_P values between parent PAHs and OPAHs were not as large as those of low molecular weight ones.

Size distributions of particle-bound oPAHs were similar to those for parent PAHs. The distribution can be classified into two groups based on the coal caking properties described by Char Residue Characteristics (CRC, ranged from 1 to 8, and the higher the CRC, the stronger the caking and swelling properties of the coal). Particulate phase OPAHs from the two briquettes and one raw chunk from Yulin (Chunk-2) with CRC of 1 or 2 were primarily found in fine particles with diameter less than 0.7 μm, while those from two other chunk coals with CRC of 5

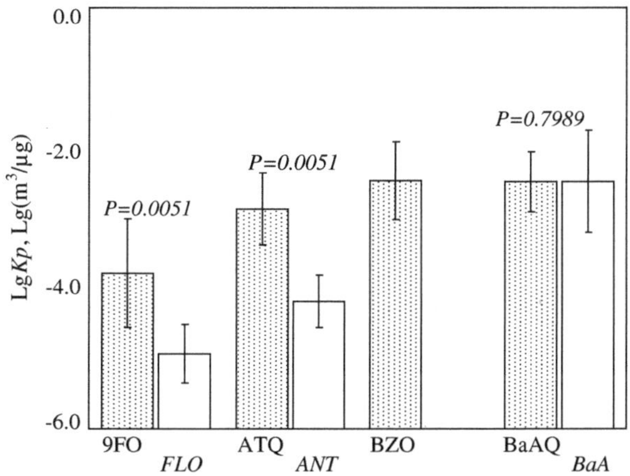

Fig. 6.4 The measured gas-particle partition coefficients (K_P) of 4 OPAHs from coal combustion. The results are compared with those of parent PAHs (except the parent PAHs for BZO which was not measured) emitted at the same time. The means and standard deviations are shown in log-scale. Reprinted from Shen et al. (2011) with permission of American Chemical Society

and 6, were dominantly in particles with Da between 0.7–1.1 and 1.1–2.1 μm, contributing 16–30 and 24–49 % of the total (Fig. 6.5).

Since oPAHs are polar organics and have low vapor pressure in comparison with non-polar parent PAHs, it is expected that oPAHs had more tendencies to be present in finer particles. Figure 6.6 compares the mass percent of oPAHs and corresponding parent PAHs in fine PM with diameter less than 0.4 μm. It is clear that for both low caking and high caking coals which had different size distributions, the mass percents of fine $PM_{0.4}$ bound oPAHs were much higher than that for the corresponding parent PAHs. For example, 17 ± 6, 19 ± 8, and 22 ± 7 % of 9FO, ATQ, and BaAQ and 10 ± 6, 13 ± 9, and 17 ± 9 % of FLO, ANT, and BaA occurred in $PM_{0.4}$ (PM with $D_a \leq 0.4$ μm) from two higher caking raw coals.

6.2 Indoor Crop Straw Burning

6.2.1 EFs and Influencing Factor

Total EFs of the 4 OPAHs for the 9 crop residues studied ranged from 2.8 ± 0.2 for soybean to 8.1 ± 2.2 mg/kg for wheat (Table 6.4) with a mean and standard deviation of 4.5 ± 2.0 mg/kg. Among the 4 compounds, 9FO was the most abundant (2.0 ± 0.8 mg/kg), followed by BZO (1.4 ± 0.7 mg/kg), ATQ (1.0 ± 0.4 mg/kg), and BaAQ (0.11 ± 0.04 mg/kg). Although EF_{OPAHs} for various crop residues were significant different, they were within the same order of

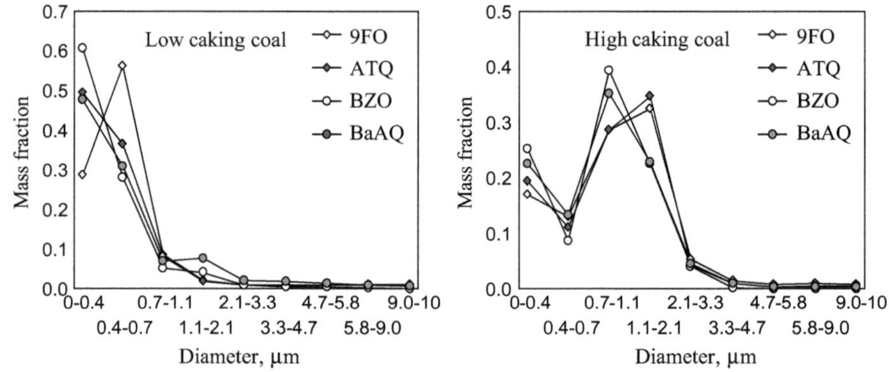

Fig. 6.5 Size distributions of particulate phase OPAHs from combustions of low CRC coals (*middle panel*), and high CRC coals (*right panel*). Modified from Shen et al. (2011) with permission of American Chemical Society

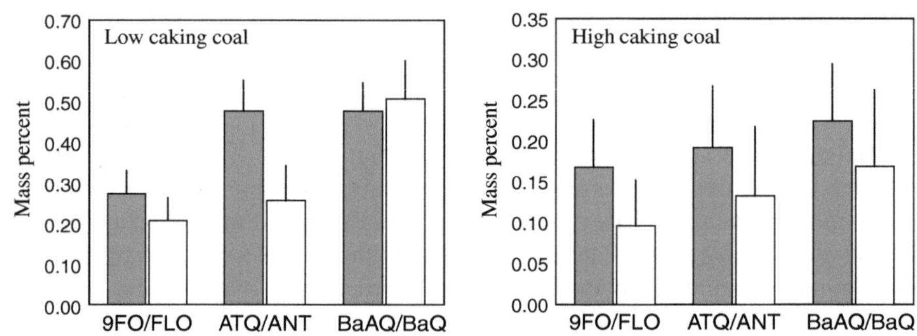

Fig. 6.6 Comparison of mass percents of fine $PM_{0.4}$-bound oPAHs and those of corresponding parent PAHs

magnitude. Hays et al. (2005) measured EF_{OPAH} from open field burning of rice and wheat residues and reported EF_{9FO}, EF_{ATQ}, and EF_{BZO} of 0.022–0.31, 0–0.042, and 0.60–0.95 mg/kg, respectively. The EF_{OPAH} for crop residue from indoor stove burning in our present study are 1–2 orders of magnitude higher than those from open field burning. This might be explained by the different amount of oxygen supply resulting in lower combustion efficiencies and relatively high temperature in the enclosed residential stoves due to low heat loss.

For oPAHs from crop residue burning, MCE and moisture negatively affected EF_{OPAH} of crop residues ($p < 0.05$). A relatively low MCE was favorable for incomplete combustion, while relatively high moistures could reduce combustion temperatures resulting in lower OPAH formations. Suppressed parent PAH formation resulted from relatively higher moistures could be also another reason for

Table 6.4 Emission factors (mg/kg) of individual OPAHs for indoor crop residue burning

Fuel type	9FO	ATQ	BZO	BaAQ
Horsebean (*Vicia faba*)	$1.6 \pm 0.5 \times 10^0$	$6.5 \pm 3.2 \times 10^{-1}$	$9.7 \pm 7.9 \times 10^{-1}$	$9.1 \pm 6.4 \times 10^{-2}$
Peanut (*Arachis hypogaea*)	2.3×10^0	1.1×10^0	1.4×10^0	1.2×10^{-1}
Soybean (*Cassia agnes*)	$1.4 \pm 0.3 \times 10^0$	$5.7 \pm 0.5 \times 10^0$	$7.7 \pm 0.1 \times 10^{-1}$	$6.2 \pm 0.5 \times 10^{-2}$
Cotton (*Anemone vitifolia*)	$1.1 \pm 0.1 \times 10^0$	$6.4 \pm 0.8 \times 10^{-1}$	$1.2 \pm 0.0 \times 10^0$	$9.4 \pm 0.4 \times 10^{-2}$
Rice (*Oryza sativa*)	$3.1 \pm 1.5 \times 10^0$	$1.2 \pm 0.3 \times 10^0$	$2.6 \pm 1.8 \times 10^0$	$1.2 \pm 0.6 \times 10^{-1}$
Wheat (*Triticum aestivum*)	$3.4 \pm 1.5 \times 10^0$	$1.8 \pm 0.4 \times 10^0$	$2.7 \pm 0.3 \times 10^0$	$2.1 \pm 0.2 \times 10^{-1}$
Rape (*Brassica napus*)	$2.2 \pm 0.4 \times 10^0$	$1.2 \pm 0.4 \times 10^0$	$1.6 \pm 0.3 \times 10^0$	$1.4 \pm 0.4 \times 10^{-1}$
Sesame (*Sesamum indicum*)	$1.5 \pm 0.5 \times 10^0$	$1.1 \pm 0.4 \times 10^0$	$4.9 \pm 0.4 \times 10^{-1}$	$7.8 \pm 2.8 \times 10^{-2}$
Corn (*Zea mays*)	$1.3 \pm 0.1 \times 10^0$	$6.7 \pm 1.6 \times 10^{-1}$	$9.2 \pm 5.2 \times 10^{-1}$	$8.2 \pm 3.1 \times 10^{-2}$

Data shown are means and standard derivation from duplicated experiments. Reprinted from Shen et al. (2011) with permission of American Chemical Society

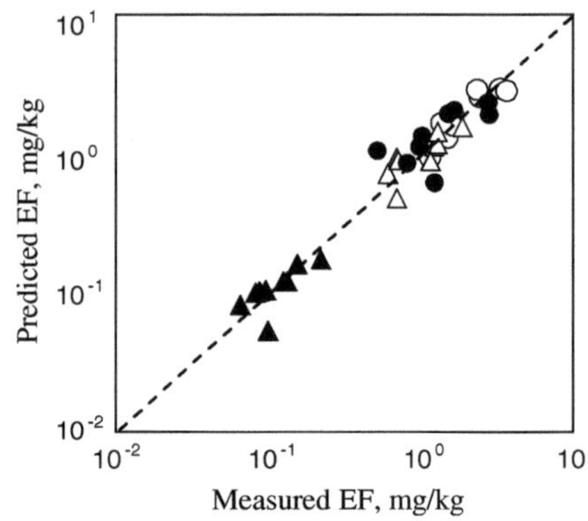

Fig. 6.7 Comparison of measured and predicted emission factors of 9-fluorenone (*open circle*), 9,10-anthraquinone (*open triangle*), benzanthrone (*filled circle*) and Benz[a]anthrane-7,12-dione (*filled triangle*) from residential crop straw burning. Reprinted from Shen et al. (2011) with permission of American Chemical Society

the subsequently lower OPAH emissions. About 56–77 % of total variations in EF_{oPAHs} can be explained by these two factors, and EF_{OPAHs} could be predicted from the following regression models with moisture (M) and MCE as independent variables (Fig. 6.7).

$$\begin{aligned} EF_{9FO} &= -43 \times M(p = 0.011) - 22 \times MCE(p = 0.0045) + 24(p = 0.0027) \\ EF_{ATQ} &= -34 \times M(p = 0.032) - 9 \times MCE(p = 0.015) + 10(p = 0.0072) \\ EF_{BZO} &= -22 \times M(p = 0.049) - 18 \times MCE(p = 0.034) + 19(p = 0.026) \\ EF_{BaAQ} &= -41 \times M(p = 0.030) - 0.9 \times MCE(p = 0.028) + 1.1(p = 0.013) \end{aligned}$$

6.2.2 Relationship Between oPAHs and Parent PAHs

In emissions from indoor crop straw burning, oPAHs also positively correlated with parent PAHs (Fig. 6.8) and other co-emitted incomplete pollutants (Table 6.5). The calculated R_o were 0.40 ± 0.18, 0.89 ± 0.41, and 0.16 ± 0.05 for 9FO/FLO, ATQ/ANT, and BaAQ/BaA, respectively. It appeared that R_o values were significantly different among OPAH compounds and the measured R_o values of crop residue burning were significantly higher than those of coal burning ($p < 0.05$). Meanwhile, it is noted that different from positive correlations between *Ro* and EF_{PAH} for coal, the dependence of R_o on EF_{PAH} was negative in crop residue burning (Fig. 6.9).

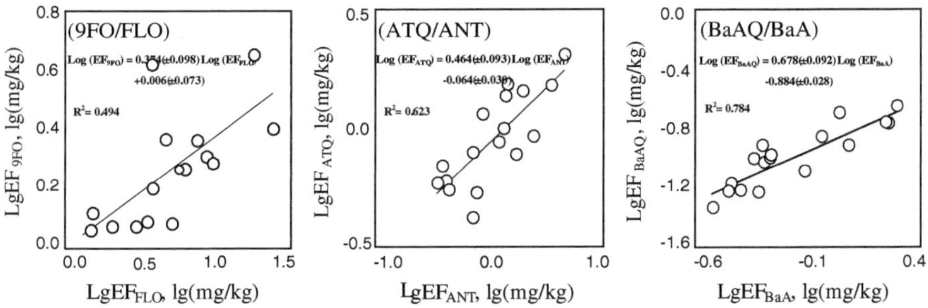

Fig. 6.8 Relationship between the log-transformed EF_{OPAH} and EF_{PAH} from residential crop residue combustions. The 3 pairs of OPAH/PAH from *left* to *right* are 9FO/FLO, ATQ/ANT, and BaAQ/BaA. Reprinted from Shen et al. (2011) with permission of American Chemical Society

Table 6.5 Correlation coefficients and *p* values between log-transformed EFs of OPAHs and those of CO, OC, PM, and parent PAHs from crop residue burning

		FLO	ANT	BaA	CO	OC	BC	PM
9FO	*r*	0.703			0.679	0.605	0.241	0.688
	p	8.2×10^{-4}			1.4×10^{-3}	5.0×10^{-3}	1.8×10^{-1}	1.1×10^{-3}
ATQ	*r*		0.789		0.527	0.553	0.274	0.674
	p		8.3×10^{-5}		1.5×10^{-2}	1.1×10^{-2}	1.4×10^{-1}	1.5×10^{-3}
BZO	*r*				0.538	0.028	0.248	0.334
	p				1.3×10^{-2}	4.5×10^{-1}	1.7×10^{-1}	9.5×10^{-2}
BaAQ	*r*			0.885	0.486	0.242	0.441	0.575
	p			1.2×10^{-6}	2.4×10^{-2}	1.7×10^{-1}	3.8×10^{-3}	7.8×10^{-3}

Reprinted from Shen et al. (2011) with permission of American Chemical Society

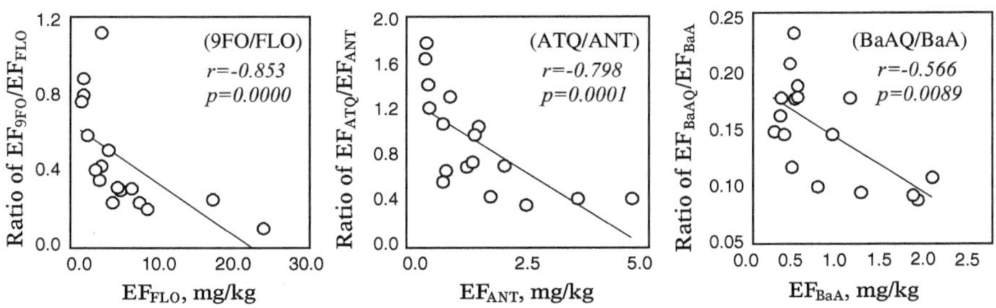

Fig. 6.9 Dependence of oxygenation rates (R_o) on EF_{PAH} from crop straw burning. The 3 pairs of OPAH/PAH from *left* to *right* are 9FO/FLO, ATQ/ANT, and BaAQ/BaA. Reprinted from Shen et al. (2011) with permission of American Chemical Society

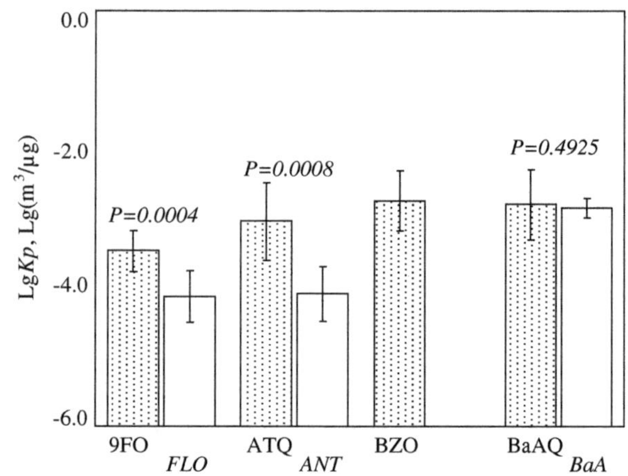

Fig. 6.10 The measured gas-particle partition coefficients (K_p) of 4 OPAHs from crop residues burning. The results are compared with those of parent PAHs (except the parent PAHs for BZO which was not measured) emitted at the same time. The means and standard deviations are shown in log-scale. Modified from Shen et al. (2011) with permission of American Chemical Society

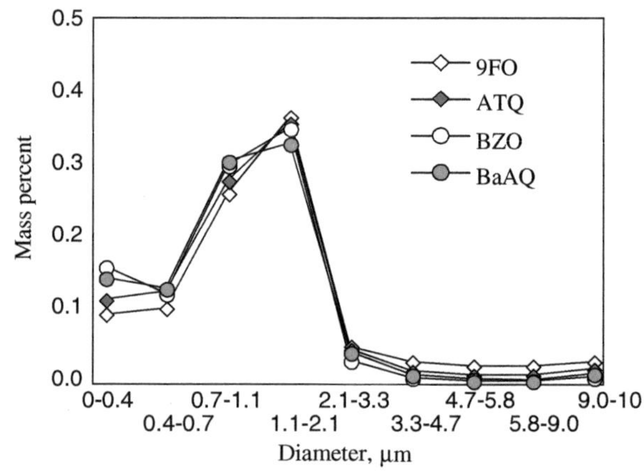

Fig. 6.11 Size distributions of particulate phase oPAHs from combustions of crop residues. Modified from Shen et al. (2011) with permission of American Chemical Society

6.2.3 Gas-Particle Partitioning and Size Distribution

Figure 6.10 shows the calculated partitioning coefficients for oPAHs in comparison with the corresponding parent PAHs. As expected, KP of organic pollutants showed a general increasing trend as molecular weight increases. Due to relatively lower vapor pressure, the oxygenated derivatives had high *Kp* values, which suggested that they are preferable to be present in particulate phase.

Size distributions of particle bound OPAHs in emissions for different crop straw were similar with a dominant fraction at 1.1–2.1 μm size range (33–37 %), followed by those between 0.7 and 1.1 μm (26–31 %) (Fig. 6.11). The distribution was similar to that in co-emitted parent PAHs. About 83, 88, 94, and 91 % of particulate phase 9FO, ATQ, BZO, and BaAQ were found in fine $PM_{2.1}$. The simultaneously measured percentages of the co-emitted FLO, ANT, and BaA associated with $PM_{2.1}$ were only 51, 65, and 88 % respectively, which again indicated that oPAHs have more tendencies to be present in finer particles in comparison with parent PAHs.

6.3 Residential Wood Combustion

6.3.1 EFs and Influencing Factor

Among the 12 nPAHs and 4 oPAHs measured, the concentrations of 7 N-BaA, 6 N-CHR, and 6 N-BaP were below the detection limit in most samples ($\sim$0.05 μg/kg). However, 5 N-ACE, 2 N-FLO, and 1 N-PYR were measured in several fuels, and 6 nPAHs (1 N-NAP, 2 N-NAP, 9 N-ANT, 9 N-PHE, 3 N-PHE, and 3 N-FLA) and 4 oPAHs (9FO, ATQ, BZO, and BaAQ) were measured in all samples. The results are listed in Table 6.6 for nPAHs and Table 6.7 for oPAHs, respectively.

For fuel wood and brushwood, the EF_{PAHn9} was 8.27 ± 5.51 and 32.2 ± 19.5 μg/kg, respectively, and EF_{PAHo4} was 1.19 ± 1.87 and 5.56 ± 4.32 mg/kg, respectively. 1 N-NAP and 2 N-NAP were the two dominant nPAHs identified. For oPAHs, the EFs for two ketones (9FO and BZO) were much higher than those for two quinones (ATQ and BaAQ). The co-emitted derivatives were positively correlated with one another (Tables 6.8 and 6.9). EF_{PAHn9} and EF_{PAHo4} from brushwood burning were significantly higher than those from fuel wood combustion (Fig. 6.12).

EFs of nPAHs were approximately 3 orders of magnitude lower than those of pPAHs, while oPAH EFs were on the same order of magnitude as pPAH EFs. In comparison with literature reported results, although limited data available, our results were comparable to those measured using woodstoves, but lower than those measured using fireplaces in general (Rogge et al. 1998; Fine et al. 2001, 2002, 2004a, b; Gullett et al. 2003). For example, average EFs for particle-bound 9FO, ATQ, and BZO were 0.841 (0.132 to 3.42), 0.612 (0.053 to 4.19), and 0.670 (0.150 to 2.09) mg/kg for wood combustion in fireplaces, respectively (Fine et al. 2001, 2002, 2004a; Gullett et al. 2003; Rogge et al. 1998), and were 0.625 (0.189 to 2.64), 0.256 (0.077 to 0.438), and 0.444 (0.091 to 1.01) mg/kg for wood burned in woodstoves, respectively (Fine et al. 2004b; Gullett et al. 2003).

Significantly negative correlations ($p < 0.05$) could be found between MCE and EFs of the derivatives, and the later were positively correlated with fuel moisture (Fig. 6.13), which might be explained by that an appreciable amount of energy is needed to vaporize the water which can reduce the combustion temperature and lead to reduced combustion efficiencies and enhanced pollutant emissions. The influence of other factors measured in this study, including density, heating value, and volatile matter content, were not statistically significant ($p > 0.05$).

Table 6.6 Average emission factors (mg/kg, dry basis) of *nitrated PAHs* from residential wood combustions of different fuel types

	1N-NAP	2N-NAP	5N-ACE	2N-FLO	9N-ANT	9N-PHE	3N-PHE	3N-FLA	1N-PYR
White Poplar	1.7×10^{-3}	1.5×10^{-3}	n.d.	n.d.	1.2×10^{-3}	3.6×10^{-4}	1.2×10^{-3}	4.3×10^{-4}	n.d.
Elm	3.0×10^{-3}	3.4×10^{-3}	n.d.	n.d.	1.1×10^{-3}	3.1×10^{-4}	2.1×10^{-4}	2.1×10^{-3}	n.d.
Locust	1.1×10^{-3}	1.3×10^{-3}	n.d.	n.d.	7.2×10^{-4}	1.8×10^{-4}	1.1×10^{-4}	2.3×10^{-4}	n.d.
Maple	1.9×10^{-3}	2.1×10^{-3}	n.d.	n.d.	8.3×10^{-4}	2.6×10^{-4}	2.9×10^{-4}	7.8×10^{-4}	3.6×10^{-4}
Fir	1.8×10^{-3}	1.7×10^{-3}	n.d.	n.d.	7.2×10^{-4}	n.d.	n.d.	4.5×10^{-4}	3.6×10^{-4}
Larch	1.7×10^{-3}	1.6×10^{-3}	n.d.	n.d.	7.7×10^{-4}	1.8×10^{-4}	n.d.	n.d.	n.d.
Water Chinese fir	2.2×10^{-3}	2.4×10^{-3}	n.d.	n.d.	8.6×10^{-4}	2.4×10^{-4}	1.9×10^{-4}	5.2×10^{-4}	n.d.
Cypress	2.3×10^{-3}	2.4×10^{-3}	6.8×10^{-4}	n.d.	1.2×10^{-3}	2.9×10^{-4}	1.6×10^{-4}	4.0×10^{-4}	2.3×10^{-4}
Oak	1.9×10^{-3}	1.9×10^{-3}	n.d.	n.d.	1.1×10^{-3}	4.9×10^{-4}	2.7×10^{-4}	5.5×10^{-4}	n.d.
Chinese Pine	1.4×10^{-3}	1.4×10^{-3}	9.2×10^{-4}	n.d.	6.3×10^{-4}	1.7×10^{-4}	1.9×10^{-4}	4.2×10^{-4}	3.5×10^{-4}
Willow	8.2×10^{-4}	1.1×10^{-3}	n.d.	n.d.	7.4×10^{-4}	2.0×10^{-4}	n.d.	n.d.	n.d.
Paulownia tomentosa	1.3×10^{-3}	1.2×10^{-3}	n.d.	n.d.	1.0×10^{-3}	3.1×10^{-4}	n.d.	3.9×10^{-4}	n.d.
Toon	1.0×10^{-3}	1.1×10^{-3}	n.d.	n.d.	5.1×10^{-4}	4.0×10^{-4}	n.d.	n.d.	n.d.
White Birch	2.9×10^{-3}	3.3×10^{-3}	3.0×10^{-3}	n.d.	1.3×10^{-3}	3.5×10^{-4}	2.8×10^{-4}	8.0×10^{-4}	n.d.
Ribbed Birch	2.5×10^{-3}	2.8×10^{-3}	5.4×10^{-4}	6.3×10^{-5}	8.4×10^{-4}	4.2×10^{-4}	3.0×10^{-4}	5.6×10^{-4}	3.6×10^{-4}
Paulownia elongata	5.1×10^{-3}	6.8×10^{-3}	1.0×10^{-3}	3.9×10^{-4}	2.2×10^{-3}	7.3×10^{-4}	4.5×10^{-4}	9.0×10^{-4}	2.7×10^{-4}
Black Poplar	5.5×10^{-3}	6.2×10^{-3}	5.2×10^{-4}	4.5×10^{-4}	2.7×10^{-3}	6.1×10^{-4}	5.2×10^{-4}	7.1×10^{-4}	1.3×10^{-3}
China Aspen	2.7×10^{-3}	2.3×10^{-3}	n.d.	1.7×10^{-4}	1.0×10^{-3}	4.2×10^{-4}	3.2×10^{-4}	7.9×10^{-4}	2.5×10^{-4}
Chinaberry	3.7×10^{-3}	4.4×10^{-3}	3.2×10^{-3}	2.7×10^{-4}	1.4×10^{-3}	6.8×10^{-4}	2.9×10^{-4}	5.4×10^{-4}	6.0×10^{-4}
Jujube tree	3.9×10^{-3}	4.4×10^{-3}	1.7×10^{-3}	1.4×10^{-4}	1.4×10^{-3}	7.4×10^{-4}	4.1×10^{-4}	9.2×10^{-4}	5.0×10^{-4}
Persimmon tree	4.0×10^{-3}	4.6×10^{-3}	5.2×10^{-3}	n.d.	1.2×10^{-3}	8.2×10^{-4}	4.8×10^{-4}	4.0×10^{-4}	3.0×10^{-4}
Mulberry tree	3.2×10^{-3}	3.6×10^{-3}	n.d.	n.d.	1.1×10^{-3}	5.2×10^{-4}	1.5×10^{-4}	1.0×10^{-3}	n.d.
Peach tree	3.0×10^{-3}	3.3×10^{-3}	n.d.	n.d.	1.2×10^{-3}	3.0×10^{-4}	1.8×10^{-4}	4.4×10^{-3}	3.7×10^{-4}
Lespedeza	1.2×10^{-2}	1.3×10^{-2}	3.4×10^{-3}	4.9×10^{-4}	1.8×10^{-3}	2.8×10^{-3}	2.4×10^{-3}	7.8×10^{-4}	2.9×10^{-3}
Buxus sinica	1.1×10^{-2}	1.2×10^{-2}	6.2×10^{-3}	3.9×10^{-4}	3.0×10^{-3}	2.3×10^{-3}	1.9×10^{-3}	1.7×10^{-3}	2.1×10^{-3}
Holly	4.4×10^{-3}	5.5×10^{-3}	2.6×10^{-3}	8.7×10^{-5}	1.2×10^{-3}	6.5×10^{-4}	5.5×10^{-4}	1.2×10^{-3}	5.1×10^{-4}
Bamboo	3.3×10^{-3}	3.5×10^{-3}	3.4×10^{-3}	n.d.	1.5×10^{-3}	1.2×10^{-3}	5.6×10^{-4}	9.3×10^{-4}	1.1×10^{-3}

For each fuel, combustion experiments were done in triplicate. Adapted from Shen et al. (2012) with permission of American Chemical Society

Table 6.7 Average emission factors (mg/kg, dry basis) of *oxygenated PAHs* from residential wood combustions of different fuel types

	9FO	ATQ	BZO	BaAQ
White Poplar	4.0×10^{-1}	6.7×10^{-2}	4.5×10^{-2}	1.4×10^{-3}
Elm	9.2×10^{-1}	2.1×10^{-1}	7.0×10^{-2}	6.0×10^{-3}
Locust	3.9×10^{-1}	1.3×10^{-1}	3.8×10^{-2}	2.8×10^{-3}
Maple	4.5×10^{-1}	1.7×10^{-1}	8.8×10^{-2}	7.5×10^{-3}
Fir	4.0×10^{-1}	1.5×10^{-1}	7.2×10^{-2}	4.7×10^{-3}
Larch	2.4×10^{-1}	6.9×10^{-2}	2.8×10^{-2}	2.6×10^{-3}
Water Chinese fir	4.3×10^{-1}	1.5×10^{-1}	5.5×10^{-2}	5.3×10^{-3}
Cypress	3.5×10^{-1}	1.4×10^{-1}	5.1×10^{-2}	1.6×10^{-1}
Oak	3.9×10^{-1}	1.5×10^{-1}	6.1×10^{-2}	9.9×10^{-3}
Chinese Pine	2.2×10^{-1}	8.7×10^{-2}	6.2×10^{-2}	5.5×10^{-3}
Willow	2.4×10^{-1}	8.3×10^{-2}	2.6×10^{-2}	1.5×10^{-2}
Paulownia tomentosa	3.7×10^{-1}	1.5×10^{-1}	9.2×10^{-2}	3.0×10^{-2}
Toon	2.4×10^{-1}	6.2×10^{-2}	3.4×10^{-2}	2.7×10^{-1}
White Birch	5.3×10^{-1}	1.7×10^{-1}	1.1×10^{-1}	7.0×10^{-3}
Ribbed Birch	5.8×10^{-1}	1.9×10^{-1}	1.3×10^{-1}	9.5×10^{-3}
Paulownia elongata	3.3×10^{0}	6.8×10^{-1}	1.7×10^{-1}	1.3×10^{-2}
Black Poplar	6.1×10^{0}	1.3×10^{0}	5.0×10^{-1}	2.5×10^{-2}
China Aspen	1.2×10^{0}	1.9×10^{-1}	1.7×10^{-1}	5.1×10^{-3}
Chinaberry	1.5×10^{0}	3.6×10^{-1}	1.3×10^{-1}	9.9×10^{-3}
Jujube tree	7.5×10^{-1}	2.0×10^{-1}	3.2×10^{-1}	3.4×10^{-2}
Persimmon tree	8.9×10^{-1}	2.2×10^{-1}	1.7×10^{-1}	9.2×10^{-3}
Mulberry tree	1.0×10^{0}	2.5×10^{-1}	7.5×10^{-2}	6.5×10^{-2}
Peach tree	7.0×10^{-1}	1.6×10^{-1}	1.0×10^{-1}	7.0×10^{-3}
Lespedeza	3.5×10^{0}	1.7×10^{0}	3.7×10^{0}	1.3×10^{-1}
Buxus sinica	2.2×10^{0}	1.2×10^{0}	2.2×10^{0}	9.6×10^{-2}
Holly	1.1×10^{0}	3.8×10^{-1}	3.1×10^{-1}	2.6×10^{-2}
Bamboo	1.0×10^{0}	4.1×10^{-1}	7.0×10^{-1}	5.1×10^{-2}

For each fuel, combustion experiments were done in triplicate. Adapted from Shen et al. (2012) with permission of American Chemical Society

6.3.2 Relationship Between oPAHs and Parent PAHs

The derivatives were significantly correlated with parent PAHs (Table 6.10 and Fig. 6.14) and co-emitted PM and CO (Fig. 6.15). The results were more or less expected since they both significantly affected by fuel moisture and MCE.

The calculated *Ro* were 2.06 ± 0.84, 0.792 ± 0.237, and $6.56 \pm 7.54 \times 10^{-2}$ for 9FO/FLO, ATQ/ANT, and BaAQ/BaA, respectively. Since EFs of nPAHs were about 3 orders of magnitude lower than the parent PAHs, the calcualetd R_N values were as low as 1.51×10^{-4} (3N-PHE/PHE) to 4.67×10^{-3} (9N-ANT/ANT), significantly lower than the *Ro* value. Since nPAHs could be formed from photochemical reactions following direct emission from combustion (Albinet et al. 2007, 2008a, b; Wang et al. 2011), the ratio is believed to increase once emitted into the ambient air. In a previous study in a rural household burning crop residue

Table 6.8 Correlation coefficient and p values between EFs of measured nitrated PAHs

		1N-NAP	2N-NAP	5N-ACE	2N-FLO	9N-ANT	9N-PHE	9N-PHE	3N-FLA
1N-NAP	r	1.000							
	p								
	n	27							
2N-NAP	r	0.991	1.000						
	p	8.5×10^{-24}							
	n	27	27						
5N-ACE	r	0.501	0.518	1.000					
	p	4.0×10^{-2}	3.5×10^{-2}						
	n	13	13	13					
2N-FLO	r	0.579	0.560	0.239	1.000				
	p	5.1×10^{-2}	5.8×10^{-2}	2.8×10^{-1}					
	n	9	9	8	9				
9N-ANT	r	0.824	0.844	0.254	0.819	1.000			
	p	6.4×10^{-8}	1.6×10^{-8}	2.0×10^{-1}	3.4×10^{-3}				
	n	27	27	13	9	27			
9N-PHE	r	0.846	0.849	0.573	0.655	0.737	1.000		
	p	2.7×10^{-8}	2.1×10^{-8}	2.0×10^{-2}	2.8×10^{-2}	8.6×10^{-6}			
	n	26	26	13	9	26	26		
9N-PHE	r	0.707	0.667	0.530	0.557	0.493	0.833	1.000	
	p	1.2×10^{-4}	3.5×10^{-4}	3.1×10^{-2}	6.0×10^{-2}	9.8×10^{-3}	7.6×10^{-7}		
	n	22	22	13	9	22	22	22	
3N-FLA	r	0.463	0.472	0.488	0.061	0.426	0.365	0.304	1.000
	p	1.1×10^{-2}	9.9×10^{-3}	4.5×10^{-2}	4.4×10^{-1}	1.9×10^{-2}	4.3×10^{-2}	8.5×10^{-2}	
	n	24	24	13	9	24	23	22	24
1N-PYR	r	0.802	0.795	0.395	0.587	0.642	0.776	0.820	0.332
	p	9.3×10^{-5}	1.2×10^{-4}	1.0×10^{-1}	4.8×10^{-2}	3.7×10^{-3}	3.3×10^{-4}	9.1×10^{-5}	1.0×10^{-1}
	n	16	16	12	9	16	15	15	16

Adapted from Shen et al. (2012) with permission of American Chemical Society

Table 6.9 Correlation coefficient and p values between EFs of measured oxygenated PAHs

		9FO	ATQ	BZO	BaAQ
9FO	r	1.000	0.940	0.821	0.637
	p		1.6×10^{-13}	7.8×10^{-8}	1.8×10^{-4}
	n	27	27	27	27
ATQ	r		1.000	0.901	0.793
	p			7.1×10^{-11}	4.0×10^{-7}
	n		27	27	27
BZO	r			1.000	0.816
	p				1.1×10^{-7}
	n			27	27
BaAQ	r				1.000
	p				
	n				27

Adapted from Shen et al. (2012) with permission of American Chemical Society

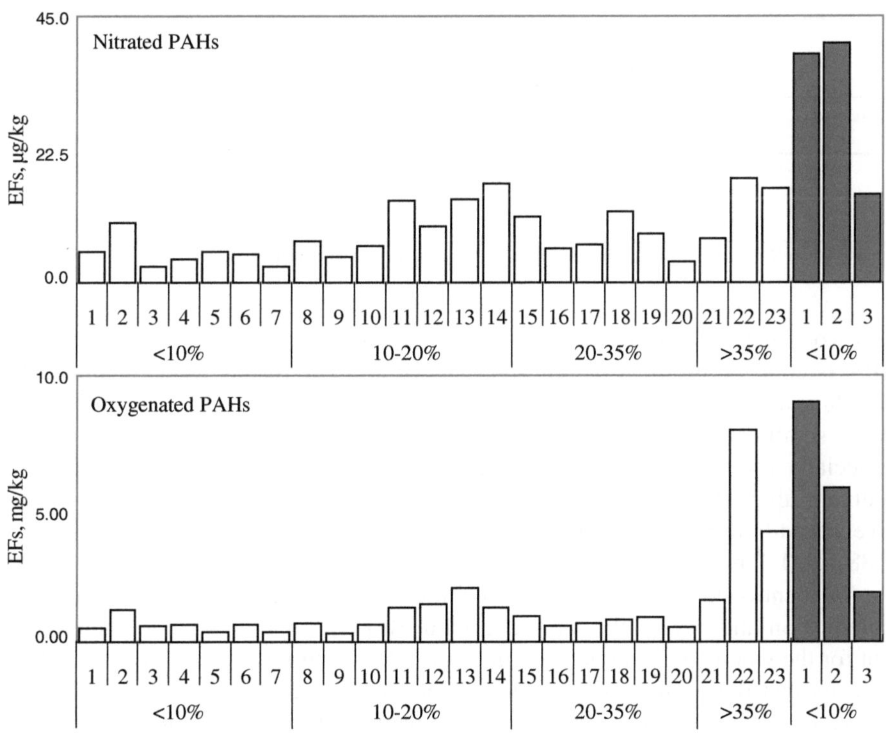

Fig. 6.12 Comparison of oPAHs and nPAHs among different wood fuel types

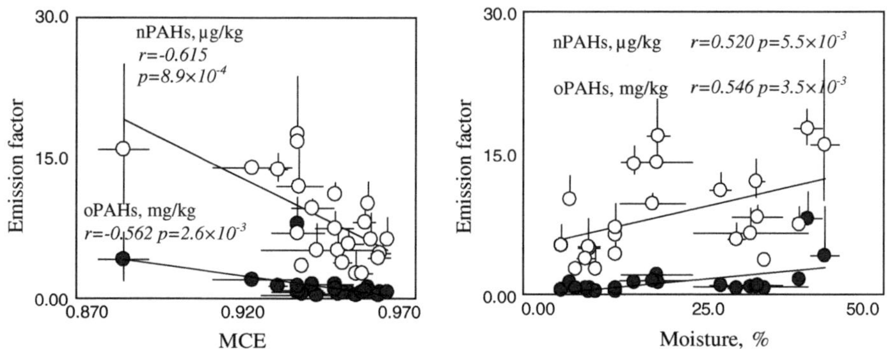

Fig. 6.13 Correlations between EFs of nPAHs and oPAHs and MCE (*left*) and fuel moisture (*right*) from residential fuel wood combustion. Modified from Shen et al. (2012) with permission of American Chemical Society

Table 6.10 Correlation coefficients between PAH derivatives and corresponding parent PAHs

Derivative Parent	1N-NAP NAP	2N-NAP NAP	5N-ACE ACE	2N-FLO FLO	9N-ANT ANT	9N-PHE PHE	3N-PHE PHE
r	0.843	0.829	0.375	0.684	0.810	0.858	0.771
p	3.4×10^{-8}	9.2×10^{-8}	2.1×10^{-1}	4.2×10^{-2}	3.1×10^{-7}	2.1×10^{-8}	2.7×10^{-5}
N	27	27	13	9	27	26	22

Derivative Parent	3N-FLA FLA	1N-PYR PYR	9FO FLO	ATQ ANT	BaAQ BaA
r	0.465	0.890	0.838	0.916	0.588
p	2.2×10^{-2}	3.9×10^{-6}	4.8×10^{-8}	2.1×10^{-11}	1.3×10^{-3}
N	24	16	27	27	27

Adapted from Shen et al. (2012) with permission of American Chemical Society

and wood for cooking, Ding et al. (2012) measured ambient pPAHs, nPAHs, and oPAHs concentrations in indoor and outdoor air. The results were calculated for the R_N and R_O. In comparison to the present study, R_N and R_O ratios in ambient air were significant higher than in emissions from wood combustion ($p < 0.05$), especially for R_N in summer. For example, the average 1 N-NAP/NAP ratios in ambient air were 1.17×10^{-2} and 2.89×10^{-1} in winter and summer, respectively, and both ambient air ratios were significantly higher than the same ratio of 7.88×10^{-4} measured in this study for primary wood combustion.

As mentioned above, R_O was positively correlated with EF_{pPAHs} for coal combustion and negatively correlated with EF_{pPAHs} for crop residue combustion, but for the wood combustion, no significant relationship was found (Table 6.11). It appears that these ratios may be useful for identifying emissions from these different combustion sources, if more field data would be available and a sound conclusion could be reached.

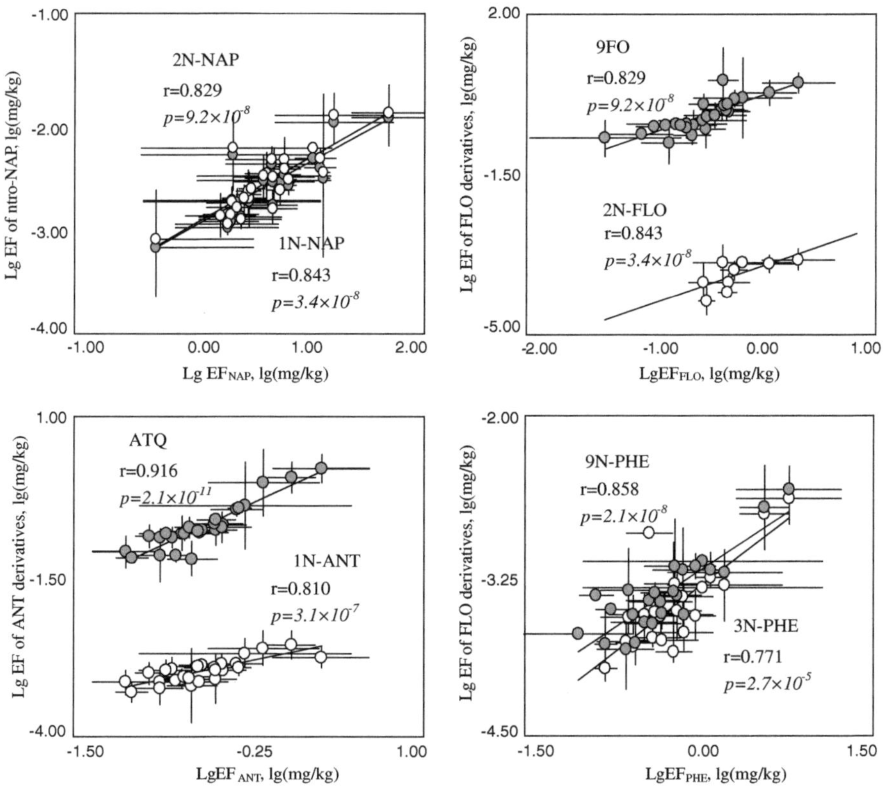

Fig. 6.14 Correlations between the measured EFs PAH derivatives and EFs of the corresponding parent PAHs. The data are log-transformed. Means and standard deviations are shown. Modified from Shen et al. (2012) with permission of American Chemical Society

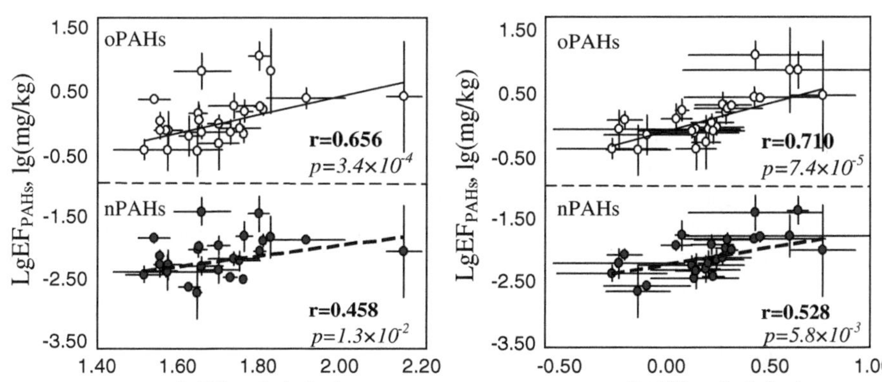

Fig. 6.15 Relationships between oPAHs/nPAHs and EF_{CO} and between oPAHs/nPAHs and EF_{PM} from residential wood combustion. Modified from Shen et al. (2012) with permission of American Chemical Society

Table 6.11 Correlation coefficient and *p* values between EF ratios of derivatives to parent PAHs (EF_{nPAHs}/EF_{pPAHs}), and EF_{pPAHs}

Ratio	1N-NAP/NAP	2N-NAP/NAP	9N-ANT/ANT	9N-PHE/PHE	3N-PHE/PHE
Parent	NAP	NAP	ANT	PHE	PHE
r	−0.481	−0.379	−0.719	0.107	−0.234
p	2.0×10^{-2}	7.4×10^{-2}	1.1×10^{-4}	6.7×10^{-1}	3.5×10^{-1}
N	23	23	23	18	18

Ratio	3N-FLA/FLA	9FO/FLO	ATQ/ANT	BaAQ/BaA
Parent	FLA	FLO	ANT	BaA
r	−0.153	−0.213	−0.321	−0.449
p	5.2×10^{-1}	3.3×10^{-1}	1.4×10^{-1}	3.1×10^{-2}
N	20	23	23	23

Adapted from Shen et al. (2012) with permission of American Chemical Society

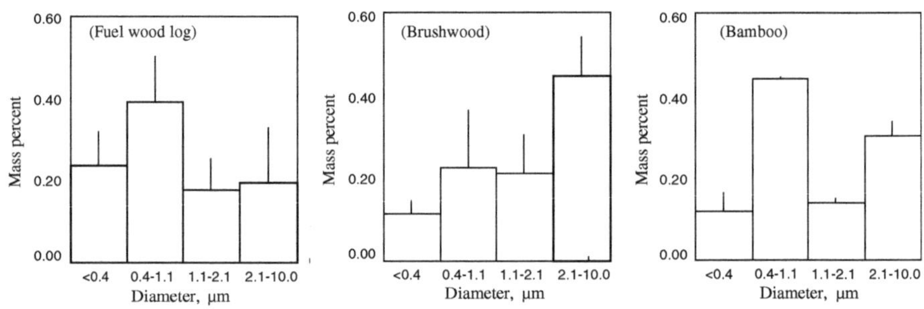

Fig. 6.16 Size distributions of oPAHs in emissions from residential fuel wood log, brushwood and bamboo combustion in a brick cooking stove

6.3.2.1 Gas-Particle Partitioning and Size Distribution

The size distributions of particle-bound oPAHs were similar between fuel wood and bamboo, which were different from that of brushwood (Fig. 6.16). oPAHs associated with $PM_{0.4-1.1}$ comprised up to 39.4 ± 10.9 % and 44.2 ± 0.3 % of the total for fuel wood and bamboo, respectively. A total of 80.6 and 70.0 % of particulate phase oPAHs were in fine $PM_{2.1}$. For brushwood, the mass percentage of oPAHs in coarse $PM_{2.1-10}$ was 44.9 ± 9.3 % of the total particle-bound oPAHs, significantly higher than those of 19.4 ± 13.5 % for fuel wood and 30.0 ± 3.6 % for bamboo. Figure 6.17 shows the distributions of four oPAH individuals. In the fine PM such as $PM_{0.4}$ and $PM_{0.4-1.1}$, the mass percents of high molecular weight BZO and BaAQ were generally higher than those for low molecular weight 9FO and ATQ.

For nitrated PAHs, due to much lower emission concentration, we only detected 1 N-NAP and 2 N-NAP in size segregated samples. The other nitrated PAHs including 9 N-ANT, 9 N-PHE, 3 N-PHE, 3 N-FLA and 1 N-PYR, were only detectable in fine $PM_{0.4}$. Figure 6.18 shows the size distributions of particulate

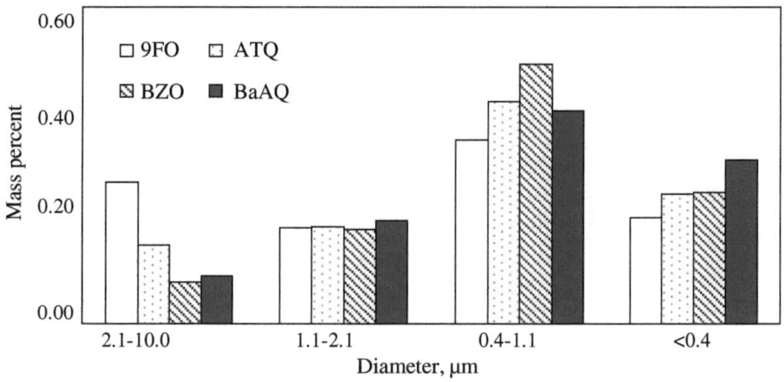

Fig. 6.17 Size distributions of oPAH individuals from residential wood combustion

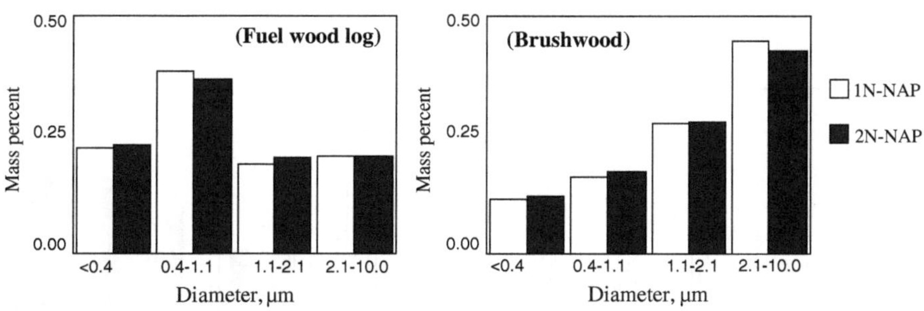

Fig. 6.18 Size distributions of particulate phase 1 N-NAP and 2 N-NAP from residential fuel wood log and brushwood combustions

1 N-NAP and 2 N-NAP. Again, distinct distributions were between the emissions from brushwood and fuel wood log combustions. The later produced much more coarse PM, and also higher mass percentages of total particulate phase PAH derivatives in coarse particle. The mass percents of 1 N-NAP and 2 N-NAP in coarse PM were 42.9 ± 1.2 and 44.9 ± 1.0 % of the total particulate phase nitrated NAP. In the emission from fuel wood log combustion, most nitrated PAHs were in fine particle, with over 80 % in fine $PM_{2.5}$. There were 36.6 ± 9.8 and 38.5 ± 11.3 % of total 1 N-NAP and 2 N-NAP present in $PM_{0.4-1.1}$.

The calculated K_p values of PAH derivatives increased in general with the increase of compound molecular weight (Fig. 6.19), indicating the preferable present in particulate phase for high molecular weight organics. Figure 6.20 compares the calculated K_p values of PAH derivatives and that for the corresponding parent PAHs. The Kp of most derivatives were significantly higher than those of pPAHs, except 3 N-FLA, 1 N-PYR, and BaAQ ($p > 0.05$). In comparison with the corresponding pPAHs, the derivatives often have lower vapor pressures and are more likely to be present in the particulate phase.

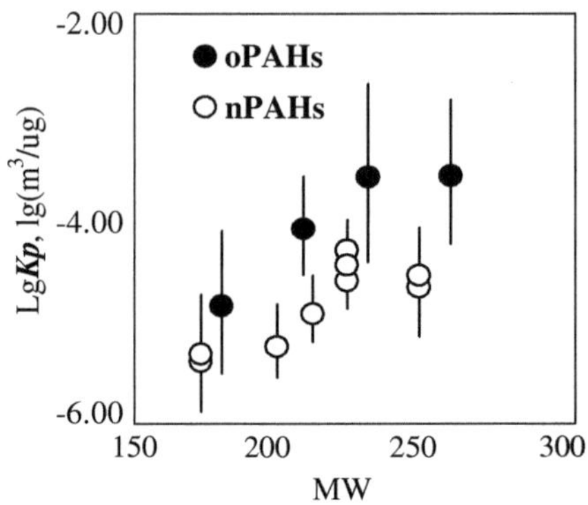

Fig. 6.19 Relationship between K_p of various PAH derivatives and their molecular weight from residential wood combustion. Modified from Shen et al. (2012) with permission of American Chemical Society

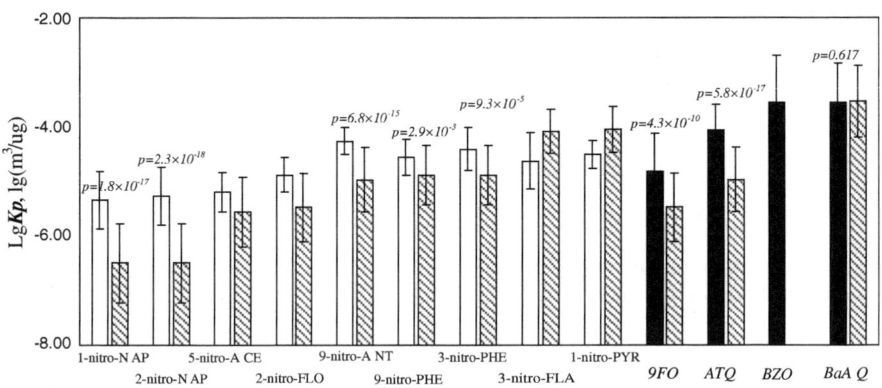

Fig. 6.20 Comparison of calculated Kp of pPAHs and their derivatives from residential wood combustion. Modified from Shen et al. (2012) with permission of American Chemical Society

Gas-particle partitioning is thought to be controlled by the relative importance of absorption into organic matter and adsorption onto the particle surface. Unfortunately, P_L^0 and K_{OA} values were not available for the nPAHs and oPAHs. However, based on the positive correlations between EFs of pPAHs and EFs of their derivatives, it might be speculated that the partitioning mechanism for nPAHs and oPAHs is similar to that for pPAHs, suggesting the governance of absorption. In fact, both pPAHs and their derivatives were positively correlated with OC ($p < 0.05$), suggesting that the absorption mechanism may be dominant for these freshly emitted PAH derivatives.

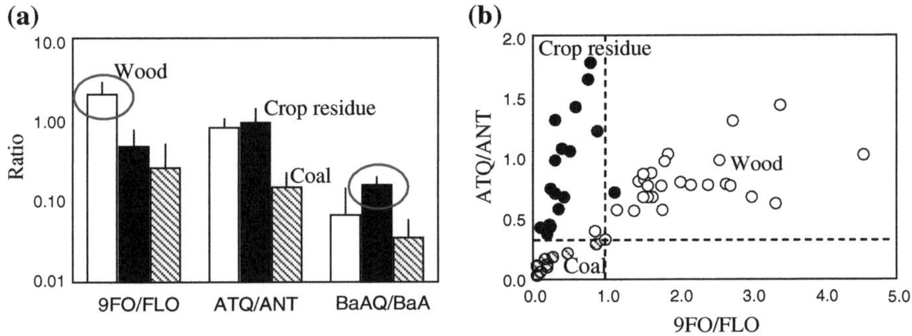

Fig. 6.21 Comparison of oxygenated-parent PAH ratio among different solid fuel types. **a** The ratios of 9FO/FLO, ATQ/ANT and BaAQ/BaA among different fuel types. **b** comparison of ATQ/ANT and 9FO/FLO

6.4 Fuel Comparison

6.4.1 Emission Factor

In regard of EF_{OPAHs} from residential coal combustion, anthracite briquette had the lowest EF of 0.049 ± 0.009 mg/kg. For bituminous coal, briquette bituminous was only 0.29 ± 0.02 mg/kg while for the other three raw bituminous chunk, EF_{OPAHs} averaged at 16.6 mg/kg. EF_{OPAHs} for fuel wood log, brushwood and crop residue were 1.2 ± 1.9, 5.6 ± 4.3 and 4.5 ± 2.0 mg/kg, respectively. In general, emission factor of oPAHs for different fuel types followed the tendency of briquette < fuel wood log < brushwood–crop residue < chunk.

6.4.2 Correlation Between Parent PAHs and Derivatives

In both coal and biomass fuel burning, emission factors of oxygenated PAHs were positively correlated with the corresponding parent PAHs. In comparison with parent PAHs, oPAHs are more preferable to partition into the particulate phase, particularly fine particles. The gas-particle partitioning of these polar organics, similar to that for parent PAHs, was thought to be mainly controlled by the absorption rather than adsorption in fresh emissions.

It is interesting to note that the calculated oxygenation ratio, *Ro*, was positively correlated with emission factor of parent PAHs in emissions from coal combustion, but negatively correlated in crop residue burning emissions, and no significant correlation was found in emissions from wood combustion. It was found that wood combustion produced relatively higher ratios of 9FO/FLO while crop residue burning had high ratios of ATQ/ANT and BaAQ/BaA, as shown in Fig. 6.21. In comparison with biomass fuel, coal combustion emitted lower *Ro* ratios. It appears that such relationship could be used for the source apportionment of oxygenated PAHs. However, it should be pointed out that the results are only based on the limited sample size in the present study, and the ratio would change obviously once emitted into the atmosphere. The use of specific ratio for oPAH source apportionment need more detailed studies in future, in field and laboratory simulation study to reach a sound conclusion.

6.5 Summary

In the present study, emission factors of oxygenated and nitrated PAHs were in the first time reported for residential solid fuel combustion in China. These PAH derivatives are believed to cause more adverse health outcome than the parent PAHs, and hence of wide and growing concern all over the world. The results showed that EFs of oPAHs were generally within the same order of magnitude of that of parent PAHs while EFs of nPAHs were about 2–3 orders of magnitude lower.

The EFs of 4 oPAHs for crop residue, fuel wood log, brushwood and coal were 4.5 ± 2.0, 1.19 ± 1.87, 5.56 ± 4.32 and 0.049–40 mg/kg, respectively. Similar to the parent PAHs, the oPAH emission varied dramatically between raw chunk and briquette coal with a range of EF in 1–2 orders of magnitude.

PAH derivatives can be also present in either gaseous or particulate phases. But they have more tendencies to be associated with particle, particularly fine particles in comparison with parent PAHs. The gas-particle partitioning behaviors of these derivatives were thought to be mainly governed by absorption of organics rather than adsorption.

PAH derivatives positively correlated with parent PAHs. The ratio in emission factor between oPAHs and corresponding parent PAHs varies among different fuel types. It is interesting in future to investigate whether the ratio can be developed into the source apportionment of PAH derivatives with more data available from field measurement, laboratory experiment and modeling studies.

References

Albinet, A., Leoz-Garziandia, E., Budzinski, H., & Villenave, E. (2007). Polycyclic aromatic hydrocarbons (PAHs), nitrated PAHs and oxygenated PAHs in ambient air of the Marseilles area (South of France): Concentrations and sources. *Science of Total Environment, 384*, 280–292.

Albinet, A., Leoz-Garziandia, E., Budzinski, H., Villenave, E., & Jaffrezo, J. (2008a). Nitrated and oxygenated derivatives of polycyclic aromatic hydrocarbons in the ambient air of two French alpine valleys Part 1: Concentrations, sources and gas/particle partitioning. *Atmospheric Environment, 42*, 43–54.

Albinet, A., Leoz-Garziandia, E., Budzinski, H., Villenave, E., & Jaffrezo, J. (2008b). Nitrated and oxygenated derivatives of polycyclic aromatic hydrocarbons in the ambient air of two French alpine valleys Part 2: Particle size distribution. *Atmospheric Environment, 42*, 55–64.

Ding, J., Zhong, J., Yang, Y., Li, B., Shen, G., & Su, Y., et al. (2012). Occurrence and exposure to polycyclic aromatic hydrocarbons and their derivatives in a rural Chinese home through biomass fuelled cooking. *Environmental Pollution*, doi:10.1016/j.envpol.2011.10.008

Fine, P. M., Cass, G. R., & Simoneit, B. R. T. (2001). Chemical characterization of fine particle emissions from fireplace combustion of woods grown in the Northeastern United States. *Environmental Science and Technology, 35*, 2665–2675.

Fine, P. M., Cass, G. R., & Simoneit, B. R. T. (2002). Chemical characterization of fine particle emissions from fireplace combustion of woods grown in the Southern United States. *Environmental Science and Technology, 36*, 1442–1451.

Fine, P. M., Cass, G. R., & Simoneit, B. R. T. (2004a). Chemical characterization of fine particle emissions from fireplace combustion of wood stove combustion of prevalent United States tree species. *Environmental Engineering and Science, 21*, 705–721.

Fine, P. M., Cass, G. R., & Simoneit, B. R. T. (2004b). Chemical characterization of fine particle emissions from fireplace combustion of wood types grown in the Midwestern and Western United States. *Environmental Engineering and Science, 21*, 387–409.

Gullett, B. K., Touati, A., & Hays, M. D. (2003). PCDD/F, PCB, HxCBz, PAH, and PM emission factors for fireplaces and woodstove combustion in the San Francisco Bay region. *Environmental Science and Technology, 37*, 1758–1765.

Hays, M. D., Fine, P. M., Geron, C. D., Kleeman, M. J., & Gullett, B. K. (2005). Open burning of agricultural biomass: physical and chemical properties of particle-phase emissions. *Atmospheric Environment, 39*, 6747–6764.

Howard, P. H., & Meylan, W. M. (Eds.). (1997). *Handbook of physical properties of organic chemicals*. Florida: CRC Press.

Mackay, D., Shiu, W. Y., Ma, K., & Lee, S. (Eds.). (1992) *Handbook of physical-chemical properties and environmental fate for organic chemicals* (2nd ed.). CRC Press: Florida.

Rogge, W. F., Hildemann, L. M., Mazurek, M. A., Cass, G. R., & Simoneit, B. R. T. (1998). Sources of fine organic aerosol 9. Pine, oak and synthetic log combustion in residential fireplaces. *Environmental Science and Technology, 32*, 13–22.

Shen, G. F., Tao, S., Wang, W., Yang, Y., Ding, H., Xue, M., Mi, Y., Zhu, C., Shen, H., Li, W., Wang, B., Wang, R., Wang, W., Wang, X., Russell, A. (2011) Emissions of Oxygenated Polycyclic Aromatic Hydrocarbons from indoor solid fuel combustion. *Environmental Science & Technology*, *45*, 3459–3465.

Shen, G. F., Tao, S., Wei, S., Zhang, Y., Wang, R., Wang, B., Li, W., Shen, H., Huang, Y., Chen, Y. C., Chen, H., Yang, Y., Wang, W., Wang, X., Liu, W., Simonich, S. (2012) Emissions of parent, nitro, and oxygenated polycyclic aromatic hydrocarbons from residential wood combustion in rural China. *Environmental Science & Technology, 46*, 8123–8130.

Walgraeve, C., Demesstere, K., Dewulf, J., Zimmermann, R., & van Langenhove, H. (2010). Oxygenated polycyclic aromatic hydrocarbons in atmospheric particulate matter: Molecular characterization and occurrence. *Atmospheric Environment, 44*, 1831–1846.

Wang, W., Jariyasopit, N., Schrlau, J., Jia, Y., Tao, S., & Yu, T., et al. (2011). Concentration and photochemistry of PAHs, NPAHs, and OPAHs and toxicity of PM2.5 during the Beijing Olympic Games. *Environmental Science and Technolgy*, *45,* 6887–6895.

Chapter 7
Field Measurement

In the built kitchen, we simulated the combustions of 9 crop residues, 26 wood materials and 5 coals, and measured emission factors of PM and PAHs. Two stoves were used. Because of large differences in fuel properties and burning conditions, the measured EFs varied dramatically, and identified influencing factors only explained the variations partly. In rural area, there are many different fuel/stove combinations in rural China. The burning process and subsequent pollutant emissions varied dramatically from site to site. It is realized that more studies are required, and more important, field studies are preferable since the field measurement can reflect the real fuel burning process and are more reliable. As a result, two field measurement campaigns were conducted in rural Shanxi and Jiangsu provinces.

7.1 Indoor Crop Residue Burning in Rural Jiangsu

7.1.1 Emission Factor

Field measurement on pollutant emissions from indoor crop residue burning was conducted in rural Jiangsu. Two stoves with same design but different usage were used. By comparing the emissions between the two stoves, it can be better to understand the historical change of pollutant emissions from indoor crop residue burning.

The measured EF_{PM}, EF_{OC}, EF_{EC}, EF_{28pPAH}, and EF_{4oPAH} are listed in Tables 7.1 and 7.2 for each fuel/stove combination in detail. For all stove-fuel combinations, arithmetic means and standard deviations of EF_{PM}, EF_{OC}, and EF_{EC} were 9.1 ± 5.7, 2.6 ± 2.9, and 1.1 ± 1.2 g/kg, respectively, which were higher than the corresponding median values, indicating right-skewed frequency distributions. In fact, log-normal distributions of EFs are often reported for residential coal and biomass burning. Similarly, EF_{28pPAH} and EF_{4oPAH} were also right-skewed with arithmetic means higher than median values. Average EF of the total

G. Shen, *Emission Factors of Carbonaceous Particulate Matter and Polycyclic Aromatic Hydrocarbons from Residential Solid Fuel Combustions*, Springer Theses,
DOI: 10.1007/978-3-642-39762-2_7,

Table 7.1 EF_{PM}, EF_{OC}, EF_{EC}, EF_{28pPAH}, and EF_{4oPAH} for indoor crop residue burning in the two stoves

Stove	Fuel	PM g/kg	OC g/kg	EC g/kg	pPAHs mg/kg	oPAHs mg/kg
Stove 1	wheat straw	9.8 ± 4.7	2.6 ± 1.8	0.46 ± 0.19	310 ± 128	17 ± 9.8
	rape straw	3.7 ± 3.0	0.58 ± 0.17	0.47 ± 0.48	561 ± 348	8.2 ± 7.2
	rice straw	5.2 ± 2.9	1.1 ± 0.9	0.51 ± 0.37	257 ± 105	22 ± 3.5
	cotton straw	5.7 ± 4.6	3.1 ± 4.5	1.2 ± 1.4	122 ± 53	2.2 ± 2.2
Stove 2	wheat straw	17 ± 7	4.1 ± 1.8	1.4 ± 0.9	122 ± 41	7.3 ± 6.3
	rape straw	13 ± 5	4.9 ± 6.5	2.7 ± 2.4	276 ± 88	14 ± 11
	rice straw	8.2 ± 2.8	1.5 ± 0.6	0.64 ± 0.19	166 ± 5	6.9 ± 6.3
	cotton straw	10 ± 4	2.5 ± 1.8	0.96 ± 0.90	188 ± 119	7.0 ± 12
Mean ± standard deviation		9.1 ± 5.7	2.6 ± 2.9	1.1 ± 1.2	259 ± 189	11 ± 14
Median		7.9	1.4	0.71	176	7.1
Geometric mean		7.3	1.6	0.67	214	3.5

Arithmetic means and standard deviations from triplicate measurements are shown for individual stove-fuel combinations and overall means (standard deviations), median, and geometric means are also listed. Reprinted from Environmental Pollution 184, Wei et al., Field measurement on the emissions of PM, OC, EC and PAHs from indoor crop straw burning in rural China, 18–24, with the permission from Elsevier

16 U.S. EPA priority PAHs, which are often reported in the literature, was 252 ± 185 mg/kg.

The measured EF_{PM}, EF_{OC}, and EF_{EC} in this study are well correlated. Whether the data for the two stoves are plotted separately or together, the correlations between EF_{PM} and EF_{OC} and between EF_{PM} and EF_{EC} are all significant ($p < 0.05$). For both new and old stoves considered, the slopes of the regression equations were 1.18 ($R^2 = 0.748$) for EF_{PM} against EF_{OC}, and 1.15 ($R^2 = 0.713$) for EF_{PM} against EF_{EC}, respectively. EFs of various parent PAHs were generally correlated with one another, especially for those with similar molecular weight, with a single exception of RET.

Similarly, the measured EFs of individual oPAH compounds correlated well with each another ($p < 0.05$), suggesting similar formation processes of them. Significant correlations between oPAHs and their corresponding parent PAHs were identified for the 1-year stove, but there was no such relationship for the old stove (Fig. 7.1). It appears that the aging of stoves affected the formation and emission of pPAHs ($p = 0.082$) and oPAHs ($p = 0.742$) differently.

It is noted that the variations of the measured EFs were much high although only two stoves and four crop residues were tested. The calculated coefficients of variations for EF_{PM}, EF_{OC}, EF_{EC}, EF_{28pPAH}, and EF_{4oPAH} were 63, 114, 113, 73, and 130 %, respectively. Such a high variation is sometimes expected since a lot of factors, such as crop type, stove design, burning temperature, air supply, fueling and fire management behaviors, can affect the EFs.

Table 7.2 Emission factors of pPAH and oPAH individuals for each fuel/stove combination

Stove	Stove 1: 1 year new stove				Stove 2: 15-year old stove			
Fuel	Wheat straw	Rape straw	Rice straw	Cotton straw	Wheat straw	Rape straw	Rice straw	Cotton straw
NAP	$1.3 \pm 0.6 \times 10^{2}$	$2.4 \pm 1.6 \times 10^{2}$	$8.0 \pm 4.3 \times 10^{1}$	$4.6 \pm 1.7 \times 10^{1}$	$4.3 \pm 2.0 \times 10^{1}$	$1.1 \pm 0.5 \times 10^{2}$	$8.0 \pm 0.5 \times 10^{1}$	$7.6 \pm 6.3 \times 10^{1}$
ACY	$4.5 \pm 1.9 \times 10^{1}$	$1.1 \pm 0.8 \times 10^{2}$	$4.2 \pm 2.2 \times 10^{1}$	$1.7 \pm 0.8 \times 10^{1}$	$1.6 \pm 0.9 \times 10^{1}$	$4.7 \pm 1.5 \times 10^{1}$	$2.6 \pm 0.6 \times 10^{1}$	$2.8 \pm 1.8 \times 10^{1}$
ACE	$3.9 \pm 1.6 \times 10^{0}$	$5.7 \pm 3.7 \times 10^{0}$	$2.1 \pm 0.9 \times 10^{0}$	$1.6 \pm 0.4 \times 10^{0}$	$1.9 \pm 0.8 \times 10^{0}$	$2.8 \pm 0.9 \times 10^{0}$	$2.4 \pm 0.8 \times 10^{0}$	$2.4 \pm 1.4 \times 10^{0}$
FLO	$1.1 \pm 0.4 \times 10^{1}$	$2.0 \pm 1.4 \times 10^{1}$	$9.2 \pm 4.9 \times 10^{0}$	$4.5 \pm 2.1 \times 10^{0}$	$4.7 \pm 2.2 \times 10^{0}$	$9.6 \pm 3.1 \times 10^{0}$	$6.4 \pm 1.8 \times 10^{0}$	$7.1 \pm 4.1 \times 10^{0}$
PHE	$4.2 \pm 1.6 \times 10^{1}$	$6.7 \pm 4.9 \times 10^{1}$	$4.2 \pm 2.0 \times 10^{1}$	$2.3 \pm 1.1 \times 10^{1}$	$1.8 \pm 0.7 \times 10^{1}$	$3.3 \pm 1.0 \times 10^{1}$	$1.9 \pm 0.1 \times 10^{1}$	$2.9 \pm 1.6 \times 10^{1}$
ANT	$7.4 \pm 2.9 \times 10^{0}$	$1.4 \pm 1.0 \times 10^{1}$	$8.6 \pm 4.4 \times 10^{0}$	$3.9 \pm 2.2 \times 10^{0}$	$3.2 \pm 1.3 \times 10^{0}$	$6.8 \pm 2.2 \times 10^{0}$	$3.3 \pm 0.2 \times 10^{0}$	$5.3 \pm 3.2 \times 10^{0}$
FLA	$2.0 \pm 0.8 \times 10^{1}$	$2.6 \pm 0.9 \times 10^{1}$	$2.0 \pm 0.9 \times 10^{1}$	$7.4 \pm 2.1 \times 10^{0}$	$9.0 \pm 1.9 \times 10^{0}$	$1.6 \pm 0.5 \times 10^{1}$	$8.7 \pm 0.4 \times 10^{0}$	$1.2 \pm 0.7 \times 10^{1}$
PYR	$2.2 \pm 0.8 \times 10^{1}$	$3.0 \pm 1.0 \times 10^{1}$	$2.3 \pm 1.2 \times 10^{1}$	$8.5 \pm 1.8 \times 10^{0}$	$1.1 \pm 0.2 \times 10^{1}$	$2.1 \pm 0.5 \times 10^{1}$	$9.7 \pm 0.1 \times 10^{0}$	$1.4 \pm 0.9 \times 10^{1}$
RET	$4.3 \pm 2.2 \times 10^{-1}$	$4.9 \pm 3.6 \times 10^{-1}$	$2.3 \pm 0.3 \times 10^{-1}$	$5.0 \pm 4.8 \times 10^{-1}$	$1.8 \pm 0.7 \times 10^{-1}$	$2.3 \pm 0.3 \times 10^{-1}$	$2.5 \pm 0.3 \times 10^{-1}$	$2.7 \pm 1.5 \times 10^{-1}$
BcP	$1.6 \pm 0.7 \times 10^{0}$	$1.9 \pm 0.3 \times 10^{0}$	$1.6 \pm 0.7 \times 10^{0}$	$6.7 \pm 2.9 \times 10^{-1}$	$8.7 \pm 1.6 \times 10^{-1}$	$1.5 \pm 0.4 \times 10^{0}$	$6.8 \pm 0.2 \times 10^{-1}$	$7.6 \pm 1.3 \times 10^{-1}$
CPP	$4.6 \pm 2.1 \times 10^{0}$	$8.7 \pm 3.2 \times 10^{0}$	$5.4 \pm 3.9 \times 10^{0}$	$1.0 \pm 0.6 \times 10^{0}$	$2.5 \pm 0.8 \times 10^{0}$	$5.9 \pm 0.5 \times 10^{0}$	$1.8 \pm 0.2 \times 10^{0}$	$1.8 \pm 0.6 \times 10^{0}$
BaA	$4.0 \pm 1.8 \times 10^{0}$	$5.3 \pm 1.1 \times 10^{0}$	$4.2 \pm 2.0 \times 10^{0}$	$1.6 \pm 0.9 \times 10^{0}$	$2.3 \pm 0.5 \times 10^{0}$	$4.1 \pm 0.8 \times 10^{0}$	$1.6 \pm 0.1 \times 10^{0}$	$1.8 \pm 0.4 \times 10^{0}$
CHR	$4.6 \pm 2.0 \times 10^{0}$	$5.5 \pm 1.0 \times 10^{0}$	$4.8 \pm 2.1 \times 10^{0}$	$2.2 \pm 1.4 \times 10^{0}$	$2.8 \pm 0.5 \times 10^{0}$	$4.4 \pm 0.8 \times 10^{0}$	$2.2 \pm 0.1 \times 10^{0}$	$2.3 \pm 0.4 \times 10^{0}$
BbF	$3.8 \pm 1.9 \times 10^{0}$	$6.5 \pm 1.8 \times 10^{0}$	$4.5 \pm 2.6 \times 10^{0}$	$1.8 \pm 1.5 \times 10^{0}$	$2.2 \pm 0.5 \times 10^{0}$	$4.7 \pm 0.8 \times 10^{0}$	$1.5 \pm 0.1 \times 10^{0}$	$1.9 \pm 0.9 \times 10^{0}$
BkF	$1.2 \pm 0.5 \times 10^{0}$	$2.1 \pm 0.5 \times 10^{0}$	$1.4 \pm 0.7 \times 10^{0}$	$6.4 \pm 4.9 \times 10^{-1}$	$7.7 \pm 1.6 \times 10^{-1}$	$1.7 \pm 0.3 \times 10^{0}$	$6.0 \pm 0.2 \times 10^{-1}$	$7.5 \pm 3.5 \times 10^{-1}$
BeP	$1.2 \pm 0.6 \times 10^{0}$	$2.2 \pm 0.6 \times 10^{0}$	$1.4 \pm 0.8 \times 10^{0}$	$6.1 \pm 5.0 \times 10^{-1}$	$7.4 \pm 1.6 \times 10^{-1}$	$1.6 \pm 0.3 \times 10^{0}$	$5.4 \pm 0.3 \times 10^{-1}$	$7.2 \pm 3.3 \times 10^{-1}$
BaP	$2.3 \pm 1.2 \times 10^{0}$	$4.6 \pm 1.4 \times 10^{0}$	$2.9 \pm 1.7 \times 10^{0}$	$9.7 \pm 8.6 \times 10^{-1}$	$1.3 \pm 0.3 \times 10^{0}$	$3.2 \pm 0.6 \times 10^{0}$	$7.9 \pm 0.7 \times 10^{-1}$	$1.1 \pm 0.6 \times 10^{0}$
PER	$3.9 \pm 2.2 \times 10^{-1}$	$8.0 \pm 2.5 \times 10^{-1}$	$4.8 \pm 3.0 \times 10^{-1}$	$1.5 \pm 1.2 \times 10^{-1}$	$2.1 \pm 0.6 \times 10^{-1}$	$5.3 \pm 1.1 \times 10^{-1}$	$1.2 \pm 0.2 \times 10^{-1}$	$1.9 \pm 1.0 \times 10^{-1}$
IcdP	$9.7 \pm 5.5 \times 10^{-1}$	$2.2 \pm 0.9 \times 10^{0}$	$1.3 \pm 0.9 \times 10^{0}$	$3.6 \pm 2.9 \times 10^{-1}$	$4.5 \pm 1.6 \times 10^{-1}$	$1.3 \pm 0.4 \times 10^{0}$	$2.9 \pm 0.2 \times 10^{-1}$	$3.8 \pm 2.0 \times 10^{-1}$
DahA	$9.3 \pm 4.8 \times 10^{-2}$	$1.8 \pm 0.6 \times 10^{-1}$	$1.2 \pm 0.7 \times 10^{-1}$	$4.1 \pm 2.9 \times 10^{-2}$	$4.1 \pm 1.2 \times 10^{-2}$	$1.1 \pm 0.3 \times 10^{-1}$	$4.5 \pm 0.8 \times 10^{-2}$	$4.6 \pm 2.1 \times 10^{-2}$
BghiP	$6.3 \pm 0.3 \times 10^{-1}$	$1.5 \pm 0.6 \times 10^{0}$	$8.8 \pm 5.9 \times 10^{-1}$	$2.5 \pm 1.9 \times 10^{-1}$	$3.3 \pm 1.2 \times 10^{-1}$	$9.1 \pm 2.4 \times 10^{-1}$	$2.1 \pm 0.1 \times 10^{-1}$	$2.9 \pm 1.6 \times 10^{-1}$
DacP	$2.7 \pm 1.6 \times 10^{-1}$	$7.4 \pm 4.5 \times 10^{-1}$	$4.0 \pm 3.0 \times 10^{-1}$	$6.4 \pm 5.1 \times 10^{-2}$	$1.0 \pm 0.4 \times 10^{-1}$	$3.5 \pm 1.2 \times 10^{-1}$	$7.8 \pm 1.8 \times 10^{-2}$	$8.6 \pm 4.7 \times 10^{-2}$
DalP	$7.5 \pm 3.2 \times 10^{-2}$	$2.0 \pm 0.8 \times 10^{-1}$	$1.1 \pm 0.6 \times 10^{-1}$	$4.2 \pm 3.9 \times 10^{-2}$	$4.7 \pm 1.7 \times 10^{-2}$	$1.1 \pm 0.4 \times 10^{-1}$	$3.6 \pm 0.0 \times 10^{-2}$	$4.3 \pm 2.0 \times 10^{-2}$
DaeF	$5.1 \pm 2.6 \times 10^{-2}$	$1.1 \pm 0.5 \times 10^{-1}$	$8.3 \pm 4.8 \times 10^{-2}$	$2.1 \pm 1.8 \times 10^{-2}$	$2.6 \pm 0.9 \times 10^{-2}$	$7.1 \pm 2.8 \times 10^{-2}$	$2.1 \pm 0.3 \times 10^{-2}$	$2.1 \pm 1.5 \times 10^{-2}$
COR	$5.7 \pm 2.8 \times 10^{-2}$	$2.0 \pm 1.0 \times 10^{-1}$	$1.0 \pm 0.7 \times 10^{-1}$	$2.1 \pm 1.3 \times 10^{-2}$	$2.7 \pm 1.1 \times 10^{-2}$	$8.3 \pm 3.0 \times 10^{-2}$	$2.0 \pm 0.3 \times 10^{-2}$	$2.3 \pm 0.8 \times 10^{-2}$
DaeP	$1.6 \pm 0.6 \times 10^{-2}$	$4.5 \pm 2.0 \times 10^{-2}$	$2.6 \pm 1.3 \times 10^{-2}$	$9.8 \pm 0.8 \times 10^{-3}$	$1.0 \pm 0.5 \times 10^{-2}$	$2.3 \pm 0.8 \times 10^{-2}$	$6.4 \pm 2.3 \times 10^{-3}$	$7.3 \pm 4.5 \times 10^{-3}$
DaiP	$2.1 \pm 0.7 \times 10^{-2}$	$5.8 \pm 2.2 \times 10^{-2}$	$3.5 \pm 2.1 \times 10^{-2}$	$1.0 \pm 1.2 \times 10^{-2}$	$1.2 \pm 0.5 \times 10^{-2}$	$3.5 \pm 0.8 \times 10^{-2}$	$4.2 \pm 6.0 \times 10^{-3}$	$9.2 \pm 6.3 \times 10^{-3}$

(continued)

Table 7.2 (continued)

Stove	Stove 1: 1 year new stove				Stove 2: 15-year old stove			
Fuel	Wheat straw	Rape straw	Rice straw	Cotton straw	Wheat straw	Rape straw	Rice straw	Cotton straw
DahP	$8.9 \pm 5.0 \times 10^{-3}$	$3.3 \pm 1.4 \times 10^{-2}$	$2.2 \pm 1.3 \times 10^{-2}$	n.d.	$5.2 \pm 3.8 \times 10^{-3}$	$1.8 \pm 1.0 \times 10^{-2}$	n.d.	$6.2 \pm 4.0 \times 10^{-3}$
Total P16	$\mathbf{3.0 \pm 1.2 \times 10^{2}}$	$\mathbf{5.5 \pm 3.4 \times 10^{2}}$	$\mathbf{2.5 \pm 1.0 \times 10^{2}}$	$\mathbf{1.2 \pm 0.5 \times 10^{2}}$	$\mathbf{1.2 \pm 0.4 \times 10^{2}}$	$\mathbf{2.7 \pm 0.9 \times 10^{2}}$	$\mathbf{1.6 \pm 0.1 \times 10^{2}}$	$\mathbf{1.8 \pm 1.2 \times 10^{2}}$
Total P28	$\mathbf{3.1 \pm 1.3 \times 10^{2}}$	$\mathbf{5.6 \pm 3.5 \times 10^{2}}$	$\mathbf{2.6 \pm 1.0 \times 10^{2}}$	$\mathbf{1.2 \pm 0.5 \times 10^{2}}$	$\mathbf{1.2 \pm 0.4 \times 10^{2}}$	$\mathbf{2.8 \pm 0.9 \times 10^{2}}$	$\mathbf{1.7 \pm 0.1 \times 10^{2}}$	$\mathbf{1.9 \pm 1.2 \times 10^{2}}$
9FO	$1.2 \pm 0.9 \times 10^{1}$	$6.1 \pm 6.4 \times 10^{0}$	$1.9 \pm 3.2 \times 10^{1}$	$3.1 \pm 2.4 \times 10^{-1}$	$4.9 \pm 4.8 \times 10^{0}$	$1.1 \pm 0.9 \times 10^{1}$	$5.1 \pm 4.5 \times 10^{0}$	$5.8 \pm 9.9 \times 10^{0}$
ATQ	$3.1 \pm 2.2 \times 10^{0}$	$1.3 \pm 1.1 \times 10^{0}$	$2.3 \pm 3.2 \times 10^{0}$	$5.3 \pm 4.6 \times 10^{-1}$	$1.4 \pm 1.1 \times 10^{0}$	$1.5 \pm 1.3 \times 10^{0}$	$1.2 \pm 1.3 \times 10^{0}$	$9.7 \pm 1.5 \times 10^{-1}$
BZO	$1.5 \pm 1.7 \times 10^{0}$	$6.0 \pm 8.9 \times 10^{-1}$	$5.4 \pm 7.2 \times 10^{-1}$	$1.3 \pm 1.4 \times 10^{0}$	$9.4 \pm 1.1 \times 10^{-1}$	$7.7 \pm 8.7 \times 10^{-1}$	$5.6 \pm 5.9 \times 10^{-1}$	$1.5 \pm 1.3 \times 10^{-1}$
BaAQ	$1.8 \pm 1.9 \times 10^{-1}$	$2.0 \pm 3.1 \times 10^{-1}$	$1.1 \pm 1.1 \times 10^{-1}$	$1.4 \pm 1.2 \times 10^{-1}$	$1.1 \pm 0.9 \times 10^{-1}$	$8.7 \pm 4.9 \times 10^{-2}$	$6.0 \pm 4.9 \times 10^{-2}$	$4.3 \pm 1.8 \times 10^{-2}$
Total oPAHs	$\mathbf{1.7 \pm 0.9 \times 10^{1}}$	$\mathbf{8.2 \pm 7.2 \times 10^{0}}$	$\mathbf{2.2 \pm 3.5 \times 10^{1}}$	$\mathbf{2.2 \pm 2.2 \times 10^{0}}$	$\mathbf{7.3 \pm 6.3 \times 10^{0}}$	$\mathbf{1.4 \pm 1.1 \times 10^{1}}$	$\mathbf{6.9 \pm 6.3 \times 10^{0}}$	$\mathbf{7.0 \pm 1.2 \times 10^{0}}$

Means and standard deviations from triplicate measurements are shown. Reprinted from Environmental Pollution 184, Wei et al., Field measurement on the emissions of PM, OC, EC and PAHs from indoor crop straw burning in rural China, 18–24, with the permission from Elsevier

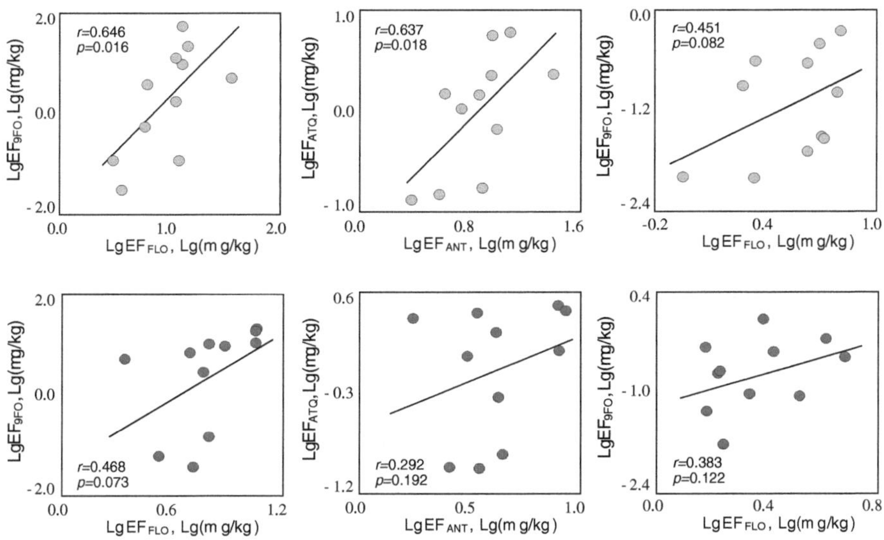

Fig. 7.1 Relationship between EFs of 9FO, ATQ, and BaAQ (from *left to right*) and EFs of their corresponding parent PAHs for crop straws burned in a 1-year old stove (*top row*) and a 15-year old stove (*bottom row*). Reprinted from Environmental Pollution 184, Wei et al., Field measurement on the emissions of PM, OC, EC and PAHs from indoor crop straw burning in rural China, 18–24, with the permission from Elsevier

7.1.2 Composition Profile and Isomer Ratios

Figure 7.2 shows compound profiles of PAHs for the two stoves. The composition profiles of the two stoves were very similar to each other, and also very similar among the crop residues. For parent PAHs, the emissions were dominated by NAP (39 ± 8 %), ACY (16 ± 3 %) and PHE (14 ± 2 %), followed by FLA (6.5 ± 1.5 %) and PYR (7.7 ± 2.0 %). The total contribution of the 20 high molecular weight parent PAHs (mw $\geq$ 228) contributed merely 9.0 % of the total.

The calculated means of several commonly used isomer ratios, including ANT/(ANT + PHE), FLA/(FLA + PYR), BaA/(BaA + CHR), IcdP/(IcdP + BghiP), BbF/(BbF + BkF), BaP/(BaP + BghiP), and BeP/(BeP + BaP), were 0.16 ± 0.01, 0.46 ± 0.02, 0.46 ± 0.03, 0.58 ± 0.01, 0.74 ± 0.02, 0.78 ± 0.02, and 0.36 ± 0.03, respectively. Insignificant differences were observed for these isomer ratio values among the four fuel types and the two stoves ($p > 0.05$). Among the 4 oxygenated PAHs, EFs of 9FO and ATQ were higher than the other two oPAHs, making up to 54 ± 28 and 26 ± 16 % of the total oPAHs, respectively.

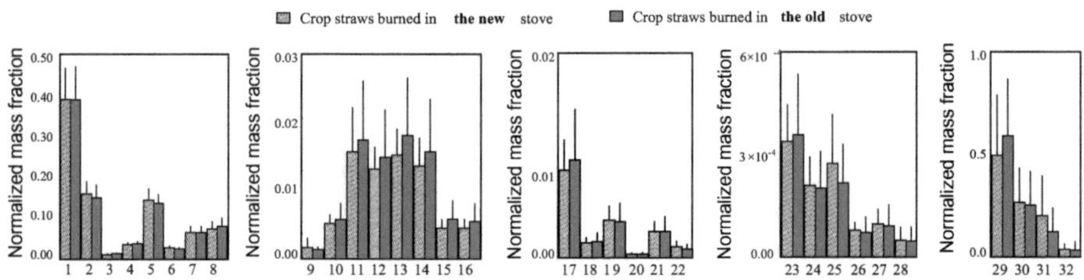

Fig. 7.2 Composition profiles of the measured pPAHs and oPAHs for crop residue burned in the new (1 year) and old (15 years) stoves. The results are arithmetic means (*bars*) and standard deviations (*sticks*) of the four crop residues. The compounds are 28 parent PAHs (1. NAP, 2. ACY, 3. ACE, 4. FLO, 5. PHE, 6. ANT, 7. FLA, 8. PYR, 9. RET, 10. BcP, 11. CPP, 12. BaA, 13. CHR, 14. BbF, 15. BkF, 16. BeP, 17. BaP, 18. PER, 19. IcdP, 20. DahA, 21. BghiP, 22. DacP, 23. DalP, 24. DaeF, 25. COR, 26. DaeP, 27. DaiP, and 28. DahP) and 4 oxygenated PAHs (29. 9FO, 30. ATQ, 31. BZO, 32. BaAQ). Reprinted from Environmental Pollution 184, Wei et al., Field measurement on the emissions of PM, OC, EC and PAHs from indoor crop straw burning in rural China, 18–24, with the permission from Elsevier

7.1.3 Influence of Fuel Type and Stove Age

Based on the measured EFs for the target pollutants, a two-way analysis of variance was conducted to test the influence of fuel type and stove usage. For all the pollutants, no interaction between the two factors (stove age and fuel type) was found ($p > 0.05$). Significant difference among the four fuel types was found for EF_{28pPAH} ($p = 0.027$). EF_{28pPAH} for the 4 tested crop residues followed the trend as rape straw ($4.2 \pm 2.8 \times 10^2$ mg/kg) > rice straw ($2.2 \pm 0.9 \times 10^2$ mg/kg) ~ wheat straw ($2.2 \pm 1.3 \times 10^2$ mg/kg) > cotton straw ($1.6 \pm 0.9 \times 10^2$ mg/kg).

The most significant difference occurred between the two stoves of different usage. With an exception of EF_{4oPAH} ($p = 0.742$), EF_{PM} ($p = 0.004$), EF_{OC} ($p = 0.090$), EF_{EC} ($p = 0.043$), and EF_{28pPAH} ($p = 0.082$) were all significant at different levels below 10 %, indicating that stove age is critical in terms of emissions of carbonaceous particles and PAHs. For example, the measured EF_{PM}, EF_{OC} and EF_{EC} for the stove used for about 1 year were 6.1 ± 4.1, 1.9 ± 2.4, and 0.66 ± 0.72 g/kg, respectively, which was some 2.5 times lower than 12 ± 5, 3.4 ± 3.4, and 1.5 ± 1.5 g/kg for the stove used for approximately 15 years. It was reported that EF_{PM} for residential wood combustion in a stove used for 1 year was 50 % higher than a new one of same type (Roden et al. 2009). The increase in EFs of carbonaceous particles in aged stoves might be explained by the stove degradation, like flue block after long time use. Such a difference in stove age adds more complexity to the variation in emissions, which should be taken into consideration in inventory development in the future.

Table 7.3 EFs of PM, OC, EC, total parent PAHs, oxy- and nitro-PAHs from residential combustions of coal briquette, coal cake, and wood

	Briquette	Coal cake	Wood
PM, g/kg	0.54–0.64	3.2–8.5	8.1–8.5
OC, g/kg	0.13–0.14	0.38–0.58	2.2–3.6
EC, g/kg	0.0040–0.0041	0.022–0.052	0.91–1.6
28 pPAHs, mg/kg	14–16	168–223	182–297
16 pPAHs, mg/kg	13–15	148–210	141–276
4 oPAHs, mg/kg	1.7–2.6	4.7–9.5	7.8–10
9 nPAHs, mg/kg	0.64–0.83	0.16–2.4	0.14–0.55

Data shown are the range from duplicate measurements. Reported values for PAHs and their derivatives are gas- and particle-phase are combined. Reprinted from Shen et al. (2013) with permission of American Chemical Society

7.2 Residential Coal and Wood Combustion in Rural Shanxi

7.2.1 Emission Factor

The measured EFs are summarized in Table 7.3 in terms of pollutant mass per burned fuel mass (g/kg). Besides a total EF of 28 pPAHs (EF_{28pPAH}), a total EF of 16 priority PAHs (EF_{16pPAH}) is also provided. EFs of individual pPAH, oPAH, and nPAH are listed in Table 7.4.

7.2.2 Coal Combustion

To the best of our knowledge, field measurements on emissions from residential coal briquette combustion in rural China have not been conducted previously. The ratios of EC/OC, EC/PM, and OC/PM for the coal briquette were 0.031 ± 0.002, 0.69 ± 0.07 % and 22 ± 4 %, respectively. These measured EFs were generally higher in comparison with those measured from combustions under laboratory conditions. For instance, EF_{PM}, EF_{OC}, and EF_{EC} of the coal briquettes (also from Shanxi) that were burned in an iron stove were 0.17 ± 0.010, 0.021 ± 0.0020, and 0.0042 ± 0.0021 g/kg, respectively. It had been also reported that EF_{PM} of honeycomb briquettes burned in residential metal stoves were in the range of 0.032–0.62 g/kg (Zhang et al. 2000). Coal properties including chemical compositions, volatile matter, and ash contents often vary widely, resulting in differences in combustion characteristics and emissions of various incomplete combustion byproducts. Therefore, the differences in EFs between this study and others could be caused by the varying properties of coal from which the briquettes were made of. In addition, stove type and configuration are also critical in emission measurements. It was estimated that the emissions of PM, OC, and EC for coals burned in improved stoves with an upper lid and chimney could be reduced by 56, 61, and

Table 7.4 EFs (mg/kg) of individual parent PAHs, oxy- and nitro-PAHs from residential briquette, coal cake, and wood combustions

	Briquette	Coal cake	Wood
Parent-PAHs			
naphthalene (NAP)	2.2–3.2	18–87	9.3–172
acenaphthylene (ACY)	0.59–0.82	32–42	15–28
acenaphthene (ACE)	0.052–0.086	1.7–2.1	1.3–2.0
fluorene (FLO)	0.29–0.42	9.5–9.6	4.1–4.9
phenanthrene (PHE)	3.0–3.5	26–29	16–22
anthracene (ANT)	0.33–0.38	8.1–9.1	3.5–4.6
fluoranthene (FLA)	2.2–2.3	11–14	14–26
pyrene (PYR)	1.6–1.7	8.6–11	13–23
retene (RET)	0.058–0.11	0.70–0.85	0.88–8.1
benzo[c]phenanthrene (BcP)	0.14–0.16	0.99–1.5	1.0–1.9
cyclopenta[c,d]pyrene (CPP)	0.022–0.11	7.8–12	12–22
benzo(a)anthracene (BaA)	0.58–0.74	4.3–6.6	4.5–8.1
chrysene (CHR)	1.2–1.3	2.8–4.2	3.6–6.6
benzo(b)fluoranthene (BbF)	0.31–0.38	2.0–2.9	2.7–4.2
benzo(k)fluoranthene (BkF)	0.24–0.30	1.8–2.5	3.0–4.5
benzo(a)pyrene (BaP)	0.20–0.22	0.95–1.3	1.7–2.6
benzo(e)pyrene (BeP)	0.068–0.077	2.4–3.4	3.8–5.9
perylene (PER)	0.0094–0.0096	0.43–0.55	0.62–0.96
indeno(l,2,3-cd)pyrene (IcdP)	0.034–0.035	1.8–2.7	2.7–3.9
dibenz(a,h)anthracene (DahA)	0.018–0.019	0.19–0.32	0.33–0.39
benzo(g,h,i)perylene (BghiP)	0.033–0.034	0.86–1.3	1.9–2.7
dibenzo[a,c]pyrene (DacP)	0.0014–0.0018	0.67–0.67	0.78–1.5
dibenzo[a,l]pyrene (DalP)	0.0050–0.0067	0.41–0.58	0.36–0.55
dibenzo[a,e]flluoranthene (DaeF)	0.0022–0.0039	0.31–0.39	0.33–0.37
Coronene(COR)	0.0055–0.0066	0.75–1.2	2.2–2.6
dibenzo[a,e]pyrene (DaeP)	0.0084–0.012	0.23–0.39	0.16–0.39
dibenzo[a,i]pyrene (DaiP)	0.0014	0.27–0.47	0.22–0.45
dibenzo[a,h]pyrene (DahP)	n.d.*	0.048–0.067	0.053–0.097
Oxy-PAHs			
9-fluorenone (9FO)	1.1–1.8	3.4–4.1	2.9–4.1
anthracene-9,10-dione (ATQ)	0.50–0.68	0.69–1.5	1.5–2.0
benzanthrone (BZO)	0.077–0.16	0.65–3.8	3.44–3.89
benzo[a]anthracene-dione(BaAQ)	0.014–0.018	0.018–0.062	0.096–0.15
Nitro-PAHs			
1-nitro-naphthalene (1N-NAP)	0.00038–0.00049	0.0037–0.0041	0.0098–0.011
2-nitro-naphthalene (2N-NAP)	0.00071–0.0012	0.0043–0.0045	0.010–0.011
5-nitro-acenaohthene (5N-ACE)	0.00032–0.00062	0.0011–0.0033	0.0054–0.0070
2-nitro-fluorene (2N-FLO)	0.00063–0.00067	0.0012–0.0033	0.0047–0.0068
9-nitro-anthracene (9N-ANT)	0.47–0.59	0.13–2.1	0.04–0.33
9-nitro-phenanthrene (9N-PHE)	0.16–0.24	0.016–0.35	0.042–0.18
3-nitro-phenanthrene (3N-PHE)	0.00022–0.00025	0.00030–0.00088	0.0015–0.0020
3-nitro-fluoranthene (3N-FLA)	0.00073–0.00075	0.00057–0.0017	0.0052–0.010
1-nitro-pyrene (1N-PYR)	0.0011–0.0017	0.0030–0.0053	0.0059

Data shown are the range from duplicate measurements. Reprinted from Shen et al. (2013) with permission of American Chemical Society

*n.d. not detected

14 % compared to emissions from a traditional stove with a lid near the bottom and without a chimney (Zhi et al. 2009).

For PAHs and their derivatives, measured EFs are again much higher than those measured in the laboratory experiment of coal briquette combustion (14 ± 3.1 and 2.9 ± 0.16 × 10^{-1} mg/kg for EF_{16pPAH} and EF_{4oPAH}, respectively, while nitro-PAHs were not quantified). Low and median molecular weight PAHs, including PHE (22 ± 0.11 %), NAP (18 ± 2.9 %), FLA (16 ± 1.3 %), and PYR (12 ± 0.91 %) dominated the pPAH composition profile from the briquette combustion. Of the 4 oPAHs, EFs of 9FO (1.1–1.8 mg/kg) and ATQ (0.50-0.68 mg/kg) were higher than the other two oPAHs. Similar to emissions from the wood burning, EF_{9FO} and EF_{ATQ} were of the same order of magnitude as EFs of their corresponding parent PAHs, while EF_{BaAQ} was roughly 2 orders of magnitude lower than EF_{BaA}. The ratios of 9FO/FLO, ATQ/ANT, and BaAQ/BaA were 4.3, 1.7, and 0.025, respectively. Also like the wood fuel combustion, EF_{nPAH} values were 2–5 orders of magnitude lower than those of their corresponding parent PAHs with a single exception of 9 N-ANT, whose EF was 0.48–0.59 mg/kg, comparable to its parent ANT (0.33–0.38 mg/kg).

For coal cake, OC and EC mass percents were 9.3 ± 3.6 and 0.65 ± 0.06 %, respectively. The calculated EC/OC ratio was 0.075 ± 0.022. There was only one previous study in the literature reporting the emissions of PM and BaP from a similar coal cake made of raw coals from rural Shanxi (Ge et al. 2004) which reported EF_{PM} and EF_{BaP} of 1.0 ± 0.06 g/kg and 2.1 ± 2.2 μg/kg, respectively. These EFs were much lower than what we found in the present study. Different types of coals that were used in the made of coal cake, in addition to different stove designs, may explain the difference between the two studies. The volatile matter and ash contents of the coal cake in our study were 6 and 87 %, respectively, while they were 11 and 30 % of the coal cake tested by Ge et al. (2004). For parent PAHs, low and median molecular weight compounds again dominated in the emissions. Approximately 25 ± 9.8, 19 ± 0.3, and 14 ± 3.7 % of the total 28 PAHs were NAP, ACY, and PHE. Of the 4 oPAHs from the coal cake burning, the EF of 9FO was the highest. The calculated ratios of 9FO/FLO, ATQ/ANT, and BaAQ/BaA were 0.39, 0.12, and 0.007, respectively. Nitro-PAHs tested were again orders of magnitude lower than those of their corresponding parent PAHs. 9N-ANT and 9N-PHE were the most abundant compounds.

7.2.3 Wood Combustion

In the previous field study conducted in China, it was reported that the overall means of EFs of $PM_{2.5}$ (PM with diameter less than 2.5 μm), OC, and EC for wood were 3.1 ± 0.82, 1.14 ± 0.40, and 1.5 ± 0.69 g/kg, respectively (Li et al. 2009). So far, most EFs were measured in laboratory conditions. It is interesting to see that the EFs reported based on laboratory tests were generally lower than those obtained in the field. In comparison with the results for residential wood combustion in our simulated kitchen study which reported EF_{PM}, EF_{OC}, and EF_{EC} values of 1.6 ± 0.32, 0.60 ± 0.35, and 0.94 ± 0.40 g/kg, respectively. The EFs were 5.2, 4.8, and 1.3 times lower than that obtained in the field test. The stove used in the built kitchen was a so-called "improved" new one with a shorter distance between the grid and cookware, a smaller firebox, and a taller chimney. The improved stoves are now commonly used in many places in China after the implementation of the National Improved Stove Program during the 1980s to 1990s. However, the simple movable wood stove without flue tested in this study is also used quite often. It is generally expected that the combustion efficiency in a stove with a chimney is higher than that without one due to an increased draft condition. In addition, mainly wood branches, instead of wood logs, were burned. This could also lead to relatively high emissions, since EFs for brushwood/branch were reported to be higher than those for fuel wood logs. Another possible reason for higher emissions is that the fire management behavior in the laboratory test was essentially "normalized", while the process observed in the field was rather random, which could include some low combustion efficiency occasions. Calculated ratios of EC/OC, EC/PM, and OC/PM from the wood combustion were 0.42 ± 0.02, 15 ± 6, and 35 ± 13 %, respectively. In comparison with the results of the previous laboratory test (2.1 ± 1.3, 49 ± 20, and 31 ± 16 %), it appears that a higher proportion of EC, but not OC, was generated while operating the improved stove with a chimney under laboratory conditions. Such a discrepancy can be explained again by the fact that under laboratory conditions, the stove performed better than that in the field test, and the former test may fail to reproduce some low-efficiency burning occasions (Chen et al. 2012). In fact, these ratios varied dramatically not only among sources, but also among burning conditions, suggesting that caution should be exercised when using these ratios.

The EFs of PAHs for residential wood combustion were tested in the field for the first time in this study. The measured EF_{28pPAH} and EF_{16pPAH} were about two orders of magnitude higher than those detected in the previous combustion experiments using an improved stove (6.4 ± 2.8 and 6.0 ± 2.7 mg/kg, respectively). The reasons for the relatively high EF_{pPAH} are similar to those for PM. However, the difference in EF_{PAH} was larger than that in EF_{PM}. A larger difference in EF_{pPAH} between two distinct combustion emissions in comparison with EF_{PM} was also mentioned in the literature (Dhammapala et al. 2007a, b; Lamberg et al. 2011), suggesting that the formation and emission of PAHs could be more sensitive to the variability in fuel/stove types and fire management behavior than PM. Though EF values for pPAHs measured in field were higher than those from the laboratory tests, the composition profiles were similar to each other, made up of approximately one third of NAP, followed by FLA, PYR, and PHE. Attention should be paid to those compounds with relatively high molecular weight PAHs (>228) that are often carcinogenic. The 7 carcinogenic compounds with high molecular weight (BaA, CHR, BbF, BkF, BaP, DahA, and IcdP) accounted for 16 % of the total emissions of the 16 U.S. EPA priority PAHs. This percentage was much higher than 6.3 % of the global total emissions from all sources. Similar to the 16 priority pollutants, EFs for 12 other parent PAHs, most of which were highly toxic, were also much higher than those measured in the previous combustion under laboratory conditions. For example, EFs of CPP and DalP in the present field measurement were 17 ± 7.0 and 0.46 ± 0.14 mg/kg, while those from the combustion under laboratory conditions were only 0.021 ± 0.0030 and 0.0020 ± 0.0018 mg/kg, respectively.

EFs of oPAHs and nPAHs were also measured for the first time in the field. For the same reason discussed above, these EFs were 1–2 orders of magnitude higher than the values reported in the previous laboratory combustion test (0.37 ± 0.27 and 0.0051 ± 0.0029 mg/kg for EF_{4oPAH} and EF_{9nPAH}, respectively). Of the 4 oxy-PAHs measured, EFs of 9FO (3.5 mg/kg) and BZO (3.7 mg/kg) were the highest, followed by ATQ (1.7 mg/kg). EFs of 9FO and ATQ were the same order of magnitude as those of corresponding parent PAHs and EF of BaAQ was around 2 orders of magnitude lower than that of BaA. The calculated EF ratios of oxy-PAHs over their corresponding parent PAHs were 0.79 (9FO/FLO), 0.42 (ATQ/ANT), and 0.020 (BaAQ/BaA), which were close to those in emissions from the previous laboratory test. Among all nitro-PAHs measured, 9N-ANT and 9N-PHE were the two most abundant species having EFs of 0.047–0.33 and 0.042–0.18 mg/kg, respectively. EFs of individual nPAHs were 2–5 orders of magnitude lower than EFs of their corresponding pPAHs. The calculated EF ratios of nPAHs over their corresponding pPAHs ranged from 0.000057 ($EF_{1N\text{-}NAP}/EF_{NAP}$) to 0.095 ($EF_{9N\text{-}ANT}/EF_{ANT}$).

7.2.4 Fuel Comparison

EFs for the target pollutants were significantly different ($p < 0.05$) among the fuels, except for EF_{9nPAH} ($p = 0.346$). The highest values were found for the burning of wood, and the lowest for the coal briquette. The EF_{28pPAH} of the coal briquette was more than an order of magnitude lower than those of the coal cake and wood. The difference in MCE may be one reason for different EFs among the three fuels. The measured MCE for the wood (92 ± 3 %) and the coal cake (86 ± 10 %) combustions were significantly lower ($p = 1.3 \times 10^{-3}$ and 1.0×10^{-4}, respectively) than that of the coal briquette combustion (94 ± 4 %).

Coal cake is often thought as a very dirty fuel causing high emissions of various air pollutants due to low burning efficiency. EF_{PM}, EF_{OC}, EF_{EC}, EF_{28pPAH}, $EF_{4o\text{-}PAH,\ and\ EF9nPAH}$ for the coal cake were 9.9, 3.6, 9.2, 14, 3.3, and 1.8 times the associated EF values for the coal briquette, respectively. In addition to the difference in MCE, ash content of the coal cake (87 %) was much higher than that of the coal briquette (47 %). It was well established that ash content can affect the fuel pyrolysis process and prevent the oxidation reaction of compounds during the combustion (Raveendran et al. 1995; Kazanc et al. 2011). Also, an increase in ash content can also reduce the heating values of coal (Wang et al. 2011). Therefore, high ash content may lead to higher emissions. In addition, incombustible ash content can remain in the exhaust smoke resulting in higher particle emissions

(Bond et al. 2004). High emissions of other pollutants such as SO_2 and NOx from coal cake combustion in comparison with coal briquette combustion were also reported (Ge et al. 2004). Stove design, fuel addition interval and fire management practice can surely affect the measured emissions as well. Unfortunately, it is very difficult to quantify these factors in practice.

Wood fuel is extensively used in rural China and many other developing countries. In this study, the EFs for the wood were found to be higher or equivalent to those for the coal cake. This is particularly true for OC and EC. EF_{OC} and EF_{EC} for the wood burning were 6.4 and 40 times of those for the coal cake, and 22 and 306 times of those for the coal briquette. Higher emissions for wood can be partly explained by relatively low combustion efficiency and high volatile matter content. Although overall average MCE for the wood was close to that for the briquette, the difference was statistically significant, most likely because EFs measured in this study are filter based, rather than real time data. Future studies focusing on the real time emissions are needed for a full understanding.

It was shown for all the three fuels that EF_{9FO} and EF_{ATQ} were comparable to their parent PAHs of FLO and ANT (within 0–1 order of magnitude difference), and EF_{BaAQ} and EF_{nPAHs} were about 2–5 orders of magnitude lower than EFs of their corresponding parent PAHs. Although the total emissions of pPAHs varied dramatically among the three fuels, the composition profiles were similar in general. For all three fuels tested, the parent PAHs were dominated by low and median molecular weight compounds especially NAP, ACY, PHE, and PHE. The only noted differences were the relatively high fractions of CPP, BaP, and PAHs with molecular weight larger than 302 from the wood and coal cake combustions. For instance, of the total 28 PAHs, mass percentages of CPP, BaP, and DalP were 5.3, 1.6, and 0.26 % in the coal cake emission, and 8.1, 2.3, and 0.21 % for the wood combustion, but only 0.48, 0.50, and 0.041 % in the emission from the coal briquette combustion. It is revealed that the coal cake and wood combustion cause not only higher emissions, but also larger fractions of highly toxic PAHs. Exposure to these highly toxic PAHs, in either indoor or outdoor environments, is believed to be associated with high risks of many diseases, including lung cancer. The composition profiles of PAH derivatives for all individual tests again were similar although EFs varied significantly. Generally, 9FO and BZO were two oxy-PAHs with higher emissions, and 9N-ANT and 9N-PHE were the two highest emitted nitrated PAHs.

The field measured emissions of various pollutants from the wood and coal combustions were higher than (or at least different from) those reported in stove combustions conducted in a laboratory under controlled conditions, which suggested that the latter used in inventory development may underestimate pollutant emissions obviously. In laboratory studies, new stoves purchased directly from market are often used, while in field many old stoves are used and the combustion efficiency goes down as the stoves wear out and the flue gets chocked. Another reason for the difference between laboratory and field tests is the way of refuel. In a laboratory, the refueling process is unintentionally normalized by the researchers, who are trained to reproduce the experimental results, while in the reality, however, fuels are often introduced randomly. Therefore, more realistic EFs, with high variations, can only be obtained in a field survey, while laboratory tests are useful for understanding the emission processes and influencing factors and conditions.

7.3 Summary

It is generally accepted that EFs measured in the field are closer to the reality and are the best data source for developing emission inventories. However, it is usually difficult to collect enough data to generate representative statistics due to extremely high variations and relatively high costs and labor intensity. In the present study, field measurement on residential solid fuel combustion was conducted in randomly selected households in rural Shanxi and Jiangsu Provinces. Much high emissions of CPM and PAHs were found from the combustion of a specific coal cake made by the residents in home in rural Shanxi. EFs of PM, EC and OC were 5.87 ± 3.77, 0.037 ± 0.021 and 0.477 ± 0.140 g/kg, respectively, and EFs of parent PAHs and oxygenated PAHs were 178 ± 43 and 7.09 ± 3.35 mg/kg, respectively. For firewood burned in an movable simple iron stove, the EFs were 8.32 ± 0.27 g/kg, 1.24 ± 0.48 g/kg, 2.91 ± 1.01 g/kg, 208 ± 62 mg/kg and 8.96 ± 1.59 mg/kg for PM, EC, OC, parent and oxygenated PAHs, respectively. In the field measurement in indoor crop residue burning measured in rural Jiangsu household, the pollutant emissions were generally higher for the burning in an old stove in comparison with those for crop residue burned in the newly built stove, indicating the analysis of pollutant emission historical change should take different EFs into consideration.

It should be noted that this study mainly focused on specific fuel/stove combinations in rural Shanxi, and included rather limited sample collections due to high labor intensity, cost, and technical difficulty. Therefore, the results of this study cannot be generalized. Instead, the results provide us valuable data and help further understanding of the difference between field and laboratory tests. It is also realized that EFs reported in the literature often vary in orders of magnitude, so the difference within one order of magnitude between the field tests and laboratory measurements found in the current study is relatively small. It is reasonable to expect that variation in the field-measured EFs should be much higher than that that measured in laboratory tests. More campaigns with relatively large sample sizes are strongly recommended before emissions can be fully understood.

References

Bond, T. C., Streets, D. G., Yarber, K. F., Nelson, S. M., Woo, J., Klimont Z. (2004). A technology-based global inventory of black and organic carbon emissions from combustion. *Journal of Geophysical Reaserch, 109*, D14203, doi:10.1029/2003JD003697.

Chen, Y., Roden, C., & Bond, T. (2012). Characterizing biofuel combustion with patterns of real-time emission data (PaRTED). *Environmental Science and Technology, 46*(11), 6110–6117.

Dhammapala, R., Claiborn, C., Simpson, C., & Jimenez, J. (2007a). Emission factors from wheat and Kentucky bluegrass stubble burning: Comparison of field and simulated burn experiments. *Atmospheric Environment, 41*, 1512–1520.

Dhammapala, R., Claiborn,C., Jimenez, J., Corkill, J., Gullett, B., & Simpson, C., et al. (2007b). Emission factors of PAHs, methoxyphenols, levoglucosan, elemental carbon and organic carbon from simulated wheat and Kentucky bluegrass stubble burns. *Atmospheric Environment, 41,* 2660–2669.

Ge, S., Xu, X., Chow, J., Watson, J., Sheng, Q., & Liu, W., et al. (2004). Emissions of air pollutants from household stoves: Honeycomb coal versus coal cake. *Environmental Science and Technology, 38*(17), 4612–4618.

Kazanc, F., Khatami, R., Crnkovic, P., & Levendis, Y. (2011). Emissions of NOx and SO_2 from coals of various ranks, bagasse, and coal-bagasse blends burning in O_2/N_2 and O_2/CO_2 environments. *Energy and Fuel, 25*(7), 2850–2861.

Lamberg, H., Nuutinen, K., Tissari, J., Ruusunen, J., Yli-Pirila, P., & Sippula, O., et al. (2011). Physicochemical characterization of fine particles from small-scale wood combustion. *Atmospheric Environment, 45*, 7635–7643.

Li, X., Wang, S., Duan, L., Hao, J., & Nie, Y. (2009). Carbonaceous aerosol emissions from household biofuel combustion in China. *Environmental Science and Technology, 43*, 6076–6081.

Raveendran, K., Ganesh, A., & Khilar, K. (1995). Influence of mineral matter on biomass pyrolysis characteristics. *Fuel, 74*(12), 1812–1822.

Roden, C. A., Bond, T. C., Conway, S., Pinel, A. B. O., MacCarty, N., & Still, D. (2009). Labloratory and field investigations of particulate and carbon monoxide emissions from traditional and improved cookstoves. *Atmospheric Environment, 43*, 1170–1181.

Wang, C., Liu, Y., Zhang, X., & Che, D. (2011). A study on coal properties and combustion characteristics of blended coals in Northwestern China. *Energy and Fuel, 25*(8), 3634–3645.

Zhang, J., Smith, K. R., Ma, Y., Ye, S., Jiang, F., & Qi, W., et al. (2000). Greenhouse gases and other airborne pollutants from household stoves in China: A database for emission factors. *Atmospheric Environment, 34*, 4537–4549.

Zhi, G., Peng, C., Chen, Y., Liu, D., Sheng, G., & Fu, J. (2009). Deployment of coal briquettes and improved stoves: possibly an option for both environment and climate. *Environmental Science and Technology, 43,* 5586–5591.

Chapter 8
Biomass Pellet

The inefficient combustions of traditional solid fuels contribute high emissions of a variety of pollutants. These emissions not only cause severe indoor air pollution and premature death of rural residents, but also contribute to regional and even global air pollution and climate forcing. Replacing traditional fuels with cleaner and more affordable fuels is a challenge in most developing countries. Among the various options, biomass pellets, especially those made of crop residues, have great potential because a large volume of crop residues is available in rural areas. The use of biomass pellets has been heavily promoted over the last several years because they are beneficial both from an environmental and social-economic standpoint. In China, promotion of biomass pellets has been included in the National Medium- and Long-Term Strategy Plan for Renewable Energy Development and the target goal is 50 million tons of annual consumption by 2020. However, due to limited experimental data on emission performance of biomass pellet burning, it is difficult to assess the real benefits and outcome of these cleaner alternatives in comparison with traditional solid fuels. In our study, two biomass pellets made from corn straw and pine wood, respectively, were burned in a modern pellet burner. By comparing the emissions from pellet burning and raw fuel combustion in a traditional cooking stove, the potential reduction in the total emissions of these pollutants from residential sector was discussed.

8.1 Combustion Temperature and MCE

Two commercial biomass pellets made of pine wood and corn straw, respectively, and a pellet burner were purchased from a local market in rural Beijing. These pellets and the burner are currently the most popular in the market. For comparison, raw biomass fuels were also collected and burned. Detailed information about the fuel property and the pellet burner can be found in the Method section. Relatively high ash content would affect the formation and emission of PM and its size distribution from the pellet burning. Although there is no direct evidence on the cause of such high ash content, additives like magnesium and calcium

G. Shen, *Emission Factors of Carbonaceous Particulate Matter and Polycyclic Aromatic Hydrocarbons from Residential Solid Fuel Combustions*, Springer Theses,
DOI: 10.1007/978-3-642-39762-2_8,

carbonates, aluminum hydroxide, kaolin and so on used in the pellet manufacture can not be ruled out (Wang et al. 2011; Yao et al. 2010; Yuan et al. 2009a, b). In many Europe countries, there are strict standards on the pellet property and use of additives (Fiedler 2004; García-Maraver et al. 2011). The limitation of additives in some regulations is 2.0 %, and in some other regulations, though no limitation value, the type of additive was required. Unfortunately, chemical compositions of the pellets or ash were not measured in this study. Future studies are recommended to address the reason causing such high ash content and the subsequent impacts on combustion and emissions.

During the pellet combustion, the kitchen room air was much less smoky in comparison to the room air when the raw materials were combusted in the traditional stove. The chamber temperatures in the pellet burner without the secondary air supply were slightly higher (mode I, 650–900 and 500–800 °C for pine wood and corn straw pellets, respectively) than those with secondary air supply (mode II, 550–750 and 400–700 °C for pine wood and corn straw pellets, respectively). However, both of these modes had higher temperatures than the traditional cooking stove during the combustions of raw pine wood (350–650 °C) and corn straw (300–500 °C). MCE for pine wood pellet burning in mode I and II were 99.4 ± 0.2 and 99.8 ± 0.0 %, respectively, and for corn straw pellet burning were 97.4 ± 1.9 and 98.5 ± 0.8 %, respectively.

8.2 Carbonaceous Particulate Matter

8.2.1 Emission Factor

The measured EF_{CO}, EF_{OC}, EF_{EC}, and EF_{PM} for the two types of biomass pellets (pine wood and corn straw) in the two different combustion modes (with and without secondary side air admission) are listed in Table 8.1 as means and standard derivations. The EFs varied obviously, even among the triplicate combustion experiments for the same fuel in the same combustion mode, indicating high variability in the emission. For example, the coefficients of variation for EF_{OC}, EF_{EC}, and EF_{PM} for pine wood pellets were 78, 84, and 27 % in mode I (without secondary side air admission) and 84, 82, and 42 % in mode II (with secondary side air admission), respectively. In general, EFs for corn straw pellets were higher than those for pine wood pellets ($p < 0.05$).

In mode II, increased excess air cooled the combustion temperature and subsequent lower the combustion efficiency (Boman et al. 2011; Johansson et al. 2004). EF_{OC} and EF_{PM} for pine wood pellet combustion in mode II were significantly higher than those in mode I ($p < 0.05$). EF_{EC} for the pine wood pellets and EF_{OC}, EF_{EC}, and EF_{PM} for corn straw pellets were not significantly different between the two modes ($p > 0.05$), though the average EFs measured in mode II were higher than those in mode I. Similarities in EFs of PCDD/F, PCBs, and HCB

Table 8.1 MCE, burning rate (R), and EFs of CO, OC, EC and PM (g/kg or mg/MJ) for raw and pellets in this study

Fuel	Pine wood	Pine wood pellet		Corn straw	Corn straw pellet	
Mode		Mode I	Mode II		Mode I	Mode II
MCE	0.941 ± 0.016	0.998 ± 0.000	0.994 ± 0.002	0.961 ± 0.004	0.985 ± 0.008	0.974 ± 0.019
R, kg/min	$7.79 \pm 1.03 \times 10^{-2}$	$2.12 \pm 0.15 \times 10^{-2}$	$1.67 \pm 0.52 \times 10^{-2}$	$4.47 \pm 0.20 \times 10^{-2}$	$1.52 \pm 0.17 \times 10^{-2}$	$1.84 \pm 0.36 \times 10^{-2}$
CO, mg/MJ	$2.30 \pm 0.71 \times 10^{3}$	$1.63 \pm 0.26 \times 10^{2}$	$3.68 \pm 1.24 \times 10^{2}$	$2.42 \pm 0.20 \times 10^{3}$	$1.10 \pm 0.60 \times 10^{3}$	$1.95 \pm 1.49 \times 10^{3}$
OC, mg/MJ	$3.61 \pm 2.09 \times 10^{1}$	$7.00 \pm 5.46 \times 10^{-1}$	$9.11 \pm 7.64 \times 10^{0}$	$5.69 \pm 2.37 \times 10^{1}$	$4.34 \pm 0.97 \times 10^{0}$	$1.30 \pm 1.55 \times 10^{1}$
EC, mg/MJ	$5.64 \pm 2.37 \times 10^{1}$	$2.07 \pm 1.73 \times 10^{0}$	$1.99 \pm 1.63 \times 10^{0}$	$2.60 \pm 0.19 \times 10^{1}$	$1.41 \pm 1.20 \times 10^{1}$	$8.35 \pm 1.86 \times 10^{0}$
PM, mg/MJ	$9.53 \pm 1.92 \times 10^{1}$	$2.96 \pm 0.81 \times 10^{1}$	$1.12 \pm 0.48 \times 10^{2}$	$3.01 \pm 0.04 \times 10^{2}$	$1.72 \pm 0.94 \times 10^{2}$	$2.04 \pm 0.84 \times 10^{2}$
CO, g/kg	$3.84 \pm 1.18 \times 10^{1}$	$2.69 \pm 0.43 \times 10^{0}$	$6.07 \pm 2.04 \times 10^{0}$	$3.74 \pm 0.31 \times 10^{1}$	$1.52 \pm 0.83 \times 10^{1}$	$2.71 \pm 2.06 \times 10^{1}$
OC, g/kg	$6.04 \pm 3.49 \times 10^{-1}$	$1.15 \pm 0.90 \times 10^{-2}$	$1.50 \pm 1.26 \times 10^{-1}$	$8.79 \pm 3.67 \times 10^{-1}$	$6.03 \pm 1.35 \times 10^{-2}$	$1.81 \pm 2.15 \times 10^{-1}$
EC, g/kg	$9.43 \pm 3.96 \times 10^{-1}$	$3.41 \pm 2.86 \times 10^{-2}$	$3.29 \pm 2.69 \times 10^{-2}$	$4.01 \pm 0.29 \times 10^{-1}$	$1.96 \pm 1.67 \times 10^{-1}$	$1.16 \pm 0.26 \times 10^{-1}$
PM, g/kg	$1.59 \pm 0.32 \times 10^{0}$	$4.88 \pm 1.34 \times 10^{-1}$	$1.85 \pm 0.78 \times 10^{0}$	$4.65 \pm 0.07 \times 10^{0}$	$2.39 \pm 1.31 \times 10^{0}$	$2.84 \pm 1.16 \times 10^{0}$

Data listed are means and standard derivations from triplicate experiments. For pellets in the burner, two different burn modes were tested, without (I) and with (II) secondary side air admission. Reprinted from Shen et al. (2012) with permission of American Chemical Society

for firewood and pellet combustions between full and reduced air supplies have been previously reported (Hedman et al. 2006).

In total PM emitted from corn straw pellet burning, total carbon mass contributed 14 % on average, and the ratio of OC/TC was 0.40 ± 0.20. In emissions from pine wood pellet burning, OC/TC and total carbon mass percent in PM were 0.61 ± 0.29 and 8.2 %, respectively. There was no significant difference in these ratios between the two combustion modes ($p > 0.05$). It was reported that in emissions from *Miscanthus* and *Triticale* pellet burning, the OC/TC ratio was 0.52 ± 0.26, very similar to our results here (Schimidl et al. 2011).

There are some reports on pollutant emissions from pellet burning in the literature although still limited compared the study on traditional solid fuels and most of the studies are in developed countries or regions (Bølling et al. 2009; Houch and Eagle 2006; Nussbaumer et al. 2008). EF_{CO} and EF_{PM} for pellets reported in the literature varied dramatically among various fuel types and burning conditions (Boman et al. 2011; Chen et al. 2010; Johansson et al. 2004; Schimidl et al. 2011; Bäfver et al. 2011; Lamberg et al. 2011). For example, for *Triticale* pellets, *Miscanthus* pellets, and wood pellets burned in a 40 kW burner, EF_{PM} were 114, 14.7, and 21.9 mg/MJ (average for start-up, full load, and part load phases), respectively (Schimidl et al. 2011). Johansson et al. (2004) reported EF_{CO} and EF_{PM} in the range of 30–1100 and 12–65 mg/MJ for wood pellets burned in two pellet burners under different burning conditions (3–22 kW). In a stove with output energy of 3–5 kW (very close to 2.6 kW in our study), EF_{CO} and EF_{PM} for wood pellets were measured at 57–270 and 19–58 mg/MJ (Bäfver et al. 2011). Boman et al. (2011) summarized the published EFs for pellets in the literature and reported that EF_{CO} and EF_{PM} were 58–2600 and 11–600 mg/MJ in the field measurements, and 24–2000 and 2–150 mg/MJ in the laboratory studies, respectively. By thoroughly reviewing reported values in the literature, it was summarized that average EF_{PM} for wood pellets in residential wood combustion were 1.91 and 3.99 g/kg in U.S. EPA certified (1988 New Source Performance Standards for woodstoves) and exempt modes, respectively (EPA 2001). In a survey for pellet heaters used in the MANE-VU region (11 states and the District of Columbia), mean EF_{PM} was estimated at 1.53 g/kg based on measurements in old, modern, EPA-certified, and EPA-exempt models and pellet heating systems (Houch and Eagle 2006).Generally, EF_{PM} for wood pellets burned under high efficiencies was about 10–50 mg/MJ (Bølling et al. 2009; Nussbaumer et al. 2008). Our results for EF_{CO} were 136–505 mg/MJ for pine wood pellets and 414–3838 mg/MJ for corn straw pellets. EF_{PM} were 17.6–176 and 65.5–332 mg/MJ for these two pellets, respectively. Average EF_{PM} for pine wood pellets was 71.0 ± 54.0 mg/MJ (1.17 ± 0.89 g/kg). Generally, the results fell within the reported range in the literature.

Measured EF_{OC} and EF_{EC} for biomass pellets were rather limited. In a 25 kW burner, EF_{EC} and EF_{OC} for wood pellets were reported at 0.1 ± 0.17 and 0.9 ± 2.1 mg/MJ (Lamberg et al. 2011). In an inventory on carbonaceous PM emission in Europe, 0.83 mg/MJ was adopted for both EF_{EC} and EF_{OC} for biomass burned in an automatic feed pellet boiler (<50 MWh) under uncontrolled condition

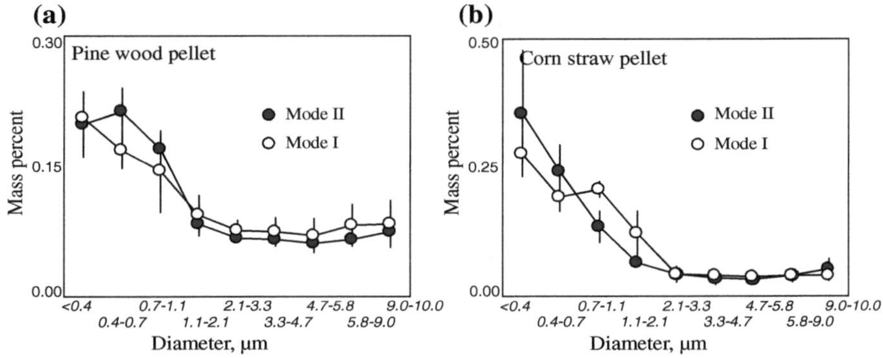

Fig. 8.1 Size distributions of PM from pine wood (**a**) and corn straw (**b**) pellet burning. Biomass pellets were burned in two combustion modes without (*mode I*) and with (*mode II*) secondary side air supply. Results shown are means and standard deviation in triplicate measurements. Reprinted from Shen et al. (2012) with permission of American Chemical Society

(Kupiainen and Klimont 2007). These values were lower than the EF_{OC} (5.74 ± 7.17, 0.335–4.60 mg/MJ) and EF_{EC} (2.02 ± 1.57, 0.223–18.5 mg/MJ) for pine wood pellets measured in our study (results from both two modes). The difference could be due to the different pellet burners and fuel types.

8.2.2 Size Distribution

PM size distributions were similar between the two combustion modes (Fig. 8.1). $PM_{0.4}$ was the most abundant, contributing 31.6 ± 9.0 and 20.3 ± 1.9 % of the total from corn straw and pine wood pellet combustions, respectively. $PM_{1.1}$ and $PM_{2.1}$ contributed 54.5 ± 5.9 and 63.4 ± 7.1 % of the total PM_{10} for pine wood pellets and 70.5 ± 6.1 and 79.9 ± 4.5 % of the total PM_{10} for corn straw pellets, respectively. Calculated MMDs of PM for corn straw pellets were 0.62 ± 0.24 (0.29–0.92) and 1.2 ± 0.2 (1.0–1.5) μm for pine wood pellets.

The prevalence of fine PM in pellet burning emission has been reported in the literature. In general, $PM_{1.0}$ contributed approximately 90 % of the total PM emitted in pellet burning (Jokiniemi et al. 2008; Bäfver et al. 2011; Lamberg et al. 2011). For example, by using a Dekati Low-Pressure Impactor (DLPI), Johansson et al. (2004) found that size distributions of particles from two pellet burners were similar with a peak size of 130 nm. Bäfver et al. (2011) reported a high mass fraction of $PM_{1.0}$ and $PM_{2.5}$ of 81–94 and 84–96 %, respectively, in particles from wood pellet burning in several pellet stoves of 3–5 kW output power. In comparison, pellet burning in our study produced more coarse particles. The difference is likely due to the distinct pellets (fuel property, e.g. moisture and ash content) and burners used (combustion condition, e.g. air supply and combustion temperature) (Boman et al. 2011; Johansson et al. 2004; Jokiniemi et al. 2008). It was found that pellet burning can also produce relative large amounts of coarse PM under given conditions. For example, as high as 15 % $PM_{1.0-10}$ was found in PM emissions from wood pellet burning in a 5–6 kW stove. And in the startup phase in a 2 kW burner under intermittent operation, $PM_{1.0}$ only made up 70 % of the total (Boman et al. 2011).

8.3 Polycyclic Aromatic Hydrocarbons

8.3.1 Emission Factors

EF_{PAH} for pine wood and corn straw pellet combustions were 0.329 ± 0.261 and 0.711 ± 0.325 mg/MJ (5.41 ± 4.30 and 9.87 ± 4.51 mg/kg) in mode I, and 0.718 ± 0.369 and 1.32 ± 0.76 mg/MJ (11.8 ± 6.1 and 18.3 ± 10.6 mg/kg) in mode II, respectively (Table 8.2). There was no statistical difference between these two types of pellets ($p > 0.05$). The sixteen U.S. EPA priority pollutant PAHs contributed more than 90 % of the total PAHs measured. No statistical difference was identified between the two combustion modes ($p > 0.05$), although the measured mean EFs in mode II were always higher than those in mode I.

There were several reports on PAH emissions from wood pellet combustion (Boman et al. 2011; EPA 2001; Houch and Eagle 2006; Johansson et al. 2004). For instance, EFs of 16 PAHs and BaP for pellet heaters in the MANE-VU region were estimated at 350 and 3.34 mg/kg, respectively (Houch and Eagle 2006). For wood pellets burned in residential burners operated under 3–22 kW, EFs of 27 particle-bound PAHs ranged from 0.06 to 8.5 mg/MJ (Johansson et al. 2004). In another study, the total EFs of 40 PAHs in both gaseous and particulate phases were only 0.002–0.34 mg/MJ (Boman et al. 2011). Our results (EFs of 28 PAHs were 0.0942–1.16 mg/MJ, with 0.0091–0.660 mg/MJ for PM-bound PAHs) also varied widely with no significant difference with the reported values.

8.3.2 Composition Profile

Although total EFs varied largely, the normalized composition profiles for two pellets in two burning modes were similar in general (Fig. 8.2). Low and median molecular weight PAHs dominated the total PAHs emitted. NAP and PHE were the most abundant, contributing 29.8 ± 11.3 and 28.4 ± 7.0 % of the total on average in the emission from pine wood pellet burning, and 47.4 ± 12.4 and 14.2 ± 5.5 % in the emission from corn straw pellet burning. The calculated isomer ratios for two pellets are provided in Table 8.3. There were no significant differences in these ratios between the pine wood and corn straw pellets or between combustion mode I and II ($p > 0.05$).

Table 8.2 EFs of 28 PAH individuals (mg/MJ) for raw and pellet fuels burned in residential stoves in this study

Fuel	Pine wood	Pine wood pellet		Corn straw	Corn straw pellet	
Mode		Mode II	Mode I		Mode II	Mode I
NAP	$2.14 \pm 1.61 \times 10^{0}$	$3.93 \pm 2.05 \times 10^{0}$	$1.28 \pm 0.94 \times 10^{0}$	$2.77 \pm 2.04 \times 10^{0}$	$9.35 \pm 7.75 \times 10^{0}$	$4.35 \pm 3.04 \times 10^{0}$
ACY	$5.47 \pm 2.98 \times 10^{-1}$	$1.01 \pm 0.49 \times 10^{0}$	$2.88 \pm 2.05 \times 10^{-1}$	$7.67 \pm 4.79 \times 10^{-1}$	$1.92 \pm 1.10 \times 10^{0}$	$7.06 \pm 2.53 \times 10^{-1}$
ACE	$4.52 \pm 2.09 \times 10^{-2}$	$1.09 \pm 0.44 \times 10^{-1}$	$5.87 \pm 0.68 \times 10^{-2}$	$7.49 \pm 5.04 \times 10^{-2}$	$1.34 \pm 0.89 \times 10^{-1}$	$3.90 \pm 2.07 \times 10^{-2}$
FLO	$1.82 \pm 0.89 \times 10^{-1}$	$4.10 \pm 1.82 \times 10^{-1}$	$2.15 \pm 2.38 \times 10^{-1}$	$2.41 \pm 1.46 \times 10^{-1}$	$4.43 \pm 2.06 \times 10^{-1}$	$1.67 \pm 0.63 \times 10^{-1}$
PHE	$1.25 \pm 0.41 \times 10^{0}$	$2.81 \pm 1.88 \times 10^{0}$	$1.58 \pm 1.68 \times 10^{0}$	$1.44 \pm 0.60 \times 10^{0}$	$2.42 \pm 1.44 \times 10^{0}$	$1.07 \pm 0.38 \times 10^{0}$
ANT	$1.38 \pm 0.46 \times 10^{-1}$	$3.51 \pm 2.67 \times 10^{-1}$	$2.85 \pm 0.33 \times 10^{-1}$	$1.68 \pm 0.82 \times 10^{-1}$	$3.18 \pm 1.46 \times 10^{-1}$	$1.37 \pm 0.38 \times 10^{-1}$
FLA	$6.57 \pm 1.98 \times 10^{-1}$	$9.31 \pm 8.00 \times 10^{-1}$	$4.21 \pm 3.08 \times 10^{-1}$	$8.12 \pm 3.14 \times 10^{-1}$	$9.59 \pm 4.95 \times 10^{-1}$	$7.17 \pm 2.91 \times 10^{-1}$
PYR	$5.53 \pm 1.08 \times 10^{-1}$	$7.40 \pm 1.86 \times 10^{-1}$	$3.35 \pm 3.92 \times 10^{-1}$	$7.19 \pm 1.42 \times 10^{-1}$	$7.78 \pm 1.00 \times 10^{-1}$	$6.84 \pm 1.64 \times 10^{-1}$
RET	$3.38 \pm 0.81 \times 10^{-1}$	$4.97 \pm 1.86 \times 10^{-1}$	$4.25 \pm 0.39 \times 10^{-1}$	$1.70 \pm 1.42 \times 10^{-1}$	$4.22 \pm 1.00 \times 10^{-1}$	$3.70 \pm 1.64 \times 10^{-1}$
BcP	$3.92 \pm 0.15 \times 10^{-2}$	$4.13 \pm 0.17 \times 10^{-2}$	$1.61 \pm 0.11 \times 10^{-2}$	$4.02 \pm 0.12 \times 10^{-2}$	$5.53 \pm 0.23 \times 10^{-2}$	$4.88 \pm 0.19 \times 10^{-2}$
CcdP	$2.13 \pm 0.35 \times 10^{-2}$	$1.08 \pm 0.87 \times 10^{-1}$	$5.57 \pm 0.60 \times 10^{-2}$	$6.65 \pm 0.63 \times 10^{-2}$	$2.15 \pm 0.16 \times 10^{-1}$	$1.75 \pm 0.97 \times 10^{-1}$
BaA	$1.14 \pm 0.63 \times 10^{-1}$	$1.45 \pm 0.78 \times 10^{-1}$	$7.06 \pm 0.82 \times 10^{-2}$	$1.43 \pm 0.45 \times 10^{-1}$	$2.08 \pm 1.10 \times 10^{-1}$	$1.86 \pm 0.94 \times 10^{-1}$
CHR	$1.37 \pm 0.75 \times 10^{-1}$	$2.10 \pm 0.85 \times 10^{-1}$	$1.53 \pm 1.27 \times 10^{-1}$	$1.62 \pm 0.38 \times 10^{-1}$	$3.36 \pm 1.13 \times 10^{-1}$	$2.60 \pm 0.98 \times 10^{-1}$
BbF	$6.92 \pm 0.53 \times 10^{-2}$	$1.12 \pm 0.57 \times 10^{-1}$	$4.69 \pm 0.41 \times 10^{-2}$	$6.71 \pm 2.46 \times 10^{-2}$	$1.51 \pm 0.95 \times 10^{-1}$	$1.62 \pm 0.62 \times 10^{-1}$
BkF	$7.09 \pm 0.53 \times 10^{-2}$	$1.01 \pm 0.34 \times 10^{-1}$	$3.58 \pm 0.32 \times 10^{-2}$	$6.07 \pm 1.80 \times 10^{-2}$	$1.17 \pm 0.67 \times 10^{-1}$	$1.42 \pm 0.53 \times 10^{-1}$
BeP	$3.78 \pm 2.61 \times 10^{-2}$	$6.65 \pm 2.61 \times 10^{-2}$	$2.93 \pm 2.51 \times 10^{-2}$	$4.86 \pm 1.50 \times 10^{-2}$	$8.74 \pm 4.25 \times 10^{-2}$	$9.41 \pm 3.12 \times 10^{-2}$
BaP	$3.73 \pm 2.64 \times 10^{-2}$	$6.85 \pm 4.69 \times 10^{-2}$	$2.40 \pm 1.99 \times 10^{-2}$	$5.56 \pm 1.83 \times 10^{-2}$	$1.03 \pm 0.67 \times 10^{-1}$	$1.22 \pm 0.65 \times 10^{-1}$
PER	$7.19 \pm 3.63 \times 10^{-3}$	$3.23 \pm 1.25 \times 10^{-2}$	$1.03 \pm 0.52 \times 10^{-2}$	$1.24 \pm 0.54 \times 10^{-2}$	$4.52 \pm 2.28 \times 10^{-2}$	$3.56 \pm 1.42 \times 10^{-2}$
IcdP	$2.27 \pm 1.63 \times 10^{-2}$	$5.27 \pm 4.27 \times 10^{-2}$	$1.93 \pm 1.85 \times 10^{-2}$	$3.96 \pm 1.57 \times 10^{-2}$	$7.67 \pm 4.58 \times 10^{-2}$	$1.18 \pm 0.61 \times 10^{-1}$
DahA	$3.10 \pm 1.92 \times 10^{-3}$	$1.13 \pm 0.87 \times 10^{-2}$	$3.23 \pm 0.29 \times 10^{-3}$	$5.12 \pm 1.96 \times 10^{-3}$	$7.75 \pm 5.72 \times 10^{-3}$	$7.39 \pm 2.78 \times 10^{-3}$
BghiP	$2.00 \pm 1.48 \times 10^{-2}$	$5.10 \pm 3.60 \times 10^{-2}$	$1.99 \pm 1.76 \times 10^{-2}$	$3.49 \pm 1.19 \times 10^{-2}$	$9.07 \pm 6.44 \times 10^{-2}$	$1.47 \pm 0.73 \times 10^{-1}$

(continued)

Table 8.2 (continued)

Fuel	Pine wood	Pine wood pellet		Corn straw	Corn straw pellet	
Mode		Mode II	Mode I		Mode II	Mode I
DacP	$1.79 \pm 0.63 \times 10^{-3}$	$5.53 \pm 0.83 \times 10^{-3}$	$3.22 \pm 2.53 \times 10^{-3}$	$4.45 \pm 2.40 \times 10^{-3}$	$1.14 \pm 0.13 \times 10^{-2}$	$1.51 \pm 1.05 \times 10^{-2}$
DalP	$1.97 \pm 1.83 \times 10^{-3}$	$8.37 \pm 0.82 \times 10^{-3}$	$4.81 \pm 3.39 \times 10^{-3}$	$5.48 \pm 2.35 \times 10^{-3}$	$8.31 \pm 0.89 \times 10^{-3}$	$1.24 \pm 0.60 \times 10^{-2}$
DaeF	$1.50 \pm 0.23 \times 10^{-3}$	$6.30 \pm 0.65 \times 10^{-3}$	$3.25 \pm 0.40 \times 10^{-3}$	$5.43 \pm 3.87 \times 10^{-3}$	$6.29 \pm 0.75 \times 10^{-3}$	$7.94 \pm 4.14 \times 10^{-3}$
Coron	$3.85 \pm 2.93 \times 10^{-3}$	$2.17 \pm 2.15 \times 10^{-2}$	$1.25 \pm 1.20 \times 10^{-2}$	$1.15 \pm 0.90 \times 10^{-2}$	$4.83 \pm 3.99 \times 10^{-2}$	$8.02 \pm 4.44 \times 10^{-2}$
DaeP	$1.48 \pm 0.23 \times 10^{-3}$	$5.23 \pm 0.51 \times 10^{-3}$	$3.99 \pm 0.39 \times 10^{-3}$	$5.29 \pm 1.10 \times 10^{-3}$	$5.15 \pm 4.88 \times 10^{-3}$	$7.90 \pm 4.14 \times 10^{-3}$
DaiP	$3.89 \pm 3.37 \times 10^{-4}$	$5.77 \pm 0.84 \times 10^{-3}$	$8.86 \pm 1.61 \times 10^{-3}$	$3.18 \pm 3.85 \times 10^{-3}$	$4.71 \pm 0.51 \times 10^{-3}$	$8.33 \pm 5.04 \times 10^{-3}$
DahP	$4.27 \pm 0.42 \times 10^{-4}$	$3.49 \pm 0.35 \times 10^{-3}$	$4.88 \pm 0.87 \times 10^{-3}$	$4.30 \pm 0.66 \times 10^{-4}$	$3.12 \pm 2.63 \times 10^{-3}$	$3.37 \pm 1.66 \times 10^{-3}$

Data shown are means and standard derivations from triplicate measurements. For pellet burning in the pellet burner, two different modes with (II) and without (I) side air admission were adopted. Reprinted from Shen et al. (2012) with permission of American Chemical Society

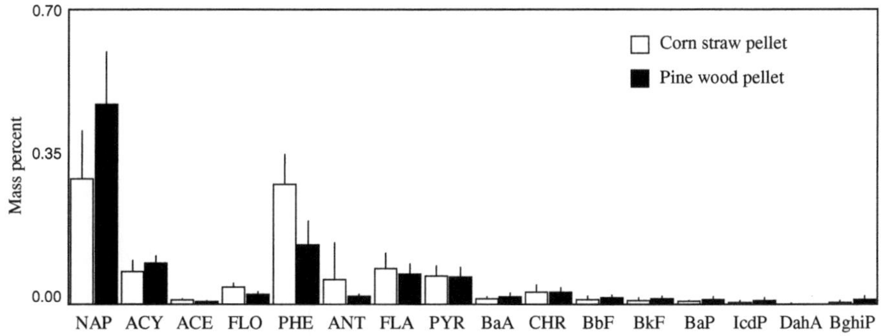

Fig. 8.2 Composition profiles of PAHs from the burning of corn straw and pine wood pellets. Data shown are means and standard deviations from all burning experiments. Reprinted from Shen et al. (2012) with permission of American Chemical Society

8.4 Fuel Comparison

8.4.1 Emission Factor

For the purpose of comparison, a relative difference (RD) was calculated as:

$$RD = \frac{EF_{pellet} - EF_{raw}}{EF_{raw}},$$

where EF_{pellet} and EF_{raw} are EFs of a given pollutant for biomass pellets (measured EFs in mode I and mode II collectively applied) and the corresponding raw fuel, respectively.

The calculated RD values for CO, OC, EC, and PM were −55 (−71 to −31, as inter-quartile range from Monte Carlo simulation), −90 (−91 to −88), −68 (−77 to −54), and −49 % (−63 to −30 %) for corn straw, and −89 (−91 to −88), −93 (−96 to −86), −97 (−98 to −56), and −43 % (−60 to −18 %) for pine wood, respectively. The reduction was statistically significant ($p < 0.05$).

For PAHs, unlike EF_{CO}, EF_{OC}, EF_{EC}, and EF_{PM}, EF_{PAH} for the pellets appeared to be higher than for the raw fuels, though the differences were not significant due to high variability ($p > 0.05$). RDs for PAHs were 64 % (56–72 %) for corn straw and 6 % (−16–33 %) for pine wood, respectively. It was reported in the literature, for some biomass pellets, EFs for PAHs and also other organics, were not significantly lower than, and sometimes slightly higher than, EFs for raw fuels (Hedman et al. 2006; Johansson et al. 2004). For example, unexpected high emissions of PCDD/Fs and PCBs were observed during combustion of wood (intermittent) and straw (continuous) pellets (Hedman et al. 2006). Relatively low moisture and high combustion temperatures are likely responsible for the slightly high EF_{PAH} of the pellet burning compared to that for raw fuels.

Table 8.3 Isomer ratios of selected PAH pairs from raw and pellet combustions in this study

Fuel	Pine wood	Pine wood Pellet		Corn straw	Corn straw pellet	
Mode		Mode II	Mode I		Mode II	Mode I
ANT/(ANT + PHE)	0.10 ± 0.00	0.11 ± 0.01	0.09 ± 0.02	0.10 ± 0.01	0.12 ± 0.02	0.12 ± 0.02
FLA/(FLA + PYR)	0.54 ± 0.03	0.52 ± 0.07	0.56 ± 0.03	0.53 ± 0.02	0.54 ± 0.05	0.51 ± 0.01
BaA/(BaA + CHR)	0.45 ± 0.00	0.39 ± 0.06	0.29 ± 0.10	0.46 ± 0.03	0.37 ± 0.06	0.40 ± 0.06
IcdP/(IcdP + BghiP)	0.54 ± 0.01	0.50 ± 0.03	0.47 ± 0.06	0.53 ± 0.01	0.47 ± 0.07	0.44 ± 0.01
BbF/(BbF + BkF)	0.49 ± 0.02	0.52 ± 0.06	0.58 ± 0.04	0.52 ± 0.02	0.56 ± 0.03	0.53 ± 0.01
BaP/(BaP + BghiP)	0.65 ± 0.06	0.58 ± 0.07	0.60 ± 0.17	0.61 ± 0.02	0.54 ± 0.08	0.44 ± 0.04
BeP/(BeP + BaP)	0.51 ± 0.01	0.52 ± 0.07	0.54 ± 0.05	0.47 ± 0.01	0.47 ± 0.05	0.47 ± 0.09

Data shown are means and standard derivations from triplicate measurements

8.4.2 PM Size Distribution

The RDs for OC and EC were much lower than those for PM, suggesting that the compressing process can reduce the emissions of OC and EC more effectively than PM. As a result, TC/PM ratios for pine wood (8 %) and corn straw (14 %) pellets were significantly lower than those for raw pine wood log (97 %) and corn straw (28 %). Relatively low organic fractions, but high inorganic fractions, in PM from pellet combustions have been reported in the literature (Bäfver et al. 2011; Boman et al. 2011).

The size distributions of PM for raw fuel and pellet were compared in Fig. 8.3. Higher percents of coarse PM were in the smoke emitted from pellet burning in comparison with those for raw fuels. The mass percents of fine PM2.1 were 63.4 ± 7.1 and 79.9 ± 4.5 % of the total PM mass in emissions from the pellet burning, which were significantly lower than those of 80.4 and 87.9 % for raw fuels. The size distribution of PM can be affected by many factors including fuel moisture, ash content, combustion temperature, excess air, and combustion efficiency (Chang et al. 2004; Lighty et al. 2000; Maguhn et al. 2003). Moreover, the impacts of these factors were complicated and sometimes interacted with one anther.

8.4.3 PAH Composition Profile

Figure 8.4 compares the normalized composition profiles between biomass pellet and raw fuels. It appears that the profiles were similar without significant difference. The calculated isomer ratios (Table 8.3) were also comparable similar between the pelletized and raw fuels. Also, PAH/PM ratios for the pellets were comparable to those for raw fuels (0.12 ± 0.02 % for pine wood and 0.07 ± 0.03 % for corn straw) ($p > 0.05$).

8.4.4 Emission Reduction

The potential reduction in total emissions of CO, OC, EC, PM, and PAHs were also estimated based on the differences in the calculated EFs between the pellets and raw fuels, reported thermal efficiencies of cooking stoves and modern burners, and emission inventories in the literature. Median R_{EF} (both results of corn straw and pine wood used collectively) of CO, OC, EC, PM, and PAHs were 0.22 (0.13–0.36, as inter-quartile range), 0.090 (0.057–0.14), 0.10 (0.067–0.15), 0.54 (0.48–0.61), and 1.3 (1.0–1.6), respectively. It was reported that the thermal efficiency of burning pellets in the modern burners was approximately 2.3 times that of raw fuels burned in the so-called improved cooking stoves and 4.5 times that of traditional cooking stoves (Chen et al. 2005).

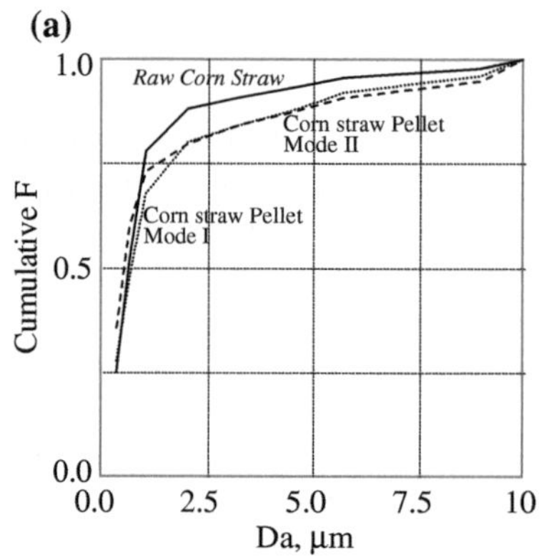

Fig. 8.3 Size distributions of PM emitted form the pellet and raw fuel burning. **a** Corn straw, **b** Pine wood. Reprinted from Shen et al. (2012) with permission of American Chemical Society

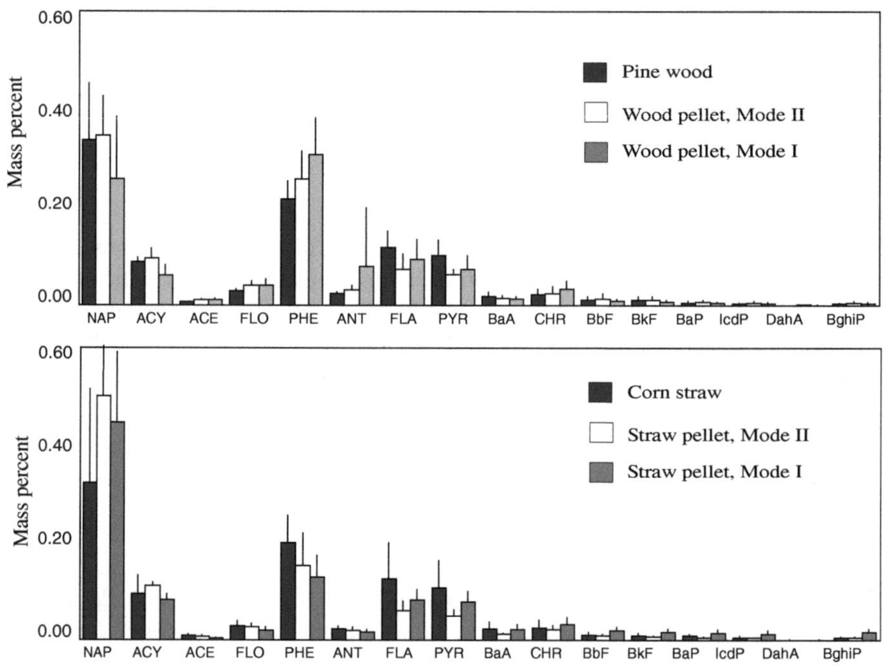

Fig. 8.4 Comparison of composition profiles of PAHs in emissions between biomass pellet and raw fuels. Reprinted from Shen et al. (2012) with permission of American Chemical Society

Taking the relatively low EFs (except for the slightly high EF_{PAHs}) and high thermal efficiency of pellet burning in the modern burner into consideration, the replacement of raw fuels with biomass pellets can reduce the emissions of most air pollutants significantly. As a rough estimation, the emission reductions of CO, OC,

EC, PM, and PAHs would be 90 (84–94), 96 (94–98), 96 (93–97), 76 (71–80), and 43 % (26–56 %), respectively, by replacing the improved cooking stoves with pellet stoves. The reduction could be as high as 95 (92–97), 98 (97–99), 98 (96–99), 88 (85–90), and 71 % (62–77 %) if the traditional cooking stoves were replaced with pellet stoves. Currently, residential biomass burning contributed approximately 39.4, 2.29, 0.59, 3.87, and 0.065 Tg of total emissions of CO, OC, EC, PM, and PAHs in China, respectively (Lei et al. 2011; Streets et al. 2006; Zhang et al. 2008). If only 20 % of the traditional cooking stoves can be replaced with pellet stoves, the total emission reductions in CO, OC, EC, PM. and PAHs would be 7366 (6262–8642), 446 (382–521), 115 (98–134), 675 (577–791), and 8.94 (7.27–10.8) Gg.

It is realized that this is a very rough estimate based on the result of a single experiment. The quantitative percents based on this study were not representative. More studies, especially field studies, are recommended for reducing the uncertainty in this prediction. Even with the uncertainties in this estimation, the environmental benefit could be significant. The quantitative reduction is strongly associated with factors including emission factor, stove thermal efficiency and the replacement ratio.

8.5 Summary

In this section, we reported emissions of CO, PM, OC, EC and PAHs from the burning of two biomass pellets burned in modern pellet burner. The results were compared with that for raw biomass fuels burned in a traditional cooking stove, and the potential reductions in these pollutant emissions were discussed. On average, the measured EF_{CO}, EF_{OC}, EF_{EC}, and EF_{PM} for straw pellet were 21.1 ± 16.2, 0.121 ± 0.158, 0.156 ± 0.121 and 2.62 ± 1.20 g/kg, respectively, and for wood pellet were 4.38 ± 2.25, 0.095 ± 0.118, 0.033 ± 0.026 and 1.17 ± 0.89 g/kg, respectively. EF_{PAHs} for these two pellets were 14.1 ± 8.9 and 8.33 ± 5.94 mg/kg, respectively. In the burning with secondary air supply, more excess air may reduce the combustion temperature, lead to lower burning efficiency and subsequent higher pollutant emissions.

In comparison with raw biomass fuels, the emission factors of CO and CPM were significantly lower for biomass pellet, however, the PAH emissions were statistically insignificant different. The compress process reduce the OC and EC emissions more effectively compared to the reduction in PM, which results in a low carbonaceous carbon fraction in PM emitted from the pellet burning. The normalized PAH composition profiles were similar in emissions between biomass pellet and raw fuels. Taking both EFs and the stove thermal efficiency into consideration, significant reductions in the total emissions of these pollutants from residential sector are expected by replacing raw fuels combusted in traditional/improved stoves with biomass pellet fuels in modern burners. As a rough estimation, The reduction in emissions of CO, OC, EC, PM, and PAHs from residential sector would be 90, 96, 96, 76, and 43 %, respectively, by replacing the improved cooking stoves with pellet stoves, and could be as high as 95, 98, 98, 88, and 71 % if the traditional cooking stoves were replaced. If only 20 % of the traditional cooking stoves can be replaced with pellet stoves, the total emission reductions in CO, OC, EC, PM. and PAHs would be 7366 (6262–8642), 446 (382–521), 115 (98–134), 675 (577–791), and 8.94 (7.27–10.8) Gg. The reduction and benefit is significant and obviously, even though the quantitative results are not representative due to limited data now. More studies in future are recommended before the reduction can be quantified. Also, factors affecting the emissions, including fuel property, stove type, and combustion condition, should be further investigated.

References

Bäfver, L. S., Leckner, B., Tullin, C., & Berntsen, M. (2011). Particle emissions from pellets stoves and modern and old-type wood stoves. *Biomass and Bioenergy, 35*, 3648–3655.

Bølling, A. K., Pagels, J., Yttri, K. E., Barregard, L., Sallsten, G., Schwarze, P. E., et al. (2009). Health effects of residential wood smoke particles: The importance of combustion conditions and physicochemical particle properties. *Particle and Fiber Toxicology, 6*, 29. doi:10.1186/1743-8977-6-29

Boman, C., Pettersson, E., Westerhol, R., Boström, D., & Nordin, A. (2011). Stove performance and emission characteristics in residential wood log and pellet combustion, Part 1: Pellet stoves. *Energy & Fuels, 25*, 307–314.

Chang, M. C., Chow, J. C., Watson, J. G., Hopke, P. K., Yi, S. M., & England, G. C. (2004). Measurement of ultrafine particle size distributions from coal-, oil-, and gas-fired stationary combustion sources. *Journal of the Air and Waste Management Association, 54*, 1494–1505.

Chen, X., Tan, Y., Wang, Y., & Yan, Y. (2005). Problems and countermeasures in application and development of the biomass pellet fuel in our country. *Renewable Energy, 119*, 41–43. (In Chinese).

Chen, Q., Zhang, X., Bradford, D., Sharifi, V., & Swithenbank, J. (2010). Comparison of emission characteristics of small-scale heating systems using biomass instead of coal. *Energy & Fuels, 24*, 4255–4265.

Fiedler, F. (2004). The state of the art of small-scale pellet-based heating systems and relevant regulations in Sweden, Austria and Germany. *Renewable and Sustainable Energy Reviews, 8*, 201–221.

García-Maraver, A., Popov, V., & Zamorano, M. (2011). A review of European standards for pellet quality. *Renewable Energy, 36*, 3537–3540.

Hedman, B., Näslund, M., & Marklund, S. (2006). Emission of PCDD/F, PCB and HCB from combustion of firewood and pellets in residential stoves and boilers. *Environmental Science and Technology, 40*, 4968–4975.

Houck, J. E., & Eagle, B. N. (2006). Control analysis and documentation for residential wood combustion in the MANE-VU region. , Beaverton, OR: OMNI Environmental Services, Inc, Dec. 2006.

Kupiainen, K., & Klimont, Z. (2007). Primary emissions of fine carbonaceous particles in Europe. *Atmospheric Environment, 41*, 2156–2170.

Johansson, L. S., Leckner, B., Gustavsson, L., Cooper, D., Tullin, C., & Potter, A. (2004). Emission characteristics of modern and old-type residential boilers fired with wood logs and wood pellets. *Atmospheric Environment, 38*, 4183–4195.

Jokiniemi, J. et al. (2008). Biomass combustion in residential heating: Particulate measurements, sampling, and physicochemcial and toxicological characterisation. ISSN 0786-4728 Final report of the project 'Biomass-PM' funded by ERA-NET Bioenergy Programme 2007–2008.;University of Kuopio, Report 1/2008; http://www.bios-bioenergy.at/uploads/media/Paper-Obernberger-Biomass-Combustion-in-Residential-Heating-2008-01-01.pdf

Lamberg, H., Nuutinen, K., Tissari, J., Ruusunen, J., Yli-Pirila, P., Sippula, O., et al. (2011). Physicochemical characterization of fine particles from small-scale wood combustion. *Atmospheric Environment, 45*, 7635–7643.

Lei, Y., Zhang, Q., He, K. B., & Streets, D. G. (2011). Primary anthropogenic aerosol emission trends for China, 1990-2005. *Atmosheric Chemistry and Physics, 11*, 931–954.

Lighty, J. S., Veranth, J. M., & Sarofim, A. F. (2000). Combustion aerosols: Factors governing their size and composition and implications to human health. *Journal of the Air and Waste Management Association, 50*, 1565–1618.

Maguhn, J., Karg, E., Kettrup, A., & Zimmermann, R. (2003). On-line analysis of the size distribution of fine and ultrafine aerosol particles in flue and stack gas of a municipal waste incineration plant: Effects of dynamic process control measures and emission reduction devices. *Environmental Science and Technology, 37*, 4761–4770.

Nussbaumer, T., Czasch, C., Klippel, N., Johansson, L. and Tullin, C. (2008). Particulate emissions from biomass combustion in IEA countries. Survey on measurements and emission factors. www.ieabcc.nl

Schimidl, C., Luisser, M., Padouvas, E., Lasselsberger, L., Rzaca, M., Curz, C. R., et al. (2011). Particulate and gaseous emissions form manually and automatically fired small scale combustion systems. *Atmospheric Environment, 45*, 7443–7454.

Shen, G. F., Tao, S., Wei, S., Zhang, Y., Wang, R., Wang, B., Li, W., Shen, H. Z., Huang, Y., Yang, Y., Wang, W., Wei, W., Wang, X., Liu, W., Wang, X., Simonich, S., (2012). Reductions in emissions of carbonaceous particulate matter and polycyclic aromatic hydrocarbons from combustion of biomass pellets in comparison with raw fuel burning. *Environmental Science & Technology, 46*, 6409–6416.

Streets, D. G., Zhang, Q., Wang, L., He, K., Hao, J., Wu, Y., et al. (2006). Revisiting China's CO emissions after the transport and chemical evolution over the Pacific (TRACE-P) mission: Synthesis of inventories, atmospheric modeling, and observations. *Journal of Geophysical Research, 111*, D14306. doi:10.1029/2006JD007118

U. S. EPA. (2001). Volume III, Chapter 2 Residential wood combustion. Boston: Eastern Research Group, Inc.

Wang, L., Skjevrak, G., Hustad, J. E., & Grønli, M. G. (2011). Effects of Sewage sludge and marble sludge addition on slag characteristics during wood waste pellets combustion. *Energy & Fuels, 25*, 5775–5785.

Yao, Z., Zhao, L., Ronnback, M., Meng, H., Luo, J., & Tian, Y. (2010). Comparison on characterization effect of biomass pellet fuels on combustion behavior. *Transactions of the Chinese Society for Agricultural Machinery, 41*, 97–102. (In Chinese).

Yuan, Y., Lin, C., Zhao, L., Tian, Y., & Meng, H. (2009a). The research progress of anti-slagging for biomass pellet fuel. *Renewable Energy, 27*, 48–51. (In Chinese).

Yuan, Y., Zhao, L., Meng, H., Lin, C., & Tian, Y. (2009b). Effects comparison on anti-slagging additives of corn straw biomass pellet fuel. *Transactions of the Chinese Society of Agricultural Engineering, 26*, 251–255. (In Chinese).

Zhang, Y., Dou, H., Chang, B., Wei Z., Qiu, W., Liu, S., Liu, W., Tao, S., (2008). Emission of polycyclic aromatic hydrocarbons form indoor straw burning and emission inventory updating in China. *Annals of the New York Academy of Science, 1140*, 218–227.

Chapter 9
Conclusion and Limitation

EF is an important input for the development of emission inventory and subsequent analysis of pollutant impacts on air quality and human health. The EFs of incomplete pollutants from residential solid fuel combustion highly depend on the fuel properties and burning conditions. In the present study, we measured EFs of CPM and PAHs from the combustions of coals, crop residues and wood materials. A number of first-hand data are obtained. The study suggested that because of oxygen deficient atmosphere formed in the small stove chamber, the pollutant emissions in the cooking stove were generally higher than those measured in laboratory chamber. The influences of fuel property and burning conditions were discussed. The study also found that the use of retene as a biomarker for conifer wood might be questionable since other fuels, particularly coal, can produce retene as well. In the present study, emission characterization of PM and PAHs for biomass pellets burned in a modern pellet burner were for the first time investigated and compared with those for raw uncompressed fuels burned in a typical brick stove.

9.1 Emission Factor

EFs of PM, OC and EC for crop residues were 8.19 ± 4.27 (3.41–16.8), 1.38 ± 0.70 (0.493–2.64) and 1.45 ± 0.62 (0.354–2.34) g/kg, respectively. They were 3.74 ± 0.80 (3.10–4.63), 0.81 ± 0.64 (0.21–1.48) and 1.53 ± 1.01 (0.48–2.49) g/kg, respectively for the brushwood, significantly higher than those of 2.04 ± 1.38 (0.71–6.23), 0.80 ± 0.85 (0.11–3.81) and 0.50 ± 0.36 (0.06–1.19) g/kg, respectively for the fuel wood log. EFs for coal had a large range from 0.065–10.8, 0.007–1.00 and 0.006–0.825 g/kg, respectively.

EFs of total 16 PAHs for crop residue ranged from 23.6 to 142 mg/kg, with a mean and standard derivation of 62.1 ± 34.6 mg/kg. For firewood material, EFs of 28 PAHs were 86.7 ± 67.6 (27.1–160 as range) and 12.7 ± 7.0 (3.2–32.7) mg/kg, respectively, and of which 90 % of the total mass was 16 priority PAHs. For coal, PAHs EFs ranged from 6.25 to 253 mg/kg. The EFs of oxygenated PAHs were

G. Shen, *Emission Factors of Carbonaceous Particulate Matter and Polycyclic Aromatic Hydrocarbons from Residential Solid Fuel Combustions*, Springer Theses, DOI: 10.1007/978-3-642-39762-2_9, © Springer-Verlag Berlin Heidelberg 2014

generally in the same order of magnitude of that of parent PAHs, while EFs of nitro-PAHs were about 2–3 orders of magnitude lower. EFs of 4 oxygenated PAHs for crop residue, fuel wood, brushwood and coal were 4.5 ± 2.0, 1.19 ± 1.87, 5.56 ± 4.32 and 0.049–40 mg/kg, respectively.

9.2 Emission Characterization

The total mass percent of carbonaceous carbon in PM from firewood burning was 61 %, significantly higher than that of 38 ± 10 % for emissions from crop residue burning. The carbon mass percent in PM emitted from coal combustion was about 32 ± 19 %. The size distribution of PM indicated that for three types of solid fuels, PM emitted from the fuel burning was fine. The mass percents of $PM_{2.5}$ were 77.5, 81.0 and 79.4 % of total PM mass. The PM size distributions in emissions from crop residue burning among different crop straw types were similar. Significant difference in PM size distribution was found in emissions between the brushwood and fuel wood log with much finer PM in the later. The most abundant particle size fraction in emissions from brushwood and fuel wood log combustions were 0.7–2.1 and <0.4 μm, respectively.

PAH composition profiles and several commonly used isomer ratios in emissions from these three fuel types were comparable in general. Slight difference was observed in the mass percentage of high molecular weight PAHs. The mass percent of high molecular weight PAHs (MW > 228) in the emission from coal combustion was higher than that for the crop residue and firewood. Because of relatively high toxic of these high molecular weight compounds, the larger fractions in coal emissions indicated that the burning of coal would result in more toxic health outcome in comparison with biomass fuel.

The gas-particle partitioning of freshly emitted PAHs and the derivatives was mainly controlled by the absorption rather than adsorption. The size distributions of particle-bound organics were similar to the size distribution of PM, with high abundance in fine particle. Because of high vaporization rate, low molecular weight PAHs are mainly emitted in the gaseous phase while high molecular weight ones are largely associated with co-emitted PM, especially fine PM. In comparison with parent PAHs, the derivatives had more tendencies to be present in particulate phase, particularly fine particles due to the compound polarity property.

Retene is often used as a marker for softwood combustion and for PAHs source apportionment. The EF of retene varied significantly among the fuels due to the differences in fuel properties and combustion conditions. EF_{RET} for pine (0.34 ± 0.08 mg/kg) and larch (0.29 ± 0.22 mg/kg) were significantly higher than those of other wood types, including fir and cypress (0.081 ± 0.058 mg/kg). However, EF_{RET} for crop residues varied from 0.048 ± 0.008 to 0.37 ± 0.14 mg/kg and were not significantly lower than those for softwood (0.074 ± 0.026 to 0.34 ± 0.08 mg/kg). The EF_{RET} for coal were very high and ranged from 2.2 ± 1.5 (anthracite briquette) to 187 ± 113 mg/kg (raw bituminous chunk).

EF_{RET} was positively correlated with EFs of co-emitted particulate matter (EF_{PM}) and phenanthrene (EF_{PHE}) for crop residue and coal, but not for wood. The study suggested that retene is not a unique PAH marker for softwood combustion and that coal combustion, in particular, should be taken into account when retene is used for PAH source apportionment.

9.3 Influencing Factor

Pollutant emission was generally low in high efficient burning. The fuel combustion efficiency can be affected by a number of factors like fuel moisture and air supply amounts. In the residential solid fuel combustion, the air supply is usually limited in the cooking stove with relatively smaller chamber, and hence oxygen deficient atmosphere resulted in incomplete fuel combustion and higher emissions of incomplete pollutants. In indoor crop residue burning, the study found that fuel moisture and MCE were two most significant factors affecting pollutant emissions. These two factors can explain 83 and 60 % of the variations in EF_{CPM} and EF_{PAHs}, respectively. In wood combustion, pollutant emission factors decreased with the increase of combustion efficiency, but the correlation between measured EFs and wood moisture was statistically insignificant. This might be due to a variety of factors affecting the pollutant emission, and these factors often interacted with each other leading to statistically insignificant dependence of EFs on a single factor. In coal combustion, as expected anthracite has lower emissions of incomplete pollutants than bituminous coal, and the emissions from the burning of honeycomb briquette were lower than that the raw coal chunk. The pollutant emission from coal combustion can be affected by coal volatile matter content, heating value and also the form.

9.4 Biomass Pellet

In the present study, two types of biomass pellet were burned and investigated for the emission characterization for the first time. A modern pellet burner purchased from the local market was used. The EFs of PM, OC and EC for corn straw pellet were 2.62 ± 1.20, 0.121 ± 0.158 and 0.156 ± 0.121 g/kg, respectively, and for the pellet made from pine wood were 1.17 ± 0.89, 0.095 ± 0.118 and 0.033 ± 0.026 g/kg, respectively. EFs of total PAHs for these two pellets were 14.1 ± 8.9 and 8.33 ± 5.94 mg/kg, respectively. The increase of air supply during the pellet burning cooled the combustion temperature and hence produced high pollutant emissions under decreased burning efficiencies. In comparison with the raw uncompressed biomass fuel, the EFs of PM, OC and EC were significantly lower for biomass pellet, but PAH EFs were not significantly different between the compressed and uncompressed fuels. Taking the emission factor (pollutant emitted from per fuel mass) and stove thermal efficiency (fuel used for cooking per task) into consideration, the reduction of pollutant emission after the deployment of biomass pellet was significant. However, the large deployment of biomass pellet required more detailed further study since there are very limited data on either monitored air quality or environment impacts of biomass pellets in China.

9.5 Field Measurement

It is widely accepted that the simulated fuel combustion differed from that in real field, and later might be more preferable in the development of emission inventory and pollution control strategy. In the study, we conducted two field measurement campaigns. This without any doubt is insufficient. It is just a trial field study. Field measured EFs had much larger variations in comparison with that measured in simulated study, but the characterization such as carbon ratio, PAHs composition profiles and isomer ratios were similar.

In field study, some specific fuel/stove combinations can be tested which can be hardly considered in simulation study. In the present study, we found that in rural Shanxi, coal cake was widely used and the pollutant emissions for coal cake were as high as 5.87 ± 3.77, 0.037 ± 0.021, 0.477 ± 0.140 g/kg, 178 ± 43 and 7.09 ± 3.35 mg/kg for PM, EC, OC, PAHs and oPAHs, respectively. For wood burned in a movable open stove, the EFs of PM, EC, OC, PAHs and oPAHs could be high as 8.32 ± 0.27, 1.24 ± 0.48, 2.91 ± 1.01 g/kg, 208 ± 62 and 8.96 ± 1.59 mg/kg, respectively. In the field study in rural Jiangsu, it was found that the pollutant emissions varied significantly for the same fuel burned in two stoves with same design but different usage. The EF_{PM} for crop residue burned in an old stove was about 2.5 times of that for that burned in a new stove.

9.6 Limitation and Future Study

In the present study, emission factors of CPM and PAHs were measured for crop residue, wood and coal burned in residential stoves in a built kitchen and in field. However, as mentioned above, the sample size in the study was still very small, particularly under the consideration of relatively larger variation in EFs and high spatial change of fuel/stove combinations in rural China. It is expected that in future more studies, especially field measurements can be conducted to get more reliable data.

In term of influencing factors, the study quantitatively analyzed the impacts of fuel properties such as elemental contents, moisture and volatile matter content on pollutant emission. It should be noted the results were under the given circumstance. The results can not be generalized simply. And some other factors like air supply amount, fire management behaviors and fuel feeding rate were not investigated. In the future study, different fuel/stove combinations and the operation behaviors of rural residents should be discussed.

The present study found the emissions from the brushwood burning were higher than those from the fuel wood log combustion. But it is realized that in most emission inventory, the difference was not accounted which may result in considerable underestimation of most pollutants. Meanwhile, the differences between the anthracite and bituminous, between the raw chunk and briquette are significant. It is fortunately to find that such differences had been taken into account in many inventory studies. However, because of limited basic measurement data available, such analysis is usually associated with relatively high uncertainties. With more confirmed results in future, it is expected that the deployment of coal briquette may reduce the pollutant emission significantly, and benefit the air quality and human health. Policy analysis and cost benefit analysis should be conducted. Similarity, the study on biomass pellet indicated that the pollutant emissions from the biomass pellet burning decreased obviously in comparison with the raw biomass fuels. Because of a small sample size in the present study, it is difficult to reach a sound conclusion now. The real benefits of the new cleaner fuels, including biomass pellets and others like biogas and natural gas, required more measurements.

MIX
Papier aus verantwortungsvollen Quellen
Paper from responsible sources
FSC® C105338

If you have any concerns about our products,
you can contact us on
ProductSafety@springernature.com

In case Publisher is established outside the EU,
the EU authorized representative is:
Springer Nature Customer Service Center GmbH
Europaplatz 3, 69115 Heidelberg, Germany

Printed by Libri Plureos GmbH
in Hamburg, Germany